Algebraic and Analytic Methods
in Representation Theory

PERSPECTIVES IN MATHEMATICS, Vol. 17
S. Helgason, editor

Algebraic and Analytic Methods in Representation Theory

Edited by

Bent Ørsted

Department of Mathematics
and Computer Science
Odense University
Odense, Denmark

Henrik Schlichtkrull

Department of Mathematics
and Physics
The Royal Veterinary and Agricultural University
Frederiksberg, Denmark

ACADEMIC PRESS

San Diego   London   Boston   New York   Sydney   Tokyo   Toronto

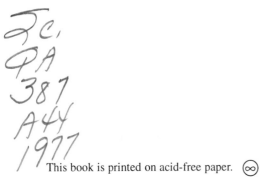

ACADEMIC PRESS, INC.
525 B Street, Suite 1900, San Diego, CA 92101-4495, USA
1300 Boylston Street, Chestnut Hill, MA 02167, USA
http://www.apnet.com

ACADEMIC PRESS LIMITED
24–28 Oval Road, London NW1 7DX, UK
http://www.hbuk.co.uk/ap/

Library of Congress Cataloging-in-Publication Data

Algebraic and analytic methods in representation theory / edited by
  Bent Ørsted, Henrik Schlichtkrull.
      p.   cm. — (Perspectives in mathematics; vol. 17)
  Main lectures from the European School of Group Theory held Aug.
  15–26, 1994, in Sønderborg, Denmark.
    Includes bibliographical references and index.
    ISBN 0-12-625440-0 (alk. paper)
    1. Lie groups. 2. Representations of groups.   I. Ørsted, Bent.
  II. Schlichtkrull, Henrik, 1954– . III. European School of Group
  Theory (1994: Sønderburg, Denmark) IV. Series.
  QA387.A44   1996
  512' .55—dc20   96-22985
                                                                  CIP

Printed in the United States of America
96 97 98 99 00 EB 9 8 7 6 5 4 3 2 1

Contents

Contributors

HENNING HAAHR ANDERSEN, Department of Mathematics, Aarhus Universitet, 8000 Århus C, Denmark

ANTHONY JOSEPH, Department of Theoretical Mathematics, The Weizmann Institute of Science, Rehovot 76100, Israel

TOSHIYUKI KOBAYASHI, Department of Mathematical Sciences, University of Tokyo, Komaba, Meguro, Tokyo 153, Japan

V. S. VARADARAJAN, Department of Mathematics, UCLA, Los Angeles, California 90095–1555, USA

DAVID A. VOGAN, JR., Department of Mathematics, MIT, Cambridge, Massachusetts 02139–4307, USA

Preface

The European School of Group Theory for the summer of 1994 was held in Sønderborg, Denmark, from August 15 to August 26 at the Sandbjerg Estate. About 50 advanced students participated, with interests quite wide but centered around algebraic and analytic aspects of Lie groups and their representations. The main lectures were delivered by five outstanding researchers, covering some of the deepest areas of the field today, yet giving introductions that were a pleasure to follow. It was a summer school with a high level of participation and interest on the part of the students, and it is our hope that the present volume will convey a similar spirit to its readers.

The theory of reductive groups is at present as active and important as ever. Building on classical tools and structure theory as well as ideas from classical and quantum physics, many problems have been solved, and new research avenues have opened up. The subject has always required knowledge in several branches of mathematics, but even more so today, where an expanding amount of algebraic, geometric, and analytic techniques are combined. Since the written material by the lecturers at the summer school went far beyond usual lecture notes and the different topics formed a unified view, albeit from perspectives quite apart, it was decided to publish the five contributions together.

Chapter 1 by Henning Haahr Andersen gives an introduction to algebraic groups and their representation theory with a clear aim at reaching recent results about modular representations and quantum groups. Indeed, one of the lessons of the past few years of quantized enveloping algebras is that one should develop a good part of the theory parallel to that of algebraic groups. For this point of view, these lectures are very valuable. In Chapter 2, Anthony Joseph gives a full account of the classification of unitary highest weight modules for semisimple Lie algebras, simplifying and expanding previously published work as well as proving results about associated varieties. In addition, several recent results are given on the algebra of differential operators on orbital varieties. Toshiyuki Kobayashi treats, in Chapter 3, discontinuous actions of discrete groups on homogeneous spaces and the corresponding problem of constructing Clifford–Klein forms. A number of classification results are given, opening up promising new areas of research on analogs of automorphic functions and locally symmetric spaces. Chapter 4 by V. S. Varadarajan develops the method of stationary phase from its very beginnings to advanced applications in geometry and in analysis on Lie groups. This fundamental technique is present in much of modern physics; in these lectures emphasis is on applications to the Plancherel formula, asymptotics of eigenvalues on locally symmetric spaces, character formulae, and related questions. David Vogan constructs, in Chapter 5, unitary representations of reductive Lie groups by the orbit method, in analogy with the quantization procedures leading from classical mechanics to quantum mechanics. Starting with the basic concepts and some motivating examples, these lectures patiently develop Lie theory, symplectic geometry, the theory of ideals in universal enveloping algebras, Dixmier algebras, and the induction techniques that enter in the orbit method. Emphasizing the distinction between semisimple and nilpotent orbits, many deep results are obtained and intriguing examples as well as conjectures are discussed.

It is a pleasure to thank the five authors for their contributions to this volume as well as for their enthusiastic lectures at the summer school. Also, we thank the participating students for their dedication to mathematics and other sports. For financial support, we acknowledge Thomas B. Thriges Fond, Odense University, Aarhus University, The Danish Natural Science Research Council, and a grant from the EC, Euroconferences in Group Theory under the program Human Capital and Mobility. Finally, we are grateful to the staff at Sandbjerg for adding to the pleasant memories of this summer school.

Djursholm, October 1995
Bent Ørsted, Henrik Schlichtkrull

Chapter 1

Modular Representations of Algebraic Groups and Relations to Quantum Groups

HENNING HAAHR ANDERSEN

University of Aarhus

CONTENTS

Introduction

The (very informal) notes in this chapter present some highlights in the "classical" representation theory for reductive algebraic groups over algebraically closed fields. The theory is developed in such a way that almost everything carries over to quantum groups. In particular, we have emphasized the similarities between the modular representation theory (the case

1

where the underlying field for the algebraic group has positive characteristic) and the representation theory for quantum groups at roots of unity. We start out by giving the basic general definitions concerning algebraic groups and their representations. One of the most important tools for us when studying representations will be the induction functor. So already in the first section we introduce this concept and give some of its first properties. In Section 2 we specialize to the case of reductive groups. Here we need some of the key results about the relations between these groups and their associated root systems. Most of what we need is collected in Facts 2.6. Our first real theorem comes in Section 3, where we present the classification of finite dimensional simple modules for reductive groups. To obtain this we rely heavily on induction from a Borel subgroup. In the following sections we explore also the higher derived functors of this induction. In particular, we prove the Borel–Weil–Bott theorem (in Section 6) and Serre duality (in Section 7). From these two results we easily deduce the complete reducibility of all finite dimensional modules for algebraic groups in characteristic zero.

Then we turn to characteristic $p > 0$. Here we shall explore the socalled Frobenius subgroups. To handle these properly, we need a more general definition of algebraic groups than the one we have been working with so far. So we begin Section 8 by introducing this more general concept (of algebraic group functors), and we have some remarks on how to generalize the previous results. Then we study (in the rest of Section 8 and in Section 9) the Frobenius subgroups and their representations in a way quite parallel to our presentation of representations of reductive groups. As a reward, we show in Section 10 how we can now very easily deduce the important Kempf vanishing theorem.

In Section 11 we sketch how the representation theory of quantum groups behaves very similarly to the representation theory for reductive algebraic groups. In fact, we leave most of the quantization to the reader, giving only some indications in a few key places. We conclude by a section on tilting modules that makes sense both for algebraic groups and quantum groups. The results here have applications to the recent work on invariants of 3-manifolds.

The material in these notes formed the major parts of my lectures at the Summer School on Group Theory at Sandbjerg in August 1994. It is a pleasure for me to thank both the organizers and all the participants for their patience and enthusiasm.

1   Algebraic groups and their representations

Throughout, k will denote an algebraically closed field.

For $n \in \mathbb{N}$ consider the general linear group $\mathrm{GL}_n(k)$ with coefficients in k, i.e.,

$$\mathrm{GL}_n(k) = \{(a_{ij})_{i,j=1,2,...,n} \mid a_{ij} \in k, \det(a_{ij}) \neq 0\}$$

Clearly, this is an open subset (in the Zariski topology) of k^{n^2} (being the complement of the zeroes of det), or alternatively, it is a closed subset of k^{n^2+1} (being identified with the zeroes of the polynomial $T\det(X_{ij}) - 1 \in k[T, (X_{ij})]$).

Definition 1.1   By *an algebraic group* (over k) we understand a Zariski closed subgroup of $\mathrm{GL}_n(k)$ for some $n \in \mathbb{N}$.

Examples 1.2   i) $\mathrm{GL}_n(k)$, $n \in \mathbb{N}$.
  ii) $\mathrm{SL}_n(k) = \{A \in \mathrm{GL}_n(k) \mid \det(A) = 1\}$, $n \in \mathbb{N}$.
  iii) *The 1-dimensional additive group* $(k, +)$. This is a subgroup of $\mathrm{GL}_2(k)$ via $x \mapsto \begin{pmatrix} 1 & x \\ 0 & 1 \end{pmatrix}, x \in k$.
  iv) *The 1-dimensional multiplicative group* $(k^\times, \cdot)$. Here $k^\times = k\backslash\{0\}$, i.e., $(k^\times, \cdot) = \mathrm{GL}_1(k)$.
  v) $T_n = T_n(k) = \left\{ \begin{pmatrix} t_1 & & 0 \\ & \ddots & \\ 0 & & t_n \end{pmatrix} \middle| \, t_i \in k^\times \right\}$. This is called *the n-dimensional torus.*
  vi) $B_n = B_n(k) = \{(a_{ij}) \in \mathrm{GL}_n(k) \mid a_{ij} = 0 \text{ for } i > j\}$, the subgroup consisting of the upper triangular matrices in $GL_n(k)$.
  vii) $U_n = U_n(k) = \{(a_{ij}) \in B_n \mid a_{ii} = 1 \text{ for } i = 1, 2, ..., n\}$. Note that U_n is a normal subgroup of B_n and that $B_n = T_n U_n$.

If V is an n-dimensional vector space over k we can (and shall) identify the group of invertible k-linear endomorphisms of V, $\mathrm{GL}(V)$ with the algebraic group $\mathrm{GL}_n(k)$.

Definition 1.3 *The coordinate ring $k[G]$* of an algebraic group G is

$$k[G] = \{f\colon G \to k \mid f \text{ is regular}\}.$$

(Here *regular* refers to the notion of "regular function on a variety," cf. [Ha]. In the following, all maps between varieties are tacitly assumed to be regular. In the case at hand, k is considered as the 1-dimensional affine space. More generally, we always consider an n-dimensional vector space as the affine n-space.)

The group structure on G gives rise to a *Hopf algebra structure*, cf. [Sw], on $k[G]$. The comultiplication $\Delta\colon k[G] \to k[G] \otimes k[G]$ comes from multiplication $G \times G \to G$, the counit $\epsilon\colon k[G] \to k$ from the unit $1 \to G$, and the antipode $S\colon k[G] \to k[G]$ from the inverse map $G \to G$.

Exercise 1.4 Show that we have the following identities:

$$k[(k,+)] = k[X], \qquad k[(k^{\times}, \cdot)] = k[X, X^{-1}],$$
$$k[T_n] = k[X_1, \ldots, X_n, X_1^{-1}, \ldots, X_n^{-1}].$$

Definition 1.5 A *homomorphism* $f\colon G_1 \to G_2$ between two algebraic groups is a map of varieties that is also a group homomorphism.

A *representation* of an algebraic group G on a finite dimensional vector space V is a homomorphism $\rho\colon G \to \mathrm{GL}(V)$. We say in short that V is a representation of G (or a *G-module*), and we often write gv instead of $\rho(g)(v)$, when $g \in G$ and $v \in V$.

If V and V' are two G-modules, then a *G-homomorphism* from V to V' is a linear map that commutes with the G-actions. The set of these is denoted $\mathrm{Hom}_G(V, V')$.

We shall need to extend the preceding definition of representations to a class of infinite dimensional ones. Namely, we say that a group homomorphism $\rho\colon G \to \mathrm{GL}(V)$ from an algebraic group G into the general linear group on some vector space V is a *a rational representation* of G if V is a union of finite dimensional representations as just defined. In the following, when we write G-module we understand such a rational representation of G.

Examples 1.6   i) The *trivial representation* of an algebraic group G is the representation on k,

$$G \to \mathrm{GL}(k) = k^\times$$

that takes any $g \in G$ into $1 \in k$. More generally, the trivial representation of G on any vector space V is given by the trivial homomorphism $G \to \mathrm{GL}(V)$.

ii) A homomorphism $G \to \mathrm{GL}_n(k)$ defines a representation of G on k^n (via the identification $\mathrm{GL}_n(k) \simeq \mathrm{GL}(k^n)$). In particular, the identity $\mathrm{GL}_n(k) \to \mathrm{GL}_n(k)$ defines the natural representation of $\mathrm{GL}_n(k)$ on k^n.

iii) A rank one representation of G is a homomorphism $G \to \mathrm{GL}(k) = k^\times$. These representations are called the *characters* of G and they constitute the set $X(G)$. We have for instance $X((k,+)) = 0$ and $X(k^\times) \simeq \mathbb{Z}$.

iv) The coordinate ring $k[G]$ of G may be equipped with several representations:

   a) *(the right regular representation)* $\rho_{\mathrm{r}} \colon G \to \mathrm{GL}(k[G])$ is defined by

$$\rho_{\mathrm{r}}(g)f \colon x \mapsto f(xg), \qquad g, x \in G, \ f \in k[G].$$

   b) *(the left regular representation)* $\rho_{\mathrm{l}} \colon G \to \mathrm{GL}(k[G])$ is defined by

$$\rho_{\mathrm{l}}(g)f \colon x \mapsto f(g^{-1}x), \qquad g, x \in G, \ f \in k[G].$$

   c) *(the conjugation representation)* $\rho \colon G \to \mathrm{GL}(k[G])$ is defined by

$$\rho(g)f \colon x \mapsto f(g^{-1}xg), \qquad g, x \in G, \ f \in k[G].$$

Using the correspondence between the algebraic group G and the Hopf algebra $k[G]$, we see that a representation of G corresponds to a comodule for $k[G]$: If V is a representation of G, then we get a k-linear map $\Delta_V \colon V \to V \otimes k[G]$ satisfying the appropriate comodule axioms, cf. [Sw, p. 30].

Definition 1.7   Let V be a representation of G. Then the *fixed points* of G in V are defined by

$$V^G = \{v \in V \mid gv = v \text{ for all } g \in G\}.$$

Note that
$$V^G \simeq \operatorname{Hom}_G(k, V).$$

Definition 1.8   Let H be a closed subgroup of the algebraic group G and let V denote a representation of H. The *induced representation* $\operatorname{Ind}_H^G V$ is defined by

$$\operatorname{Ind}_H^G V = \{f \colon G \to V \mid f(gh) = h^{-1}f(g),\ g \in G,\ h \in H\}$$

with G-action given by

$$(xf)(g) = f(x^{-1}g), \quad x, g \in G,\ f \in \operatorname{Ind}_H^G V.$$

In the case where V is infinite dimensional, we require that the elements $f \in \operatorname{Ind}_H^G V$ are maps of varieties in the sense that $f(G)$ is contained in a finite dimensional G-invariant subspace W of V such that $f \colon G \to W$ is a map of varieties.

The map $\operatorname{Ev} \colon \operatorname{Ind}_H^G V \to V$ that takes $f \in \operatorname{Ind}_H^G V$ into $f(1)$ is called the *evaluation map*. It is clearly an H-homomorphism.

Examples 1.9   i) Consider the trivial subgroup $H = 1$, and let k denote the 1-dimensional trivial representation of H. Then

$$\operatorname{Ind}_1^G k = k[G].$$

The action is the left regular representation of G on $k[G]$, cf. Example 1.6(iv).

ii) The other extreme is the case $H = G$. In that case, we have

$$\operatorname{Ind}_G^G V \simeq V$$

for all G-modules V.

iii) Let $G = \operatorname{SL}_2(k)$ and take $H = T_2 \cap G$. Then $\operatorname{Ind}_H^G k \simeq S(V)$, the symmetric algebra on the natural 2-dimensional representation of G. (To see this takes a little work, cf. also Section 5.)

Proposition 1.10 (Frobenius reciprocity)   *Let H be a closed subgroup of G. Suppose V is an H-module and M is a G-module. Then the map*

$$\operatorname{Hom}_G(M, \operatorname{Ind}_H^G V) \to \operatorname{Hom}_H(M, V)$$

that takes ϕ to $\operatorname{Ev} \circ \phi$ is an isomorphism.

Proof: The inverse map takes $\psi \in \operatorname{Hom}_H(M, V)$ into the homomorphism that maps $m \in M$ to $g \mapsto \psi(g^{-1}m)$. $\square$

Proposition 1.10 has several consequences. We mention a few of the more immediate ones:

Corollary 1.11   *Let H denote a closed subgroup of G.*
  i) The functor Ind_H^G is left exact (being the right adjoint of restriction).
  ii) Induction is transitive; i.e., if K is a closed subgroup of H, then $\operatorname{Ind}_K^G \simeq \operatorname{Ind}_H^G \circ \operatorname{Ind}_K^H$.
  iii) Ind_H^G takes injective H-modules into injective G-modules.
  (An injective G-module is by definition a G-module that is injective in the category of all rational representations of G.)

By taking $H = 1$ in (iii) we get for any G-module V an embedding of V into $\operatorname{Ind}_1^G V$. The latter is injective by (iii), and hence

iv) The category of rational representations of G has enough injectives.

We shall also need the following useful result (the proof of which takes a little more work, see [Ja, Proposition I.3.6])

Proposition 1.12 (The tensor identity)   *Let V be a G-module and W an H-module. Then there is a natural isomorphism of G-modules*

$$\operatorname{Ind}_H^G(V \otimes W) \simeq V \otimes \operatorname{Ind}_H^G W.$$

2   Reductive algebraic groups

From here on, we shall concentrate on reductive algebraic groups. In this section, we give the appropriate definitions, prove a key result on representations of tori, and list some facts about reductive groups that we shall need in developing their representation theory.

Definition 2.1 i) An algebraic group T is called a *torus* if there exists $n \in \mathbb{N}$ such that $T \simeq T_n$ (cf. Example 1.2(v)).

ii) An algebraic group G over k is called *reductive* (resp., *semisimple*) if the maximal connected solvable normal closed subgroup in G is a torus (resp., trivial).

Examples 2.2 i) A torus is clearly reductive.

ii) $\mathrm{GL}_n(k)$ is reductive: This is a consequence of Lie–Kolchin's theorem, [Hu2, 17.6], which says that a connected solvable closed subgroup N of $\mathrm{GL}_n(k)$ is conjugate to an upper triangular subgroup. If N is also normal, it follows that N is upper triangular. By symmetry, N is therefore diagonal. But if E_{ij} denotes the $n \times n$ matrix with 1 at the (i, j)-th entry and 0 elsewhere, then

$$(1 + E_{ij}) \begin{pmatrix} t_1 & & 0 \\ & \ddots & \\ 0 & & t_n \end{pmatrix} (1 + E_{ij})^{-1} = \begin{pmatrix} t_1 & & 0 \\ & \ddots & \\ 0 & & t_n \end{pmatrix} + (t_j - t_i) E_{ij}$$

for all $t_1, \ldots, t_n \in k$. Hence, we see that

$$N \leq \left\{ \begin{pmatrix} t & & 0 \\ & \ddots & \\ 0 & & t \end{pmatrix} \;\middle|\; t \in k^\times \right\} = Z(\mathrm{GL}_n(k))$$

and the claim follows (the only connected subgroups of $k^\times$ are 1 and $k^\times$).

iii) $\mathrm{SL}_n(k)$ is semisimple: This follows from the same arguments as in (ii).

Let now T be a torus. If $T \simeq T_n$, then $X(T) \simeq \mathbb{Z}^n$ (in the following we shall identify $(\lambda_1, \ldots, \lambda_n) \in \mathbb{Z}^n$ with the character $\lambda \colon T_n \to k^\times$ that takes $\begin{pmatrix} t_1 & & 0 \\ & \ddots & \\ 0 & & t_n \end{pmatrix}$ into $t_1^{\lambda_1} \cdots t_n^{\lambda_n}$, $t_1, \ldots, t_n \in k^\times$). As T is abelian, we know that all irreducible representations are 1-dimensional, i.e., belong to $X(T)$. The following result says that all finite dimensional representations split into a direct sum of 1-dimensional representations (i.e., are completely reducible).

Proposition 2.3   *Let V be a finite dimensional representation of the torus T. Then*
$$V = \bigoplus_{\lambda \in X(T)} V_\lambda,$$
where for each $\lambda \in X(T)$,
$$V_\lambda = \{v \in V \mid tv = \lambda(t)v, \ t \in T\}.$$

Proof: Choose first arbitrarily a basis $\{v_1, \ldots, v_r\}$ of V. Then we can write
$$tv_i = \sum_j f_{ij}(t)v_j, \qquad t \in T,$$
for some $f_{ij} \in k[T]$. Identifying T with T_n and recalling that $k[T_n] = k[X_1, X_1^{-1}, \ldots, X_n, X_n^{-1}]$ (cf. Exercise 1.4)), we have
$$f_{ij}\begin{pmatrix} t_1 & & 0 \\ & \ddots & \\ 0 & & t_n \end{pmatrix} = \sum_{\lambda \in X(T)} r_{ij\lambda} t_1^{\lambda_1} \cdots t_n^{\lambda_n}.$$

If we denote our representation of T on V by $\rho{:}\,T \to \mathrm{GL}(V)$, then the preceding tells us that we may write
$$\rho(t) = \sum_{\lambda \in X(T)} \lambda(t)\rho_\lambda,$$

where $\rho_\lambda \in \mathrm{End}_k(V)$ is the endomorphism that in the basis $v_1, \ldots, v_n$ is given by the matrix $(r_{ij\lambda})_{i,j=1,\ldots,n}$. Now $X(T)$ is a linearly independent subset of $k[T]$. This together with the fact that $\rho(tt') = \rho(t)\rho(t')$, $t, t' \in T$, i.e.,
$$\sum_{\lambda \in X(T)} \lambda(t)\lambda(t')\rho_\lambda = \sum_{\nu,\mu \in X(T)} \nu(t)\mu(t')\rho_\nu\rho_\mu \qquad \text{for all } t, t' \in T,$$

means first that

$$\lambda(t')\rho_\lambda = \sum_{\mu \in X(T)} \mu(t')\rho_\lambda\rho_\mu \qquad \text{for all } t' \in T,$$

and then

$$\rho_\lambda\rho_\mu = \begin{cases} \rho_\lambda, & \text{if } \lambda = \mu, \\ 0, & \text{otherwise.} \end{cases} \tag{2.1}$$

Note that we have also

$$\text{Id} = \rho \begin{pmatrix} 1 & & 0 \\ & \ddots & \\ 0 & & 1 \end{pmatrix} = \sum_{\lambda \in X(T)} \rho_\lambda. \tag{2.2}$$

From (2.1) and (2.2) it follows clearly that $V = \bigoplus_\lambda \rho_\lambda(V)$ and that $\rho_\lambda(V) = V_\lambda$. $\square$

Terminology 2.4 Let V be as in Proposition 2.3. If $\lambda \in X(T)$, then we call V_λ the λ-*weight space* in V, and we say that λ is a *weight* in V if $V_\lambda \neq 0$.

Let G be a connected reductive algebraic group, and fix a maximal torus $T \leq G$ (i.e., T is a torus, $T \leq G$ and T is maximal with these properties). If $g \in G$, the conjugation action $x \mapsto gxg^{-1}$, $x \in G$, gives rise to an endomorphism $\text{Ad}(g): \mathfrak{g} \to \mathfrak{g}$, where $\mathfrak{g}$ is the Lie algebra of G. In this way we obtain a representation $g \mapsto \text{Ad}(g)$ of G on $\mathfrak{g}$, i.e.,

$$\text{Ad}: G \to \text{GL}(\mathfrak{g})$$

is a homomorphism of algebraic groups, called the *adjoint representation* of G.

When we restrict Ad to T we have a finite dimensional representation of T, and by Proposition 2.3 it splits into a sum of weight spaces.

Definition 2.5 Let G and T be as before. The nonzero weights of T in $\mathfrak{g}$ are called the roots of G (with respect to T). The set of roots will be denoted by R.

We collect some basic facts about the correspondence between R and G; see, e.g., [Hu2].

Facts 2.6   *Let G be a connected reductive algebraic group with a maximal torus $T \leq G$.*

i) *The set of roots R of G w.r.t. T is a root system in the euclidian space $\mathrm{span}_\mathbb{R} R \subseteq X(T) \otimes_\mathbb{Z} \mathbb{R}$.*

ii) *For each $\alpha \in R$, there exists a (unique up to scalar) homomorphism $x_\alpha \colon (k, +) \to G$ such that $t x_\alpha(z) t^{-1} = x_\alpha(\alpha(t)z)$, $t \in T$, $z \in k$. The image of x_α is called the root subgroup in G associated with α. It is denoted U_α.*

iii) *Choose a set $R_+ \subset R$ of positive roots, and let U (resp., U') denote the subgroup of G generated by $\{U_\alpha \mid \alpha \in -R_+\}$ (resp., $\{U_\alpha \mid \alpha \in R_+\}$). Then $U = \prod_{\alpha \in R_+} U_{-\alpha}$ (resp., $U' = \prod_{\alpha \in R_+} U_\alpha$), with $\prod$ denoting direct product. Moreover, $U \cap T = 1 = T \cap U'$, and if we set $B = TU$ then $U'B$ is an open subset of G.*

iv) *$W = N_G(T)/T$ is a finite group, which via its natural action on $X(T)$ becomes identified with the Weyl group for R.*

Exercise 2.7   Check the facts (i)–(iv) for $G = \mathrm{GL}_n(k)$.

3   Classification of simple G-modules

In this and the next sections we consider a connected reductive algebraic group G over k with a fixed maximal torus T. We will use the notation $X(T)$, R, R_+, B, U, U', U_α, and W from Section 2.

Lemma 3.1   *Let V be a finite dimensional representation of G. Then the weights in V (w.r.t. T) form a W-stable subset of $X(T)$.*

Proof: Let $w = nT \in W$, $n \in N_G(T)$. Take $v \in V_\lambda \backslash \{0\}$. Then $t(nv) = n(n^{-1}tn)v = n(\lambda(n^{-1}tn)v) = (w\lambda)(t)nv$, for all $t \in T$; i.e., $nv \in V_{w\lambda} \backslash \{0\}$. $\square$

Lemma 3.2   *Let E be a finite dimensional representation of B. Suppose $e \in E_\lambda \backslash \{0\}$ for some $\lambda \in X(T)$. Then, for all $\alpha \in R_+$, we have*

$$U_{-\alpha} e \subseteq e + \sum_{n>0} E_{\lambda - n\alpha}.$$

12   *H. H. Andersen*

Proof: Since by Proposition 2.3 we have $E = \bigoplus_{\mu \in X(T)} E_\mu$, we may for each $\alpha \in R_+$ write

$$x_{-\alpha}(z)e = \sum_{\mu \in X(T)} f_\mu(z)e_\mu, \qquad z \in k,$$

where $e_\mu \in E_\mu$ and $f_\mu \in k[(k,+)] = k[X]$. If $a_{\mu,n} \in k$ are determined by $f_\mu(z) = \sum_{n \geq 0} a_{\mu,n} z^n$, $z \in k$, then by Fact 2.6(ii) we get

$$tx_{-\alpha}(z)t^{-1}e = x_{-\alpha}(\alpha(t)^{-1}z)e.$$

The left hand side is equal to $\lambda(t)^{-1} \sum_\mu f_\mu(z)\mu(t)e_\mu$, whereas the right hand side is $\sum_\mu f_\mu(\alpha(t)^{-1}z)e_\mu$. This means

$$\lambda(t)^{-1}f_\mu(z)\mu(t) = f_\mu(\alpha(t)^{-1}z), \qquad \text{for all } \mu, t, z,$$

i.e.,

$$\lambda(t)^{-1}\mu(t)a_{\mu,n} = a_{\mu,n}\alpha(t)^{-n}, \qquad \text{for all } \mu, t, n.$$

Hence, $f_\mu = 0$ unless $\mu = \lambda - n\alpha$ for some $n \geq 0$ and $f_{\lambda-n\alpha}(z) = a_n z^n$ for some $a_n \in k$. Using also that $x_{-\alpha}(0) = 1$, we see that $f_\lambda = 1$ and the lemma is proved. □

The choice of R_+ induces an ordering on $X(T)$, namely $\lambda \leq \mu$ if $\mu - \lambda = \sum_{\alpha \in R_+} n_\alpha \alpha$ for some $n_\alpha \in \mathbb{N}$.

Proposition 3.3 *Let E be a finite dimensional B-module. Then we have, for any $\lambda \in X(T)$ and $e \in E_\lambda \backslash \{0\}$,*

$$Ue \subseteq e + \sum_{\mu < \lambda} E_\mu.$$

In particular, $E_\lambda \subseteq E^U$ if λ is minimal among the weights in E.

Proof: Immediate from Lemma 3.2 and Fact 2.6(iii). □

Recall that by Facts 2.6 we have $U \lhd B$ and $B/U \simeq T$. Each $\lambda \in X(T)$ therefore gives a character (also denoted by λ) of B, namely,

$$\lambda(tu) = \lambda(t), \qquad t \in T, \ u \in U.$$

Now $X(k, +) = 0$, cf. Example 1.6(iii), and hence also $X(U_\alpha) = 0$ for all $\alpha \in R$. It follows that $X(U) = 0$. In other words, all characters of B come from T (as before) or

$$X(B) \simeq X(T). \tag{3.1}$$

We may therefore consider (in a unique way) every $\lambda \in X(T)$ as a 1-dimensional representation of B.

Corollary 3.4 *Any finite dimensional representation E of B has a filtration*

$$0 = F^{r+1} \subset F^r \subset \cdots \subset F^0 = E$$

by B-invariant subspaces F^i with $F^i/F^{i+1} \simeq \lambda_i$, $i = 0, 1, \ldots, r$, where $\{\lambda_0, \ldots, \lambda_r\}$ is the set of weights in E. Moreover, the filtration may be chosen such that $\lambda_i < \lambda_j$ only if $i > j$.

Notation 3.5 If E is a representation of B, then we write $H^0(E) = \operatorname{Ind}_B^G E$.

A weight $\lambda \in X(T)$ is called *dominant* if $\langle \lambda, \alpha^\vee \rangle \geq 0$ for all $\alpha \in R_+$. (Here, $\alpha^\vee$ denotes the coroot associated with α. Recall that $s_\alpha \lambda = \lambda - \langle \lambda, \alpha^\vee \rangle \alpha$ defines the reflection s_α in W associated to $\alpha \in R_+$.) Clearly, λ is dominant iff $s_\alpha \lambda \leq \lambda$ for all $\alpha \in R_+$.

The set of dominant weights is denoted $X(T)^+$.

Theorem 3.6 *Let $\lambda \in X(T)$ and suppose $H^0(\lambda) \neq 0$. Then*

i) $\lambda \in X(T)^+$.

ii) The restriction map $\phi \colon H^0(\lambda) \to k[U']$, which takes $f \in H^0(\lambda)$ into $f|_{U'}$, is injective.

iii) If we let T act on $k[U']$ via

$$t \cdot h \colon u \mapsto \lambda(t) h(t^{-1} u t), \qquad t \in T, \ u \in U', \ h \in k[U'],$$

then ϕ becomes a T-homomorphism.

iv) The weight spaces $H^0(\lambda)_\mu$ are all finite dimensional; and if $H^0(\lambda)_\mu \neq 0$, then $\mu \leq \lambda$.

v) $\dim H^0(\lambda)_\lambda = 1$.

vi) $\dim H^0(\lambda) < \infty$.

Proof: Consider first (ii): If $f \in H^0(\lambda)$ restricts to 0 on U', then $f(ub) = \lambda(b)^{-1}f(u) = 0$ for all $u \in U'$, $b \in B$; i.e., f is also zero on the open set $U'B$ (cf. Fact 2.6(iii)). Now a connected group is an irreducible variety, and hence $U'B$ is dense in G. It follows that $f = 0$.

Next we deal with (iii): Take $f \in H^0(\lambda)$, $t \in T$. Then $\phi(tf)(u) = f(t^{-1}u) = f(t^{-1}utt^{-1}) = \lambda(t)f(t^{-1}ut) = t\phi(f)(u)$ for all $t \in T$, $u \in U'$.

Let us then look at (iv): Suppose $h \in k[U']_\mu \backslash \{0\}$ for some $\mu \in X(T)$ (we consider the action of T on $k[U']$ defined in (iii)). Then we have

$$\lambda(t)h(t^{-1}ut) = \mu(t)h(u), \quad t \in T, \ u \in U'. \tag{3.2}$$

By Fact 2.6(iii) we have $u = \prod_{\alpha \in R_+} x_\alpha(z_\alpha)$ for some $z_\alpha \in k$, and we may write

$$h(u) = \sum_{\underline{r}} a_{\underline{r}} \left(\prod_{\alpha \in R_+} z_\alpha^{r_\alpha} \right)$$

for some $a_{\underline{r}} \in k$, $\underline{r} = (r_\alpha)_{\alpha \in R_+} \in \mathbb{N}^{|R_+|}$. Inserting this in (3.2) we get (using $t^{-1}ut = \prod_\alpha x_\alpha(\alpha(t)^{-1}z_\alpha)$)

$$\lambda(t) \sum_{\underline{r}} a_{\underline{r}} \left(\prod_\alpha \alpha(t)^{-r_\alpha} z_\alpha^{r_\alpha} \right) = \mu(t) \sum_{\underline{r}} a_{\underline{r}} \left(\prod_\alpha z_\alpha^{r_\alpha} \right).$$

We conclude that $\mu = \lambda - \sum_{\alpha \in R_+} r_\alpha \alpha$ for some $\underline{r} \in \mathbb{N}^{|R_+|}$ and (iv) now follows from (ii) and (iii). In fact, we get $\dim H^0(\lambda)_\mu \leq \#\{\underline{r} \in \mathbb{N}^{|R_+|} \mid \lambda - \mu = \sum_{\alpha \in R_+} r_\alpha \alpha\}$.

To prove (v) we see by the preceding that we only have left to check that $H^0(\lambda)_\lambda \neq 0$. But if $H^0(\lambda)_\lambda = 0$, then $\mathrm{Ev}_\lambda : H^0(\lambda) \to \lambda$ would be the zero map, which, by the universality (Proposition 1.10) of the induced module, would imply $H^0(\lambda) = 0$.

Now we have only left to observe that (i) follows from (iv) and (v) (by Lemma 3.1, we see that $s_\alpha \lambda$ is a weight of $H^0(\lambda)$ for all $\alpha \in R_+$) and that (vi) is a consequence of (iv) (it is well known that any weight is W-conjugate to a dominant weight and that there are only finitely many $\mu \in X(T)^+$ with $\mu \leq \lambda$). $\square$

Corollary 3.7 *If E is a finite dimensional representation of B, then $H^0(E)$ is a finite dimensional representation of G.*

Proof: Apply H^0 to a filtration of E as in Corollary 3.4 and use Theorem 3.6(vi). (Recall that H^0 is left exact). $\square$

It turns out that the converse of the implication $H^0(\lambda) \neq 0 \Rightarrow \lambda \in X(T)^+$ (see Theorem 3.6(i)) is also true, i.e.,

$$\text{Let } \lambda \in X(T). \text{ Then } H^0(\lambda) \neq 0 \text{ iff } \lambda \in X(T)^+. \tag{3.3}$$

(This involves a little algebraic geometry; see, e.g., [Hu2. 31.4].)

Theorem 3.8 *Let $\lambda \in X(T)^+$.*
i) There is a unique irreducible G-invariant subspace $L(\lambda)$ in $H^0(\lambda)$.
ii) $\dim L(\lambda)_\lambda = 1$.

Proof: We claim

$$H^0(\lambda)^{U'} = H^0(\lambda)_\lambda. \tag{3.4}$$

Note that the containment $\supseteq$ follows from Proposition 3.3 (with U replaced by U'), since $H^0(\lambda)_\mu = 0$ for all $\mu > \lambda$. To get the other inclusion, take $f \in H^0(\lambda)^{U'}$ nonzero and note that $f(u'b) = \lambda(b)^{-1}f(1)$ for all $u'b \in U'B$. Since $U'B$ is dense in G (Fact 2.6(iii)) we conclude that $\mathrm{Ev}_\lambda f = f(1)$ is nonzero. But this means that $\mathrm{Ev}_\lambda|_{H^0(\lambda)^{U'}}$ is injective, and hence $\dim H^0(\lambda)^{U'} \leq 1$.

Let us now prove (i). Suppose S and S' are two different irreducible G-invariant subspaces of $H^0(\lambda)$. Then $S \cap S' = 0$. On the other hand, using Proposition 3.3 we have $0 \neq S^{U'} \subseteq H^0(\lambda)^{U'} = H^0(\lambda)_\lambda$; and since the latter space is 1-dimensional (Theorem 3.6(v)), we conclude $S^{U'} = H^0(\lambda)_\lambda$. The same is, however, true for S', and we get a contradiction.

The same reasoning proves $L(\lambda)^{U'} = H^0(\lambda)_\lambda$; i.e., $L(\lambda)_\lambda \neq 0$ and (ii) follows. $\square$

Theorem 3.9 *If S is a finite dimensional irreducible representation of G, then there exists a unique $\lambda \in X(T)^+$ with $S \simeq L(\lambda)$.*

Proof: The uniqueness we get from Theorem 3.8(ii). To get the existence, we pick $\lambda \in X(T)$ such that λ is maximal among all weights of S. By Corollary 3.4, we can then find a B-invariant subspace $F \subseteq S$ with $S/F \simeq$

λ. In particular, $0 \neq \mathrm{Hom}_B(S, \lambda) \simeq \mathrm{Hom}_G(S, H^0(\lambda))$; see Proposition 1.10. But this means that $\lambda \in X(T)^+$ (by (3.3)) and that S may be identified with a G-invariant subspace of $H^0(\lambda)$. The conclusion follows by Theorem 3.8(i). $\square$

Example 3.10 i) Let $\lambda_0 \in X(T)$ denote the trivial character, i.e., $\lambda_0(t) = 1$ for all $t \in T$. (Sometimes we write 0 for this character, but this leads to ambiguities in the present context.) Then $H^0(\lambda_0) = k$, the trivial representation of G. (By Theorem 3.6, we see that λ_0 is the only weight of $H^0(\lambda_0)$.)

ii) Let $\omega \in X(T)^+$, and suppose ω is minimal among dominant weights (i.e., no $\lambda \in X(T)^+$ satisfies $\lambda < \omega$). Then $H^0(\omega) = L(\omega)$. (One checks that the only weights of $H^0(\omega)$ are $\{w(\omega) \mid w \in W\}$).

Let V be a finite dimensional representation of T. Then we set (cf. Proposition 2.5)

$$\mathrm{ch}\, V = \sum (\dim V_\lambda) e^\lambda \in \mathbb{Z}[X(T)], \qquad (3.5)$$

and we call this the (formal) character of V.

Clearly, ch is additive on short exact sequences. Also, we see from Lemma 3.1 that

$$\text{If } V \text{ is a representation of } G, \text{ then } \mathrm{ch}\, V \in \mathbb{Z}[X(T)]^W. \qquad (3.6)$$

Here, W acts on $\mathbb{Z}[X(T)]$ via $w\left(\sum_\lambda a_\lambda e^\lambda\right) = \sum_\lambda a_\lambda e^{w\lambda}$, $w \in W$.

Problem 3.11 *Determine* $\mathrm{ch}\, L(\lambda)$ *for all* $\lambda \in X(T)^+$.

This problem is in some sense the very first problem one encounters when studying finite dimensional representations of reductive groups. When $\mathrm{char}(k) = p > 0$, it is still open in general. (We will have more to say about it later on; see Problems 8.9, 9.7, and 11.7 and Proposition 11.8.)

4   Derived functors of induction

We still keep the general assumptions from Section 3. Consider two closed subgroups $H_1 \leq H_2 \leq G$.

Recall that $\mathrm{Ind}_{H_1}^{H_2}$ is a left exact functor from the category of H_1-modules to that of H_2-modules and that these categories have enough injectives (Corollary 1.11(iv)). Hence, we may consider the right derived functors $R^i\mathrm{Ind}_{H_1}^{H_2}$ of $\mathrm{Ind}_{H_1}^{H_2}$. If V is an H_1-module, then $R^i\mathrm{Ind}_{H_1}^{H_2}V$ is obtained by choosing an injective resolution

$$0 \to V \to I_0 \to I_1 \to I_2 \to \cdots$$

and applying $\mathrm{Ind}_{H_1}^{H_2}$. Then $R^i\mathrm{Ind}_{H_1}^{H_2}V$ is the ith cohomology of the resulting complex, $\mathrm{Ind}_{H_1}^{H_2}I_\bullet$.

In the case where $H_1 = B$ and $H_2 = G$, we write

$$H^i = R^i\mathrm{Ind}_B^G. \tag{4.1}$$

(Note that this is consistent with the use of H^0 in the previous section, see Notation 3.5). In the general case, we sometimes write $H^i(H_2/H_1, -)$ instead of $R^i\mathrm{Ind}_{H_1}^{H_2}$.

If $H_1 = 1$, then $\mathrm{Ind}_1^{H_2}(E) = k[H_2] \otimes E$, for any k-vector space E (this follows, e.g., from the tensor identity of Proposition 1.12). In particular, $\mathrm{Ind}_1^{H_2}$ is an exact functor, and hence $R^i\mathrm{Ind}_1^{H_2} = 0$ for $i > 0$. Similarly, $\mathrm{Ind}_T^{H_2}$ is exact for all $H_2 \geq T$ (because all T-modules are injective by Proposition 2.3). Note in particular that

$$\mathrm{Ind}_T^B E = k[U] \otimes E \tag{4.2}$$

for all T-representations E. Here, U acts on $k[U]$ via the left regular representation and T acts by $tf\colon u \mapsto f(t^{-1}ut)$, $t \in T$, $u \in U$, $f \in k[U]$.

Proposition 4.1   *Let E be an H_1-module and V an H_2-module. Then we have for all i an H_2-isomorphism,*

$$R^i\mathrm{Ind}_{H_1}^{H_2}(V \otimes E) \simeq V \otimes R^i\mathrm{Ind}_{H_1}^{H_2}E.$$

Proof: For $i = 0$, this is the tensor identity from Proposition 1.12. It follows for $i > 0$ by noting that if $E \to I_\bullet$ is an injective H_1-resolution of

E, then $V \otimes E \to V \otimes I_{\bullet}$ is an injective H_1-resolution of $V \otimes E$ (injectivity is preserved when tensoring by an arbitrary representation). $\square$

The preceding derived functors of induction relate to sheaf cohomology of bundles on the homogeneous space H_2/H_1 as follows:

Let V be an H_1-module. Then V induces a bundle $\mathcal{L}(V)$ on H_2/H_1, the sections of which over an open subset $U \subset H_2/H_1$ are

$$\Gamma(U, \mathcal{L}(V)) = \{f \colon \pi^{-1}(U) \to V \mid f(xb) = b^{-1}f(x), \ x \in \pi^{-1}(U), \ b \in B\}$$

(here, $\pi \colon H_2 \to H_2/H_1$ denotes the canonical map).

Note that $\Gamma(H_2/H_1, \mathcal{L}(V)) = \mathrm{Ind}_{H_1}^{H_2} V$. We claim that in fact

$$H^i(H_2/H_1, \mathcal{L}(V)) \simeq R^i \mathrm{Ind}_{H_1}^{H_2} V \qquad (4.3)$$

for all i.

Since we already observed (4.3) for $i = 0$, it follows for all i by standard dimension shift once we check that an injective H_1-module I induces a bundle $\mathcal{L}(I)$ that is acyclic for the global section functor. This in turn reduces to the standard fact that on the affine variety H_2 any bundle has vanishing higher cohomology, see [Ja, I.5.11].

Via (4.3) we are now able to carry over properties of sheaf cohomology to our derived functors of induction. Although we could make much more heavy use of this (e.g., in Theorem 3.6(vi) or later in Theorem 7.1), we shall prefer to give pure representation theoretic arguments whenever possible. However, we shall need the following vanishing theorem (see [Ha, Theorem III.2.7]):

$$H^i(H_2/H_1, \mathcal{F}) = 0 \qquad \text{for all } i > \dim H_2/H_1, \qquad (4.4)$$

for all sheaves $\mathcal{F}$ on H_2/H_1.

Let us use this opportunity to also introduce Ext-groups as derived functors:

If $H \leq G$ is a closed subgroup and if E is a representation of H we may consider the right derived functors of the left exact functor $\mathrm{Hom}_H(E, -)$ from H-modules into k-vector spaces. As usual, we write

$$\mathrm{Ext}_H^i(E, -) = R^i \mathrm{Hom}_H(E, -).$$

5   Representations of $\mathrm{SL}_2(k)$

In this section, we set $G = \mathrm{SL}_2(k)$ and

$$T = \left\{ \begin{pmatrix} t & 0 \\ 0 & t^{-1} \end{pmatrix} \;\middle|\; t \in k^\times \right\},$$

$$U = \left\{ \begin{pmatrix} 1 & u \\ 0 & 1 \end{pmatrix} \;\middle|\; u \in k \right\},$$

$$B = TU = \left\{ \begin{pmatrix} t & u \\ 0 & t^{-1} \end{pmatrix} \;\middle|\; t \in k^\times,\; u \in k \right\}.$$

It is then straightforward to check that

$$G = B \cup U \begin{pmatrix} 0 & -1 \\ 1 & 0 \end{pmatrix} B \qquad \text{(disjoint union)}. \tag{5.1}$$

Set $V = k[X, Y]$, and let G act on V by

$$\begin{pmatrix} a & b \\ c & d \end{pmatrix} f(X, Y) = f(aX + cY, bX + dY),$$

for all $\begin{pmatrix} a & b \\ c & d \end{pmatrix} \in G$, $f(X, Y) \in k[X, Y]$. Then V is a rational G-module.
In fact, $V = \bigoplus_{n \geq 0} V_n$, where

$$V_n = \mathrm{span}_k \{ X^i Y^j \mid i + j = n \}.$$

Note that V_n is G-stable, and that we have

$$\begin{pmatrix} t & 0 \\ 0 & t^{-1} \end{pmatrix} X^i Y^j = t^{i-j} X^i Y^j,$$

$$\begin{pmatrix} 1 & u \\ 0 & 1 \end{pmatrix} X^i Y^j = \sum_{m=0}^{j} \binom{j}{m} u^m X^{i+m} Y^{j-m}, \tag{5.2}$$

$$\begin{pmatrix} 0 & -1 \\ 1 & 0 \end{pmatrix} X^i Y^j = (-1)^j X^j Y^i.$$

By (5.1), these formulas tell everything.

In this case, $X(T) = \mathbb{Z}$. For $n \in \mathbb{Z}$, let $\lambda_n : T \to k^\times$ denote the character $\begin{pmatrix} t & 0 \\ 0 & t^{-1} \end{pmatrix} \mapsto t^{-n}$, $t \in k^\times$.

Proposition 5.1 *For all $n \geq 0$, there is a natural G-isomorphism*

$$\phi_n \colon V_n \simeq H^0(\lambda_n).$$

Proof: The projection $V_n \to k$ onto the basis element $Y^n \in V_n$ is a B-homomorphism if we let B act on k via λ_n. By Frobenius reciprocity (Proposition 1.10), we get a G-homomorphism $\phi \colon V_n \to H^0(\lambda_n)$. One checks that this is an isomorphism. $\square$

Corollary 5.2 *Let $n \geq 0$. Then $H^0(\lambda_n)$ has a basis $\{v_0, \ldots, v_n\}$ such that, for all $i = 0, \ldots, n$,*

i) $tv_i = \lambda_{2i-n}(t)v_i, \quad t \in T.$

ii) $\begin{pmatrix} 1 & u \\ 0 & 1 \end{pmatrix} v_i = \sum_{m=0}^{i} \binom{i}{m} u^{i-m} v_m, \quad u \in k.$

Proof: Set $v_i = \phi_n(X^{n-i}Y^i)$ and compare with (5.2). $\square$

Remark It is easy from this corollary to check that in characteristic zero the $H^0(\lambda_n)$ are all irreducible. This is a special case of a general fact; see Theorem 7.6.

We now want to compute the higher cohomology groups $H^i(\lambda_n)$, $i > 0$, $n \in \mathbb{Z}$. Here we use the exact sequence

$$0 \to \lambda_n \to k[U] \otimes \lambda_n \to Q \to 0, \tag{5.3}$$

where the first homomorphism is the natural B-homomorphism $\lambda_n \to \operatorname{Ind}_T^B \lambda_n$ (see (4.2)) and Q is the quotient. A careful examination of the long exact sequences arising from (5.3) for various $n \in \mathbb{Z}$ gives (cf. [Do, 12.2]):

Proposition 5.3 *Let $n \in \mathbb{Z}$.*

i) $H^i(\lambda_n) = 0$ for $i > 1$.

ii) If $n \geq -1$, then $H^1(\lambda_n) = 0$.

iii) If $n \leq -1$, then $H^0(\lambda_n) = 0$. Moreover, $H^1(\lambda_n)$ has a basis of elements $w_0, w_1, \ldots, w_{-n-2}$ such that

$$tw_i = \lambda_{-n-2-2i}(t)w_i, \qquad t \in T,$$

$$\begin{pmatrix} 1 & u \\ 0 & 1 \end{pmatrix} w_i = \sum_{j=0}^{-n-2} \binom{j}{i} u^{j-i} w_j.$$

If we compare the result in Proposition 5.3(iii) with Corollary 5.2 we obtain

Corollary 5.4   *Let $n \geq 0$. Then*

$$\mathrm{Hom}_G(H^1(\lambda_m), H^0(\lambda_n)) = \begin{cases} k, & \text{if } m = -n - 2, \\ 0, & \text{otherwise.} \end{cases}$$

A nonzero homomorphism $\Phi_n \colon H^1(\lambda_{-n-2}) \to H^0(\lambda_n)$ is given by

$$\Phi_n(w_i) = \binom{n}{i} v_{n-i}, \qquad i = 0, \ldots, n.$$

Corollary 5.5   *Let $n \in \mathbb{N}$. If $\Phi \in \mathrm{Hom}_G(H^1(\lambda_{-n-2}), H^0(\lambda_n))$ is nonzero, then*

$$\mathrm{Im}\,\Phi = L(\lambda_n).$$

Remark   Similar computations show that we have for each $n \geq 0$ a G-homomorphism $\Psi_n \colon H^0(\lambda_n) \to H^1(\lambda_{-n-2})$ given by

$$\Psi_n(v_i) = (n - i)!\, i!\, w_{n-i}.$$

The composite (in either order) of Ψ_n and Φ_n is multiplication by $n!$.

6   The Borel–Weil–Bott theorem

Let α be a simple root. Then we define a subgroup $P_\alpha \leq G$ by

$$P_\alpha = B \cup Bs_\alpha B, \qquad\qquad (6.1)$$

where (by abuse of notation) we have written s_α also for a representative for $s_\alpha \in W$ in $N_G(T)$ (note that $s_\alpha B$ is independent of which representative we choose).

Now $\mathrm{Ind}_B^{P_\alpha}$ behaves exactly as Ind_B^G in the case $G = \mathrm{SL}_2(k)$ considered in the previous section. Writing H_α^i instead of $R^i \mathrm{Ind}_B^{P_\alpha}$, we have thus

Proposition 6.1   *Let $\lambda \in X(T)$. Then*

i) If $i > 1$. then $H_\alpha^i(\lambda) = 0$.

ii) If $\langle \lambda, \alpha^\vee \rangle = n \geq -1$, then $H_\alpha^1(\lambda) = 0$. and $H_\alpha^0(\lambda)$ has a basis of elements $v_0, v_1, \ldots, v_n$ with

$$tv_i = (\lambda - (n - i)\alpha)(t)v_i, \qquad t \in T,$$

$$x_{-\alpha}(z)v_i = \sum_{j=0}^{i} \binom{i}{j} z^{i-j} v_j, \qquad z \in k.$$

iii) If $\langle \lambda, \alpha^\vee \rangle = n \leq -1$, then $H_\alpha^0(\lambda) = 0$. and $H_\alpha^1(\lambda)$ has a basis of elements $w_0, w_1 \ldots, w_{-n-2}$ with

$$tw_i = (\lambda - (n + i + 1)\alpha)(t)w_i, \qquad t \in T,$$

$$x_{-\alpha}(z)w_i = \sum_{j=i}^{-n-2} \binom{j}{i} z^{j-i} w_j.$$

Notation 6.2   Set $\rho = \frac{1}{2}\sum_{\alpha \in R_+} \alpha$. We have then $\langle \rho, \alpha^\vee \rangle = 1$ for all simple roots α, and we assume $\rho \in X(T)$. Then the *dot action* of W on $X(T)$ is defined by

$$w.\lambda = w(\lambda + \rho) - \rho, \qquad w \in W, \ \lambda \in X(T).$$

If $w \in W$, then we let $l(w)$ denote the length of w, i.e., the length of a reduced expression for w in terms of simple reflections. The aim of this section is to prove the following result, cf. [Bo], [De], or [An1].

Theorem 6.3   *Suppose $\mathrm{char}(k) = 0$, and let $\lambda \in X(T)$. If $w \in W$ is chosen such that $w(\lambda + \rho) \in X(T)^+$, then we have G-isomorphisms*

$$H^i(\lambda) \simeq \begin{cases} H^0(w.\lambda), & \text{if } i = l(w), \\ 0, & \text{otherwise.} \end{cases}$$

Remark   We say that λ is *singular* if there exists $\alpha \in R$ with $\langle \lambda + \rho, \alpha^\vee \rangle = 0$. In this case, $w.\lambda \notin X(T)^+$ and Theorem 6.3 therefore says $H^i(\lambda) = 0$ for all i. On the other hand, if λ is regular (i.e., nonsingular), then the $w \in W$ with $w(\lambda + \rho) \in X(T)^+$ is unique and $w.\lambda \in X(T)^+$. So in this case there is a unique nonvanishing cohomology group, namely the one in degree $l(w)$.

To prove Theorem 6.3 we need two lemmas.

Lemma 6.4   *Let $\lambda \in X(T)$, and suppose $\langle \lambda, \alpha^\vee \rangle \geq -1$ for some simple root α. Then we have a G-isomorphism*

$$H^i(\lambda) \simeq H^i(G/P_\alpha, H^0_\alpha(\lambda))$$

for all $i \geq 0$.

Proof: Let $0 \to \lambda \to I_\bullet$ be an injective B-resolution of λ. Then, since $H^i_\alpha(\lambda) = 0$ for $i > 0$ (Proposition 6.1), we have that $0 \to H^0_\alpha(\lambda) \to H^0_\alpha(I_\bullet)$ is an injective P_α-resolution. The injectivity comes from Corollary 1.11(iii). Hence, the right hand side in the lemma is the ith cohomology of the complex $H^0(G/P_\alpha, H^0_\alpha(I_\bullet))$. By transitivity (Corollary 1.11(ii)), this complex is isomorphic to $H^0(I_\bullet)$, which has $H^i(\lambda)$ as its ith cohomology. $\square$

Lemma 6.5   *Let $\lambda \in X(T)$, and suppose $\langle \lambda, \alpha^\vee \rangle \leq -1$ for some simple root α. Then we have G-isomorphisms*

$$H^i(\lambda) \simeq H^{i-1}(G/P_\alpha, H^1_\alpha(\lambda))$$

for all $i \geq 1$.

Proof: Let again $0 \to \lambda \to I_\bullet$ be an injective B-resolution. By Proposition 6.1 the complex $0 \to H^0_\alpha(I_0) \to H^0_\alpha(I_1) \to \cdots$ is exact except in degree 1, where its cohomology is $H^1_\alpha(\lambda)$. If we set $Q = I_0/\lambda \simeq \mathrm{Ker}(I_1 \to I_2)$, this means that we have exact sequences

$$0 \to H^0_\alpha(I_0) \to H^0_\alpha(Q) \to H^1_\alpha(\lambda) \to 0$$

and

$$0 \to H^0_\alpha(Q) \to H^0_\alpha(I_1) \to H^1_\alpha(I_2) \to \cdots .$$

The long exact sequence arising from the first of these gives

$$H^i(G/P_\alpha, H^1_\alpha(\lambda)) \simeq H^i(G/P_\alpha, H^0_\alpha(Q))$$

for $i \geq 1$. Arguing as in Lemma 6.4, we get

$$H^i(G/P_\alpha, H^0_\alpha(Q)) \simeq H^{i+1}(\lambda)$$

for all $i \geq 0$, and we are done. $\square$

Proof of Theorem 6.3: From Proposition 6.1 it follows that if $\lambda \in X(T)$ has $\langle \lambda, \alpha^\vee \rangle \geq -1$ for some simple root α, then $H^0_\alpha(\lambda) \simeq H^1_\alpha(s_\alpha.\lambda)$ (cf. Corollary 5.4 — it is at this point that we use char$(k) = 0$; otherwise, the binomial coefficients $\binom{n}{i}$ are not necessarily all nonzero, i.e., the Φ_n are not necessarily isomorphisms). By Lemmas 6.4 and 6.5, we get from this $H^i(\lambda) \simeq H^{i+1}(s_\alpha.\lambda)$ for all i. Using this repeatedly, we get (with w as stated)

$$H^i(\lambda) \simeq H^{i-l(w)}(w.\lambda) \qquad \text{for all } i.$$

If $i < l(w)$, we see that $H^i(\lambda) = 0$. A symmetric argument gives

$$H^i(\lambda) \simeq H^{i+N-l(w)}(w_0 w.\lambda),$$

where $N = |R_+|$ and $w_0 \in W$ is the maximal element in W (of length $l(w_0) = N$). Hence if $i > l(w)$ we get $H^i(\lambda) = 0$ by appealing to (4.3) (note that dim $G/B = N$). $\square$

Remarks i) Theorem 6.3 remains true when char$(k) = p > 0$ for those λ which satisfy $\langle \lambda + \rho, \alpha^\vee \rangle \leq p$ for all $\alpha \in R$. The proof goes through unchanged because for such λ the relevant binomial coefficients are all prime to p. If $p \geq h$ (the Coxeter number of R), we see in particular that

$$\dim H^i(w.0) = \delta_{i,l(w)} \qquad \text{for all } i \geq 0, \ w \in W. \qquad (6.2)$$

ii) Suppose again char$(k) = p \geq h$, and let $r \geq 0$. Then we have, for all $w \in W$,

$$H^i(p^r w(\rho) - \rho) \simeq \begin{cases} H^0((p^r - 1)\rho), & \text{if } i = l(w), \\ 0, & \text{otherwise.} \end{cases} \qquad (6.3)$$

Again the proof of Theorem 6.3 goes through (this time the point is that $\binom{ap^r-1}{i} \not\equiv 0 \pmod{p}$ for all $i = 0, 1, \ldots, ap^r - 1, \ a = 1, \ldots p$).

iii) Actually (6.2) and (6.3) hold also for $p < h$, cf. [An2, An3]

7   Serre duality and complete reducibility

In this section, we shall prove

Theorem 7.1   *Let $\lambda \in X(T)$.   Then, for each $i \geq 0$, there is a G-isomorphism*

$$H^i(\lambda) \simeq H^{N-i}(-\lambda - 2\rho)^*.$$

This theorem is a consequence of the Serre duality theorem for locally free sheaves on projective varieties [Ha, III.7]. However, we shall deduce Theorem 7.1 directly from the results of the previous sections. Combining it with Theorem 6.3 we obtain the complete reducibility of all finite dimensional G-modules in characteristic zero.

First, let us make sure we agree about dual representations: If V is a representation of G, then G acts on $V^* = \mathrm{Hom}_k(V, k)$ via

$$gf: v \mapsto f(g^{-1}v), \qquad g \in G, \ v \in V, \ f \in V^*.$$

Lemma 7.2 $H^N(-2\rho) \simeq k.$
Proof: This is the $w = w_0$ case of (6.2). $\square$

Lemma 7.3   *Let E and F be two B-modules. Then we have natural G-homomorphisms*

$$H^i(E) \otimes H^j(F) \to H^{i+j}(E \otimes F)$$

for all $i, j \geq 0$.
Proof: We go by induction on $i + j$. For $i = j = 0$, we have the map $H^0(E) \otimes H^0(F) \to H^0(E \otimes F)$ given by

$$f_1 \otimes f_2 \mapsto (g \mapsto f_1(g) \otimes f_2(g), \ g \in G),$$

$f_1 \in H^0(E)$, $f_2 \in H^0(F)$.
Suppose $j > 0$, and embed $F \to I$ with I an injective B-module. Then

$$0 \to H^0(F) \to H^0(I) \to H^0(I/F) \to H^1(F) \to 0 \qquad (7.1)$$

is exact, and for $j > 1$ we have $H^j(F) \simeq H^{j-1}(I/F)$. The last statement clearly gives the lemma for $j > 1$ via the induction hypothesis. To handle the $j = 1$ case, tensor (7.1) by $H^i(E)$ to obtain the commutative diagram

$$
\begin{array}{ccc}
& 0 & \\
& \downarrow & \downarrow \\
H^i(E) \otimes H^0(F) & \longrightarrow & H^i(E \otimes F) \\
\downarrow & & \downarrow \\
H^i(E) \otimes H^0(I) & \longrightarrow & H^i(E \otimes I) \\
\downarrow & & \downarrow \\
H^i(E) \otimes H^0(I/F) & \longrightarrow & H^i(E \otimes I/F) \\
\downarrow & & \downarrow \\
H^i(E) \otimes H^1(F) & & H^{i+1}(E \otimes F) \\
\downarrow & & \downarrow \\
& 0 &
\end{array}
$$

where the columns are exact and the horizontal maps exist by induction hypothesis. We get the desired map $H^i(E) \otimes H^1(F) \to H^{i+1}(E \otimes F)$. $\square$

Let E be a finite dimensional B-module. The natural map $E \otimes E^* \to k$ that takes $e \otimes f$ into $f(e)$ is clearly a B-homomorphism. By the previous two lemmas, we get G-homomorphisms

$$\phi_E^i : H^i(E) \otimes H^{N-i}(E^* \otimes -2\rho) \to H^N(-2\rho) \simeq k$$

for all $i \geq 0$. This gives us a G-homomorphism

$$
\begin{aligned}
\Psi_E^i : H^i(E) &\to H^{N-i}(E^* \otimes -2\rho)^*, \\
x &\mapsto (h \mapsto \phi_E^i(x \otimes h)).
\end{aligned}
\tag{7.2}
$$

Now Theorem 7.1 is a special case of

Claim 7.4 Ψ^i_E *is an isomorphism for each $i \geq 0$.*

Proof: Consider first the case $i = 0$ and $E = \lambda \in X(T)^+$. Set $R = \mathrm{Ker}(H^0(\lambda) \overset{\mathrm{Ev}_\lambda}{\to} \lambda)$. If we tensor the diagram

$$0 \;\to\; R \cap L(\lambda) \;\to\; L(\lambda) \;\to\; \lambda \;\to\; 0$$

$$\cap \qquad\qquad \cap \qquad\qquad \|$$

$$0 \;\to\; R \;\to\; H^0(\lambda) \;\to\; \lambda \;\to\; 0$$

by $-\lambda - 2\rho$ and take cohomology, we get

$$L(\lambda) \otimes H^N(-\lambda - 2\rho) \;\to\; H^N(-2\rho) \simeq k \;\to\; 0$$

$$\cap \qquad\qquad\qquad\qquad \|$$

$$H^0(\lambda) \otimes H^N(-\lambda - 2\rho) \;\to\; H^N(-2\rho) \simeq k \;\to\; 0$$

where we have used Proposition 4.1 to rewrite the terms on the left and (4.4) to get the 0's on the right.

This and a little weight argument (note that all weights of the quotient $H^0(\lambda)/L(\lambda)$ are strictly less than λ) prove that Ψ^0_λ is nonzero when restricted to $L(\lambda)$. Hence, Ψ^0_λ is injective (because by Theorem 3.8 we know that $L(\lambda)$ is the unique irreducible G-invariant subspace of $H^0(\lambda)$).

It is now an easy induction on $\dim E$ to check that Ψ^0_E is injective for all E.

We prove next that Ψ^0_λ is surjective. For $\lambda = (p^r - 1)\rho$, this follows from (6.3). For arbitrary $\lambda \in X(T)^+$, we choose $r \in \mathbb{N}$ such that $(p^r - 1)\rho + w_0\lambda \in X(T)^+$. Note that $w_0((p^r - 1)\rho + w_0\lambda) = -(p^r - 1)\rho + \lambda$ is the minimal weight in $H^0((p^r - 1)\rho + w_0\lambda)$, so that we have an exact B-sequence

$$0 \to \lambda \to H^0((p^r - 1)\rho + w_0\lambda) \otimes (p^r - 1)\rho \to Q \to 0.$$

This gives us the diagram

$$
\begin{array}{ccc}
0 & & 0 \\
\downarrow & & \downarrow \\
H^0(\lambda) & \longrightarrow & H^N(-\lambda - 2\rho)^* \\
\downarrow & & \downarrow \\
\begin{array}{c} H^0((p^r - 1)\rho + w_0\lambda) \\ \otimes H^0((p^r - 1)\rho) \end{array} & \longrightarrow & \begin{array}{c} H^0((p^r - 1)\rho + w_0\lambda) \\ \otimes H^N((-p^r - 1)\rho)^* \end{array} \\
\downarrow & & \downarrow \\
H^0(Q) & \longrightarrow & H^N(Q^* \otimes -2\rho)^*
\end{array}
$$

As we have observed, the middle horizontal map $1 \otimes \Psi_\lambda^0$ is an isomorphism. Since we also know that bottom horizontal map Ψ_Q^0 is injective, it follows that Ψ_λ^0 is surjective.

Again, an easy induction gives that Ψ_E^0 is surjective for all E. Hence, Ψ_E^0 is an isomorphism. Standard degree shift arguments show that so are all Ψ_E^i, $i \geq 0$. $\square$

Corollary 7.5 *Let $\lambda, \mu \in X(T)^+$. Any short exact sequence of G-modules*

$$0 \to H^0(\lambda) \to E \to H^0(\mu)^* \to 0$$

splits, i.e.,

$$\mathrm{Ext}_G^1(H^0(\mu)^*, H^0(\lambda)) = 0 \qquad \text{for all } \lambda, \mu \in X(T)^+.$$

Proof: Suppose first that $\lambda \not< -w_0(\mu)$. Then λ is maximal among the weights of E, and this implies that we have a B-homomorphism $E \to \lambda$, see Corollary 3.4. By Frobenius reciprocity (Proposition 1.10) this is the same as a G-homomorphism $E \to H^0(\lambda)$; i.e., we have found a retraction $H^0(\lambda) \to E$ of the given embedding.

If $\lambda < -w_0(\mu)$, we dualize the sequence and repeat the previous argument. $\square$

Combining Serre duality and the Borel–Weil–Bott theorem, we obtain the following classical result.

Theorem 7.6 *Suppose $\mathrm{char}(k) = 0$.*

i) $H^0(\lambda)$ is an irreducible representation of G for all $\lambda \in X(T)^+$.

ii) Any finite dimensional representation of G is completely reducible.

Proof: i) By Theorems 6.3 and 7.1, we have

$$H^0(\lambda) \simeq H^N(w_0.\lambda) \simeq H^0(-w_0\lambda)^*.$$

Now we know already that $H^0(\lambda)$ has a unique irreducible G-invariant submodule $L(\lambda)$ (Theorem 3.8(i)). The displayed isomorphism shows that $L(\lambda)$ is also the only irreducible G-quotient of $H^0(\lambda)$. It follows that $H^0(\lambda) = L(\lambda)$.

ii) follows from (i) and Corollary 7.5. $\square$

Remark   The results in Theorem 7.6 remain true in char(k) = $p > 0$ if we require $\langle \lambda + \rho, \alpha^\vee \rangle \leq p$ for all $\alpha \in R_+$ in (i), resp., if the same condition holds for all weights in the representation in (ii). On the other hand, the results fail without this bound: For $SL_2(k)$ we have, for instance, $L(\lambda_p) \neq H^0(\lambda_p)$, $p = \text{char}(k)$.

8   Frobenius kernels and their representations

In the previous sections we have been working with the "pedestrian" definition of an algebraic group given in Definition 1.1. When dealing with algebraic groups over more general fields and rings (e.g., $\mathbb{Z}$) or when considering infinitesimal groups (like the ones we are going to introduce shortly), we need a better definition:

Definition 8.1   Let R be a commutative ring. By an *algebraic group* over R, we understand a functor G from the category of R-algebras into the category of groups such that there exists an R-algebra $R[G]$ of finite type over R (i.e., $R[G]$ is a quotient of a polynomial ring in finitely many variable over R) for which we have

$$G(A) = \text{Hom}_{R\text{-alg}}(R[G], A)$$

for all R-algebras A.

We shall call $R[G]$ the *coordinate ring* of G. Then the definition says that an algebraic group is a representable functor from the category of R-algebras into the category of groups (with a coordinate ring of finite type). The coordinate ring is a Hopf algebra over R. It clearly determines G uniquely.

When $R = k$ is an algebraically closed field, an algebraic group over k is determined by its k-points, $G(k) = \text{Hom}_{k\text{-alg}}(k[G], k)$. In fact, $k[G]$ may be identified with the set of regular functions on $G(k)$, since $G(k)$ is nothing but (the closed points in) the affine variety defined by $k[G]$. In the previous sections we have just been working with these k-points.

When S is an R-algebra and G is an algebraic group over R, we can define an algebraic group G_S over S by extensions of scalars as follows:

$$G_S(A) = G(A)$$

for all S-algebras A. Note that $S[G_S] = R[G] \otimes_R S$.

On the other hand we say that an algebraic group H over S is *defined over* R if there exists an algebraic group H_0 over R with $H = H_{0S}$.

Examples 8.2   i) The *additive group* $\mathbb{G}_a$ over R is defined by

$$\mathbb{G}_a(A) = (A, +)$$

for any R-algebra A. Note that $R[G] = R[T]$, where T is an indeterminate.

ii) The multiplicative group $\mathbb{G}_m$ over R is defined by

$$\mathbb{G}_m(A) = (A^\times, \cdot)$$

for any R-algebra A. (Here $A^\times$ denotes the set of units in A.) We have $R[\mathbb{G}_m] = R[T, T^{-1}]$.

iii) The general linear group over R is given by

$$\mathrm{GL}_n(A) = \{(a_{ij})_{i,j=1,\dots,n} \in M_n(A) \mid \det(a_{ij}) \in A^\times\}$$

for any R-algebra A. For this algebraic group, we have $R[\mathrm{GL}_n] = R[X_{ij}, T]/(T \det(X_{ij}) - 1)$.

We leave it as an exercise for the reader to generalize the other algebraic groups we have encountered in the previous sections and to compute their coordinate rings (as well as to verify that the notation $k[G]$ in Definition 1.3 is consistent with the conventions just given).

Likewise, the various definitions and constructions found in the previous sections also generalize. Let us give the details in the case of the construction of induced modules:

Let G be an algebraic group over k, and suppose V is a vector space over k. Then a representation of G on V is a morphism of functors between the two algebraic groups G and $\mathrm{GL}(V)$. Here the latter is the algebraic group whose value on a k-algebra A is the set of A-automorphisms of $V \otimes_k A$.

If now H is a subgroup of G (i.e., H is itself an algebraic group, and $H(A) \leq G(A)$ for all k-algebras A), we shall construct a functor

$$\mathrm{Ind}_H^G : \{\text{Representations of } H\} \to \{\text{Representations of } G\},$$

which is a right adjoint of the restriction functor.

Suppose M is a representation of H. Let M_a denote the functor

$$M_a: \{k\text{-algebras}\} \to \{k\text{-modules}\},$$
$$A \mapsto M \otimes A.$$

We then define

$$\text{Ind}_H^G(M) = \{f \in \text{Mor}(G, M_a) \mid f_A(gh) = h^{-1}f_A(g),$$
$$\text{for all } g \in G(A), \ h \in H(A), \text{ and all } k\text{-algebras } A\}. \tag{8.1}$$

Here, $f_A: G(A) \to M_a(A)$ denotes the A-component of f.

Note that $\text{Ind}_H^G(M)$ is a representation of G via

$$(gf)_R: x \mapsto f_R(g^{-1}x), \qquad x \in G(R), \ g \in G(A), \ f \in \text{Mor}(G, M_a) \otimes A,$$

for any k-algebra A and any A-algebra R (we identify $\text{Mor}(G, M_a) \otimes A$ with $\text{Mor}(G_A, M \otimes A)$).

We have

$$M \otimes k[G] = M_a(k[G]) \simeq \text{Mor}(G, M_a),$$

and using this we get

$$\text{Ind}_H^G(M) = (M \otimes k[G])^H, \tag{8.2}$$

where H acts on $M \otimes k[G]$ via the given action on M and the right regular representation on $k[G]$, see Example 1.6(iv)(a) (properly generalized).

Our purpose in this and the next two sections is to explore the representations of Frobenius kernels. These are certain subgroup functors of G whose only point over the field k is 1:

We continue to assume that k is an algebraically closed field and that G is a connected reductive algebraic group over k, but now this is considered as an algebraic group in the new sense (whose k-points form a connected reductive algebraic group in the old sense). From now on, we assume $p = \text{char}(k)$ to be positive.

Suppose A is a k-algebra. Let $k \xrightarrow{i} A$ be the structure map. Then, for each $r \in \mathbb{Z}$, we define a new k-algebra $A^{(r)}$ by setting $A^{(r)} = A$ as a ring and choosing the composite $k \xrightarrow{F^{-r}} k \xrightarrow{i} A$ as the new structure map. Here, F is the Frobenius homomorphism on k, i.e., $F(x) = x^p$, $x \in k$.

If X is a functor from $\{k\text{-algebras}\}$ into $\{\text{Sets}\}$, then we denote by $X^{(r)}$ the functor given by

$$X^{(r)}(A) = X(A^{(-r)}), \qquad A \in \{k\text{-algebras}\}.$$

The rth Frobenius morphism on X is the morphism

$$F_X^r : X \to X^{(r)}$$

given by

$$F_X^r(A): X(A) \xrightarrow{X(\gamma_r)} X(A^{(-r)})$$

where $\gamma_r: A \to A^{(-r)}$ is the k-algebra homomorphism that takes a into a^{p^r}, $a \in A$.

When we apply this to G, we obtain in particular the Frobenius homomorphism $F_G^r: G \to G^{(r)}$. This is clearly a homomorphism of algebraic groups.

Definition 8.3

$$G_r = \mathrm{Ker}(F_G^r).$$

Similarly, we have Frobenius subgroups T_r, B_r, U_r, etc., of the subgroups T, B, U of G.

From now on, we shall assume that all our groups are defined over $\mathbb{F}_p \subset k$. Since F is the identity on $\mathbb{F}_p$, this assumption implies that $G^{(r)} = G$ for all r. We thus have an exact sequence of algebraic groups:

$$1 \to G_r \to G \xrightarrow{F_G^r} G \to 1. \tag{8.3}$$

It turns out that the representation theory for G_r is quite similar to the one for G. Without worrying too much about details and proofs (the reader is referred to [Ja] for these), we now give some of the highlights.

Definition 8.4   For any B_r-module E, we set

$$Z_r(E) = \operatorname{Ind}_{B_r}^{G_r} E.$$

One big difference between the induction functor Z_r and the "global" version H^0 (induction from B to G) is that

$$Z_r \text{ is an exact functor.} \tag{8.4}$$

The reason is that G_r/B_r is affine, namely $G_r/B_r \simeq U_r'$, so that all higher sheaf cohomology of bundles on G_r/B_r vanishes. This fact also gives

$$\dim Z_r(E) = p^{Nr} \dim E, \tag{8.5}$$

because $\dim k[U_r'] = p^{Nr}$ (cf. Fact 2.6(iii)).

In analogy with Theorems 3.8 and 3.9, we have

Theorem 8.5   *Let $\lambda \in X(T)$.*

i) $Z_r(\lambda)$ contains a unique irreducible G_r-submodule, which we denote $L_r(\lambda)$.

ii) Any finite dimensional simple G_r-module is isomorphic to some such $L_r(\lambda)$.

iii) If also $\mu \in X(T)$, then $L_r(\lambda) \simeq L_r(\mu)$ iff $\lambda \equiv \mu \pmod{pX(T)}$.

Definition 8.6   Set

$$X_{p^r}(T) = \{\lambda \in X(T) \mid 0 \le \langle \lambda, \alpha^\vee \rangle < p^r \text{ for all simple roots } \alpha\}.$$

The elements in this set are called the p^r-*restricted weights*.

Note that any $\lambda \in X(T)$ can be written uniquely

$$\lambda = \lambda^0 + p^r \lambda^1 \qquad \text{with } \lambda^0 \in X_{p^r}(T),\ \lambda^1 \in X(T). \tag{8.6}$$

Note that $p^r \lambda^1$ is trivial as a G_r-module. Hence $Z_r(\lambda) \simeq Z_r(\lambda^0)$ and $L_r(\lambda) \simeq L_r(\lambda^0)$. This explains the if-part of Theorem 8.5(iii), which can be rephrased

The finite dimensional simple G_r-modules are
parametrized by the set of p^r-restricted weights. $\qquad$ (8.7)

Now comes the surprise:

Theorem 8.7 *If $\lambda \in X_{p^r}(T)$, then $L_r(\lambda) \simeq L(\lambda)|_{G_r}$.*

For $r = 1$ this result goes back to Curtis [Cu]. It may be proved by closely examining the natural G_r-homomorphism $H^0(\lambda) \to Z_r(\lambda)$ (this idea is due to G. Kempf [Ke2]). Alternatively, see [Ja, II.3.15].

For many purposes it is convenient to develop an analogous theory for the subgroup schemes $G_r T$ and $G_r B$. Note that

$$G_r T = (F_G^r)^{-1}(T)$$

and

$$G_r B = (F_G^r)^{-1}(B).$$

We shall use the notation $\hat{Z}_r$ (resp., $\tilde{Z}_r$) for the induction functor from $\{B_r T\text{-modules}\}$ to $\{G_r T\text{-modules}\}$ (resp., from $\{B\text{-modules}\}$ to $\{G_r B\text{-modules}\}$). Then we have

Theorem 8.8 *Let $\lambda \in X(T)$.*

i) $\hat{Z}_r(\lambda)$ (resp., $\tilde{Z}_r(\lambda)$) contains a unique simple $G_r T$-submodule (resp., $G_r B$-submodule), which we denote $\hat{L}_r(\lambda)$ (resp., $\tilde{L}_r(\lambda)$).

ii) Any finite dimensional simple $G_r T$-module (resp., $G_r B$-module) is isomorphic to some such (unique!) $\hat{L}_r(\lambda)$ (resp., $\tilde{L}_r(\lambda)$).

iii) $\hat{L}_r(\lambda)|_{G_r} \simeq L_r(\lambda) \simeq \tilde{L}_r(\lambda)|_{G_r}$.

One advantage of $G_r T$-modules over G_r-modules is that we may consider their (formal) characters. In particular, we can therefore pose the

Problem 8.9 *Determine ch $\hat{L}_r(\lambda)$ for $\lambda \in X(T)$, $r \in \mathbb{N}$.*

Note that by the preceding, $\hat{L}_r(\lambda) \simeq L(\lambda^0) \otimes p^r \lambda^1$, and hence this problem is equivalent to Problem 3.11. It turns out that the crucial case is the case $r = 1$; see [Ja].

9   Injective $G_r T$-modules

We keep the assumptions from Section 8.

Proposition 9.1   *The category of finite dimensional G_rT-modules has enough injectives.*

Proof: Let M be a finite dimensional G_rT-module. We have to find a finite dimensional injective G_rT-module I containing M.

Recall that M is injective as a T-module (this follows immediately from Proposition 2.3). Hence $I = \operatorname{Ind}_T^{G_rT} M$ is an injective G_rT-module, see Corollary 1.11(iii). Moreover, I is finite dimensional (the dimension being given by a formula analogous to (8.5)). By Frobenius reciprocity, the identity $M \to M$ induces a G_rT-injection $M \to I$. $\square$

Definition 9.2   Let $\lambda \in X(T)$. Then we set $\hat{Q}_r(\lambda)$ equal to the injective hull of $\hat{L}_r(\lambda)$.

Note that by Proposition 9.1 (or rather its proof) we have $\hat{Q}_r(\lambda) \subset \operatorname{Ind}_T^{G_rT} \hat{L}_r(\lambda)$. Hence, by injectivity,

$$\hat{Q}_r(\lambda) \text{ is a direct summand of } \operatorname{Ind}_T^{G_rT} \hat{L}_r(\lambda). \tag{9.1}$$

Definition 9.3   A G_rT-module M is said to have a $\hat{Z}_r$-*filtration* if it can be filtered by G_rT-submodules

$$0 = F_0 \subset F_1 \subset \cdots \subset F_m = M$$

in such a way that $F_i/F_{i-1} \simeq \hat{Z}_r(\lambda_i)$ for certain $\lambda_i \in X(T)$.

When M has a $\hat{Z}_r$-filtration we denote by $[M : \hat{Z}_r(\mu)]$ the number of times $\hat{Z}_r(\mu)$ occurs in such a filtration; i.e., in the preceding notation

$$[M : \hat{Z}_r(\mu)] = \#\{i \mid \lambda_i = \mu\}.$$

Example 9.4   Let V be a finite dimensional B_rT-module. Then V has a B_rT-filtration

$$0 = V_0 \subset V_1 \subset \cdots \subset V_t = V$$

with $V_i/V_{i-1} \simeq \lambda_i$ for certain $\lambda_i \in X(T)$ (this is the same argument as in Corollary 3.4). Applying the exact functor $\hat{Z}_r$, we obtain a $\hat{Z}_r$-filtration of $\hat{Z}_r V$. In other words, any G_rT-module that is induced from a finite dimensional B_rT-module has a $\hat{Z}_r$-filtration. Moreover,

$$[\hat{Z}_r V : \hat{Z}_r(\mu)] = \dim V_\mu.$$

Proposition 9.5   Let $\lambda \in X(T)$. Then $\hat{Q}_r(\lambda)$ has a $\hat{Z}_r$-filtration, and for each $\mu \in X(T)$ we have

$$[\hat{Q}_r(\lambda) : \hat{Z}_r(\mu)] = [\hat{Z}_r(\mu) : \hat{L}_r(\lambda)]$$

(the latter symbol denoting composition factor multiplicity).

Proof: Note that $\mathrm{Ind}_T^{G_rT} = \hat{Z}_r \circ \mathrm{Ind}_T^{B_rT}$. Hence the first statement is a consequence of (9.1) and Example 9.4.

To prove the second half, we first need some preparation:

Let B' denote the opposite Borel subgroup; i.e., $B' = w_0 B w_0$, where w_0 is (a representative in G of) the longest element in W. Set

$$\hat{Z}'_r = \mathrm{Ind}_{B'_rT}^{G_rT}.$$

Then, clearly, $\hat{Z}'_r$ has properties completely analogous to $\hat{Z}_r$. Elementary weight considerations show

$$\mathrm{ch}\ \hat{Z}_r(\lambda) = \mathrm{ch}\ \hat{Z}'_r(\lambda - 2(p^r - 1)\rho) \tag{9.2}$$

for all $\lambda \in X(T)$. We now have

Lemma 9.6   Let $\lambda, \mu \in X(T)$. Then

i) $\mathrm{Hom}_{G_rT}(\hat{Z}'_r(\lambda), \hat{Z}_r(\mu)) \simeq \begin{cases} k, & \text{if } \mu = \lambda + 2(p^r - 1)\rho, \\ 0, & \text{otherwise.} \end{cases}$

ii) $\mathrm{Ext}^1_{G_rT}(\hat{Z}'_r(\lambda), \hat{Z}_r(\mu)) = 0$.

Proof: (ii) is proved just as Corollary 7.5 noticing that the (contravariant) dual of $\hat{Z}'_r(\lambda)$ is isomorphic to $\hat{Z}_r(\lambda + 2(p^r - 1)\rho)$. This also implies (i) via Frobenius reciprocity and some easy weight considerations. $\square$

Proof of Proposition 9.5 (continued): From Lemma 9.6, it follows that

$$[M : \hat{Z}_r(\mu)] = \dim \mathrm{Hom}_{G_rT}(\hat{Z}'_r(\mu - 2(p^r - 1)\rho), M)$$

for any G_rT-module M that has a $\hat{Z}_r$-filtration. Combining this with the equation

$$[V : \hat{L}_r(\lambda)] = \dim \mathrm{Hom}_{G_rT}(V, \hat{Q}_r(\lambda)),$$

valid for any finite dimensional G_rT-module V, we are done because of (9.2). $\square$

The equality in Proposition 9.5 is known as the *Brauer–Humphreys reciprocity law*. It shows that Problem 8.9 (or 3.11) is equivalent to the following:

Problem 9.7 *Determine* ch $\hat{Q}_r(\lambda)$ *for all* $\lambda \in X(T)$.

Corollary 9.8 $L((p^r - 1)\rho) = \hat{Z}_r((p^r - 1)\rho) = \hat{Q}_r((p^r - 1)\rho)$.

Proof: By Lemma 9.6(i) we have a nontrivial homomorphism from the module $\hat{Z}'_r(-(p^r - 1)\rho)^*$ to the module $\hat{Z}_r((p^r - 1)\rho)$. But the first (resp., second) module has $L((p^r - 1)\rho)$ as its unique simple quotient (resp., submodule), and so the first equality follows. The last equality is then a consequence of Proposition 9.5 (combined with the fact that all $\hat{Z}_r(\lambda)$ have the same dimension). $\square$

The irreducible G-module $L((p^r - 1)\rho)$ is known as the rth Steinberg module and also denoted St_r. By Corollary 9.8, it is an irreducible injective and induced (from $B_r T$) $G_r T$-module.

The standard argument shows that injectivity is preserved when tensoring by an arbitrary finite dimensional module. It is then easy to see that for any $\lambda \in X_r(T)$ we have an isomorphism of $G_r T$-modules,

$$\text{St}_r \otimes L((p^r - 1)\rho + w_0\lambda) \simeq \hat{Q}_r(\lambda) \oplus \left(\bigoplus_{\mu > \lambda} \hat{Q}_r(\mu)^{m_\mu} \right) \qquad (9.3)$$

for certain $m_\mu \geq 0$. In other words, $\hat{Q}_r(\lambda)$ appears as a distinguished $G_r T$-summand of this tensor product. For $p \geq 2(h - 1)$ it is known also to be a G-summand, whereas for smaller p the question of a G-structure on $\hat{Q}_r(\lambda)$ is still open; see [Ja, 11.11].

10 Kempf's vanishing theorem

The Frobenius morphisms F^r_G and the Steinberg modules St_r, $r \geq 0$, from the last sections are used here to prove the following.

Theorem 10.1 ([Ke1]) *Let* $\lambda \in X(T)^+$. *Then*

$$H^i(\lambda) = 0 \quad \text{for } i > 0.$$

Note that in characteristic zero this result is a special case of the Borel–Weil–Bott theorem 6.3. So, as in the previous two sections, we shall assume $\text{char}(k) = p > 0$.

Definition 10.2 Let V be a representation of G. Then we define $V^{(r)}$ to be the representation of G on the same vector space but with G-action given by $gv = F_G^r(g)v$, $g \in G$, $v \in V$.

We shall derive Theorem 10.1 from

Theorem 10.3 *Let E be a B-module. Then, for any $r \geq 0$, we have an isomorphism of G-modules*

$$H^i(E^{(r)} \otimes (p^r - 1)\rho) \simeq H^i(E)^{(r)} \otimes \mathrm{St}_r.$$

Proof: Since $H^0 = \mathrm{Ind}_B^G = \mathrm{Ind}_{G_r B}^G \circ \tilde{Z}_r$ and since $\tilde{Z}_r$ is an exact functor (8.2), which takes injective B-modules into acyclic modules for $\mathrm{Ind}_{G_r B}^G$ (in fact, into injective $G_r B$-modules, see Corollary 1.11(iii)), we have

$$H^i = H^i(G/G_r B, \tilde{Z}_r(-)).$$

Now $\tilde{Z}_r(E^{(r)} \otimes (p^r - 1)\rho) = E^{(r)} \otimes \tilde{Z}_r((p^r - 1)\rho) = E^{(r)} \otimes \mathrm{St}_r$, the first equality being the tensor identity for $\tilde{Z}_r$ and the second coming from Corollary 9.8. Hence, the tensor identity for $H^i(G/G_r B, -)$ gives

$$H^i(E^{(r)} \otimes (p^r - 1)\rho) \simeq H^i(G/G_r B, E^{(r)}) \otimes \mathrm{St}_r,$$

and we are done if we show that

$$H^i(G/G_r B, E^{(r)}) \simeq H^i(E)^{(r)}. \tag{10.1}$$

One checks this easily for $i = 0$. To do the obvious dimension shift, we need to check that $H^i(G/G_r B, I^{(r)}) = 0$ for $i > 0$ whenever I is an injective B-module. In fact, it is enough to consider $I = k[B]$, and since $k[B]^{(r)} \simeq \mathrm{Ind}_{G_r}^{G_r B}(k)$ we have $H^i(G/G_r B, k[B]^{(r)}) \simeq H^i(G/G_r, k)$. Induction from G_r to G is exact (because $G/G_r \simeq G$ is affine), and we are done. $\square$

Proof of Theorem 10.1: By Theorem 10.3 we have, for all $\lambda \in X(T)$,

$$H^i(\lambda)^{(r)} \otimes \mathrm{St}_r \simeq H^i(p^r(\lambda + \rho) - \rho),$$

and for $r \gg 0$, $\lambda \in X(T)^+$ the right hand side vanishes whenever $i > 0$. This is so because the line bundle $\mathcal{L}(\lambda + \rho)$ on G/B is ample when $\lambda \in X(T)^+$; see [Ha, III.5.2]. $\square$

Corollary 10.4 (Weyl's character formula)   *Let* $\lambda \in X(T)^+$. *Then*

$$\operatorname{ch} H^0(\lambda) = J(e^{\lambda+\rho})/J(e^\rho),$$

where J is the $\mathbb{Z}$-endomorphism on $\mathbb{Z}[X(T)]$ given by

$$J(e^\mu) = \sum_{w \in W} (-1)^{l(w)} e^{w(\mu)}, \qquad \mu \in X(T).$$

Proof: By Theorem 10.1 we have $\operatorname{ch} H^0(\lambda) = \sum_i (-1)^i \operatorname{ch} H^i(\lambda)$. The rest is standard; see, e.g., [Ja, II.5.10]. $\square$

11   Quantum groups

In this section we introduce quantum groups and briefly sketch how many of the preceding results in the representation theory for algebraic groups have quantized counterparts.

Let $(a_{ij})_{i,j=1,\dots,n}$ be a Cartan matrix, and choose a set of relatively prime integers $d_1,\dots,d_n$ such that $(d_i a_{ij})$ is symmetric. Fix also an indeterminate v, and set $v_i = v^{d_i}$, $i = 1,\dots,n$. If $m \in \mathbb{N}$ we write $[m] = (v^m - v^{-m})/(v - v^{-1})$ and $[m]! = [m][m-1]\cdots[2][1]$. For $t \in \mathbb{N}$, we set $\begin{bmatrix} m \\ t \end{bmatrix} = [m][m-1]\cdots[m-t+1]/[t]!$. We write $[m]_i$, $[m]_i!$, and $\begin{bmatrix} m \\ t \end{bmatrix}_i$ for the corresponding expressions with v replaced by v_i.

Definition 11.1   The quantum group attached to (a_{ij}) is the $\mathbb{Q}(r)$-algebra U with generators E_i, F_i, K_i, K_i^{-1}, $i = 1,\dots,n$, and relations
  i) $K_i K_i^{-1} = 1 = K_i^{-1} K_i$ and $K_i K_j = K_j K_i$
  ii) $K_i E_j K_i^{-1} = v_i^{a_{ij}} E_j, \quad K_i F_j K_i^{-1} = v_i^{-a_{ij}} F_j$
  iii) $E_i F_j - F_j E_i = \delta_{ij} \dfrac{K_i - K_i^{-1}}{v_i - v_i^{-1}}$
  iv) $\displaystyle\sum_{r+s=1-a_{ij}} (-1)^s \begin{bmatrix} 1 - a_{ij} \\ s \end{bmatrix}_i E_i^r E_j E_i^s = 0, \quad i \neq j$
  v) $\displaystyle\sum_{r+s=1-a_{ij}} (-1)^s \begin{bmatrix} 1 - a_{ij} \\ s \end{bmatrix}_i F_i^r F_j F_i^s = 0, \quad i \neq j$

It turns out that U is a Hopf algebra with comultiplication Δ given by

$$\Delta(K_i) = K_i \otimes K_i, \quad \Delta(E_i) = E_i \otimes 1 + K_i \otimes E_i, \quad \Delta(F_i) = F_i \otimes K_i^{-1} + 1 \otimes F_i,$$

coidentity ϵ given by

$$\epsilon(K_i) = 1, \qquad \epsilon(E_i) = \epsilon(F_i) = 0$$

and antipode S given by

$$S(K_i) = K_i^{-1}, \qquad S(E_i) = -K_i^{-1}E_i, \quad S(F_i) = -F_iK_i.$$

Set now $A = \mathbb{Z}[v, v^{-1}]$, and let U_A denote the A-subalgebra of U generated by $K_i, K_i^{-1}, E_i^{(m)}, F_i^{(m)}, \ i = 1, \ldots, n, \ m \in \mathbb{N}$, where $E_i^{(m)} = E_i^m/[m]_i!$ and similarly for $F_i^{(m)}$.

If q is a nonzero element in some field F, then F becomes a A-algebra by specializing v to q, and we define

$$U_{F,q} = U_A \otimes_A F.$$

Sometimes we write just U_F or U_q instead of $U_{F,q}$. By the preceding, U_F is a Hopf algebra. It is called the quantum group over F associated with (a_{ij}).

Of course, this construction makes sense also for any commutative ring F and $q \in F$ a unit. However, we shall here always assume that F is a field, and in fact for simplicity we only consider fields of characteristic zero. For instance, we may take $F = \mathbb{C}$ and let q be a nonzero complex number. By abuse of notation, we shall write $F_i^{(m)}$, etc., also for the elements in U_q corresponding to $F_i^{(m)}$, etc. $\in U_A$.

We have

$$U_q = U_q^- U_q^0 U_q^+,$$

where U_q^- (resp., U_q^+) is the subalgebra of U_q generated by $F_i^{(m)}$ (resp., $E_i^{(m)}$), $i = 1, \ldots, n, \ m \in \mathbb{N}$, and U_q^0 is the subalgebra generated by $K_i, K_i^{-1}, \left[\begin{smallmatrix} K_i; c \\ t \end{smallmatrix} \right], \ i = 1, \ldots, n, \ c, t \in \mathbb{N}$. Here, $\left[\begin{smallmatrix} K_i; c \\ t \end{smallmatrix} \right]$ is the element

$$\begin{bmatrix} K_i; c \\ t \end{bmatrix} = \prod_{j=1}^{t} \frac{K_i v_i^{c+1-j} - K_i^{-1} v_i^{-c-1+j}}{v_i^j - v_i^{-j}}.$$

Let R be the root system associated with (a_{ij}), and let $\alpha_1, \ldots, \alpha_n$ be the set of simple roots. Set X equal to the set of integral weights, i.e., $X = \mathbb{Z}^n$ where $\lambda \in X$ is an n-tuple $\lambda = (\lambda_1, \ldots, \lambda_n)$ with $\lambda_i = \langle \lambda, \alpha_i \rangle, \ i =$

$1, \ldots, n$. Such a weight gives rise to a character of U^0_A (and hence of U^0_q) via

$$\lambda(K_i) = v_i^{\lambda_i}, \qquad \lambda\left(\begin{bmatrix} K_i; c \\ t \end{bmatrix}\right) = \begin{bmatrix} \lambda_i + c \\ t \end{bmatrix}_i.$$

If V is a U^0_q-module and $\lambda \in U^0_q$, we set

$$V_\lambda = \{v \in V \mid uv = \lambda(u)v, u \in U^0_q\}.$$

It turns out (but this is a nontrivial fact; see, e.g., [APW, Section 9]) that any finite dimensional U_q-module V (of type **1**) is a sum of weight spaces (note that replacing v_i by $-v_i$ in the foregoing formulas defines an equally good character λ and type **1** simply means that we work with all signs positive). We shall consider a slightly more general category of U_q-modules. Namely, define $\mathfrak{C}_q$ to be the category consisting of all U_q-modules V that satisfy

i) $V = \bigoplus_{\lambda \in X} V_\lambda$.

ii) For every $v \in V$, we have $F_i^{(m)}v = E_i^{(m)}v = 0$ for all $m \gg 0$, $i = 1, \ldots, n$.

Note that if V is arbitrary, then

$$F(V) = \left\{ v \in \bigoplus_{\lambda \in X} V_\lambda \ \middle|\ E_i^{(m)}v = F_i^{(m)}v = 0, \ m \gg 0, \ i = 1, \ldots, n \right\}$$

is a U_q-submodule of V. By definition, $F(V) \in \mathfrak{C}_q$.

Let $\mathfrak{C}_q^-$ (resp., $\mathfrak{C}_q^0$) be the category obtained in the same way when we replace U_q by $U_q^- U_q^0$ (resp., U_q^0).

Definition 11.2   The induction functor

$$H_q^0 \colon \mathfrak{C}_q^- \to \mathfrak{C}_q$$

is defined by

$$H_q^0(V) = F(\mathrm{Hom}_{U_q^- U_q^0}(U_q, V)), \qquad V \in \mathfrak{C}_q^-.$$

Here, U_q is considered a $U_q^- U_q^0$-module via left multiplication, and the U_q-structure on $H_q^0(V)$ comes via right multiplication on U_q:

$$u \cdot f \colon x \mapsto f(xu), \qquad x, u \in U_q, \ f \in H_q^0(V).$$

It is straightforward to check that H_q^0 satisfies Frobenius reciprocity, etc. We leave to the reader the task of verifying that the results from Section 4 on remain true when we add index q.

One result, which is not clear how to quantize, is (4.4). We shall now indicate how one may derive the special case of this that we need from the classical case:

Let $\mathbb{Z}$ be the A-algebra obtained by specializing v to 1. Choose also $q \in k$ (where k is an algebraically closed field of characteristic $p > 0$ as in previous sections) such that q is a primitive lth root of unity (so $p \nmid l$). Specializing v to q, we have made k into an A-algebra.

Proposition 11.3   *There exists a Hopf algebra homomorphism* $\mathcal{F}\colon U_k \to U_{\mathbb{Z}} \otimes k$ *determined by*

$$\mathcal{F}(E_i^{(m)}) = \begin{cases} E_i^{(m/l)} \otimes 1, & \text{if } l | m, \\ 0, & \text{otherwise}, \end{cases}$$

$$\mathcal{F}(F_i^{(m)}) = \begin{cases} F_i^{(m/l)} \otimes 1, & \text{if } l | m, \\ 0, & \text{otherwise}, \end{cases}$$

$$\mathcal{F}(K_i^{\pm 1}) = K_i^{\pm 1} \otimes 1,$$

$i = 1, \ldots, n$.

Proof: [Lu1, Lu2]. $\square$

Let now $\bar{U}$ denote the enveloping algebra over $\mathbb{Q}$ for the semisimple Lie algebra corresponding to (a_{ij}). Denote the standard generators of $\bar{U}$ by $\bar{E}_i, \bar{F}_i$, $i = 1, \ldots, n$, and set $\bar{H}_i = [\bar{E}_i, \bar{F}_i]$. The Kostant $\mathbb{Z}$-form of $\bar{U}$ is the $\mathbb{Z}$-subalgebra of $\bar{U}$ generated by $\bar{E}_i^{(r)}, \bar{F}_i^{(r)}$, $i = 1, \ldots, n$, where $\bar{E}_i^{(r)} = \bar{E}_i^r / r!$ and similarly for $\bar{F}_i^{(r)}$; see [Hu1, Section 26].

Proposition 11.4   *There is a Hopf algebra isomorphism*

$$\bar{U}_{\mathbb{Z}} \xrightarrow{\sim} U_{\mathbb{Z}} / (\{K_i - 1 \mid i = 1, \ldots, n\})$$

given by

$$\bar{E}_i^{(r)} \mapsto \text{image of } E_i^{(r)},$$

$$\bar{F}_i^{(r)} \mapsto \text{image of } F_i^{(r)},$$

$$\bar{H}_i \mapsto \text{image of } \begin{bmatrix} K_i; 0 \\ 1 \end{bmatrix}.$$

Proof: [Lu1, Lu2]. □

Let now $\bar{\mathcal{F}}\colon U_k \to \bar{U}_{\mathbb{Z}} \otimes k$ be the composite obtained from Propositions 11.3 and 11.4. If we then start by a representation V of the algebraic group G over k corresponding to (a_{ij}), then this representation is also a module for the so-called hyperalgebra $\bar{U}_{\mathbb{Z}} \otimes k$. Hence V gives rise via $\bar{\mathcal{F}}$ to a module for U_k. We denote this U_k-module by $V^{[l]}$.

Imitating the arguments for (10.1) we obtain

Proposition 11.5 $H_q^i(V^{[l]}) \simeq H^i(V)^{[l]}$, $i \geq 0$.

We can now prove the quantized version of the special case $H_1 = B$, $H_2 = G$ of (4.4):

Theorem 11.6   *Let* $\lambda \in X(T)$. *Then* $H_q^i(\lambda) = 0$ *for* $i > N$.

Proof: Let u_k be the subalgebra of U_k generated by $E_i, F_i, K_i^{\pm 1}$, $i = 1, \ldots, n$. (This corresponds to $G_1 \subset G$.) Set $\tilde{Z}_q = \mathrm{Ind}_{U_k^- U_k^0}^{U_k^- U_k^G u_k} $. Then $\tilde{Z}_q$ is an exact functor, and we have

$$H_q^i(\lambda) = H^i(U_k/U_k^- U_k^0 u_k, \tilde{Z}_q(\lambda)).$$

Let $\tilde{L}_q(\mu)$ be a $U_k^- U_k^0 u_k$-composition factor of $Z_q(\lambda)$. Then $\tilde{L}_q(\mu) = L_q(\mu^0) \otimes l\mu^1$, where $\mu = \mu^0 + l\mu^1$ with $\mu^0 \in X_l$ (i.e., $0 \leq \mu_i^0 < l$, $i = 1, \ldots, n$). Hence, by Proposition 11.5, we have

$$H^i(U_k/U_k^- U_k^0 u_k, \tilde{L}_q(\mu)) \simeq L_q(\mu^0) \otimes H^i(\mu^1)^{[l]}.$$

Now (4.4) gives $H^i(\mu^1) = 0$ for $i > N$, and we are done. □

Remark   From Theorem 11.6, it is a standard base change argument to derive

$$H_A^i(\lambda) = 0 \qquad \text{for all } i > N.$$

It follows that the vanishing $H_q^i(\lambda) = 0$ for $i > N$ also holds for q any nonzero element in any field F.

Let us return to the case where $F = \mathbb{C}$ and $q \in \mathbb{C}\backslash\{0\}$. If q is not a root of unity, the Borel–Weil–Bott theorem holds for U_q and hence $L_q(\lambda) =$

$H_q^0(\lambda)$ for all $\lambda \in X(T)^+$. Moreover, in this situation $\mathfrak{C}_q$ is semisimple (as it follows from that theorem combined with quantized Serre duality, cf. Theorem 7.6).

So suppose q is a primitive lth root of unity. Then we have as an analog of Problem 3.11

Problem 11.7   *Determine* ch $L_q(\lambda)$, $\lambda \in X^+$.

Now Lusztig has conjectured that if $l = p \geq h$ (h denoting the Coxeter number of R), then

$$\text{ch } L_q(\lambda) = \text{ch } L(\lambda) \qquad \text{for all } \lambda \in X_p(T). \tag{11.1}$$

This conjecture has recently been proved ([AJS]) for $p \gg 0$. Moreover, work of Kazhdan and Lusztig [KL] relates Problem 11.7 to a similar problem for affine Lie algebras. Early in 1994 Kashiwara and Tanisaki announced a solution of this latter problem [KT], and hence Problem 11.7 is also solved (although there still seems to be some problem if (a_{ij}) is not symmetric) for all l.

The proof of (11.1) in [AJS] (for $p \gg 0$) involves the "dual" approach of injective modules, see Problem 9.7. We conclude this section with a few remarks on injective modules for U_q.

Set $\text{St}_l = L_q((l-1)\rho)$ (We are again in the situation where $q \in \mathbb{C}$ is an lth root of 1. For simplicity, we shall assume that l is odd.) Then St_l is injective in $\mathfrak{C}_q$ (this is in contrast with the situation for representations of G in characteristic $p > 0$, where no finite dimensional representation is ever injective). Considering $\text{St}_l \otimes L_q(\lambda)$, $\lambda \in X_l$, we obtain (compare Corollary 9.8)

Proposition 11.8   *For each $\lambda \in X_l$, the injective hull $Q_q(\lambda)$ of $L_q(\lambda)$ in $\mathfrak{C}_q$ is finite dimensional. In fact, $Q_q(\lambda)$ is a distinguished summand of $\text{St}_q \otimes L_q(w_0\lambda + (l-1)\rho)$.*

We also have a Brauer–Humphreys reciprocity law for the $Q_q(\lambda)$:

Definition 11.9   A U_q-module M is said to have a *good filtration* if there exists a filtration by U_q-submodules

$$0 = F_0 \subset F_1 \subset \cdots \subset F_s = M$$

with $F_i/F_{j-1} \simeq H_q^0(\lambda_i)$ for certain $\lambda_i \in X^+$, $i = 1, \ldots, s$. The number of times a given $H_q^0(\mu)$ occurs in such a filtration is denoted $[M : H_q^0(\mu)]$.

Proposition 11.10   *Let $\lambda \in X_l$. Then $Q_q(\lambda)$ has a good filtration and*

$$[Q_q(\lambda) : H_q^0(\mu)] = [H_q^0(\mu) : L_q(\lambda)].$$

12 Tilting modules

In this section we shall give a short account of the theory of tilting modules for algebraic groups, as well as for quantum groups. We will use the language of algebraic groups leaving it in most cases to the reader to quantize the results by adding the index q. Whenever this quantization is not straightforward, we have added a note or a remark.

We shall assume that G is a connected reductive group over an algebraically closed field k. The corresponding quantum group is U_q, where $q \in \mathbb{C}$ is an lth root of unity, l odd.

Definition 12.1   A finite dimensional representation M of G is called a tilting module if both M and M^* have good filtrations (see Definition 11.9).

(The dual module M^* in the quantum case has U_q-action given by $uf(m) = f(S(u)m)$, $u \in U_q$, $f \in M^*$, $m \in M$.)

Examples 12.2   Since $H^0(0) = k$, the trivial module is a tilting module. The same is true more generally whenever $H^0(\lambda)$ is irreducible (because then $H^0(\lambda)^* = L(\lambda)^* \simeq L(-w_0\lambda) = H^0(-w_0\lambda)$). Elementary weight considerations show that $H^0(\lambda)$ is irreducible for all p whenever λ is minuscule. By the linkage principle [An1], the same is true, e.g., for $\lambda = (p-1)\rho$ and for λ in the upper closure of the fundamental alcove $C = \{\nu \in X^+ \mid \langle \nu + \rho, \alpha^\vee \rangle < p \text{ for all } \alpha \in R_+\}$.

Let us denote by $\mathfrak{C}_t$ the category of tilting modules. It is easy to see that $\mathfrak{C}_t$ is closed under finite direct sums and under extensions (but not under formation of submodules and quotients). It is not trivial to see that it is also closed under tensor products:

Theorem 12.3 *If $M_1, M_2 \in \mathfrak{C}_t$, then also $M_1 \otimes M_2 \in \mathfrak{C}_t$. (The U_q-action on tensor products is obtained via the comultiplication $\Delta \colon U_q \to U_q \otimes U_q$.)*

This theorem was proved by Wang [Wa], Donkin [Do], and Mathieu [Ma]. A proof (taking care of also the quantum case) using Lusztig's canonical bases was written down by Paradowski [Pa].

Theorem 12.4 *i) For each $\lambda \in X(T)^+$, there exists an indecomposable tilting module $D(\lambda)$ with highest weight λ. Moreover, $\dim D(\lambda)_\lambda = 1$.*
ii) The modules $\{D(\lambda) \mid \lambda \in X(T)^+\}$ form a complete set of inequivalent indecomposable tilting modules.

Proof: (This theorem is due to Ringel [Ri] in a somewhat more general context.) We give a proof which only works for p not too small, namely we assume that $H^0(\omega) \in \mathfrak{C}_t$ for all fundamental weights ω. For $p > h + 6$ this is OK by Example 12.2, because then all fundamental weights ω belong to the fundamental alcove C).

(i) Write $\lambda = \sum_{i=1}^n a_i \omega_i$ where $\omega_1, \ldots, \omega_n$ are the fundamental weights. By our assumption and Theorem 12.3, $H^0(\omega_1)^{\otimes a_1} \otimes \cdots \otimes H^0(\omega_n)^{\otimes a_n}$ belongs to $\mathfrak{C}_t$, and since this module has highest weight λ and this occurs with multiplicity 1, we find $D(\lambda)$ as an indecomposable summand.

(ii) Let $M \in \mathfrak{C}_t$, and pick a maximal weight λ of M. Thus, we have a B-homomorphism $M \to \lambda$ (see Corollary 3.4), and by Frobenius reciprocity (Proposition 1.10) this gives a G-homomorphism $M \xrightarrow{\pi} H^0(\lambda)$. Since M has a good filtration, this must be a surjection. On the other hand, we also have a surjection $D(\lambda) \xrightarrow{\pi'} H^0(\lambda)$. Since $\mathrm{Ext}^1_G(H^0(\mu)^*, H^0(\lambda)) = 0$ (see Corollary 7.5), we obtain G-homomorphisms $\phi \colon M \to D(\lambda)$ and $\phi' \colon D(\lambda) \to M$ making the diagrams

$$
\begin{array}{ccc}
M & \xrightarrow{\;\pi\;} & H^0(\lambda) \\
{\scriptstyle \phi'}\big\uparrow\big\downarrow{\scriptstyle \phi} & & \big\| \\
D(\lambda) & \xrightarrow{\;\pi'\;} & H^0(\lambda)
\end{array}
$$

commutative. Since $\phi \circ \phi'$ is nonzero on the 1-dimensional space $D(\lambda)_\lambda$, we conclude that $D(\lambda)$ is a summand of M, and we are done. □

Examples 12.5   i) $D(0) = k$.

ii) $D((p-1)\rho) = \mathrm{St}_1$.

iii) $D(\lambda) = H^0(\lambda)$ for all $\lambda \in C$.

iv) Let $\lambda' \in C'$, where C' is the alcove obtained from reflecting C in the wall $H = \{\nu \in X \mid \langle \nu + \rho, \alpha_0^\vee \rangle = p\}$ (here, $\alpha_0 \in R_+$ is the highest short root — we are assuming that R is indecomposable). Then we have an exact sequence

$$0 \to H^0(\lambda) \to D(\lambda') \to H^0(\lambda') \to 0,$$

where $\lambda = s_{\alpha_0}.\lambda' + p\alpha$.

Problem 12.6   *Determine* ch $D(\lambda)$ *for all* $\lambda \in X(T)^+$.

To convince you that Problem 12.6 is an interesting (and hard) problem, we prove

Proposition 12.7   *Let* $\lambda \in X_l$. *Then* $Q_q(\lambda) = D_q(2(l-1)\rho + w_0\lambda)$.

(Contrary to our usual convention we have stated this result in its quantized version. The reason is that we have no finite dimensional injective G-modules — although it is possible to define G-modules that behave like the $Q_q(\lambda)$ by putting an upper bound on the weights allowed.)

Proof: Proposition 11.10 says that $Q_q(\lambda)$ has a good filtration. Since St_l is self-dual, we see that $Q_q(\lambda)^*$ is also injective (cf. the construction of $Q_q(\lambda)$ in Proposition 11.8), and hence $Q_q(\lambda)^*$ has also a good filtration. Thus, $Q_q(\lambda) \in \mathfrak{C}_t$. By construction, we furthermore see that $Q_q(\lambda)$ has highest weight $2(l-1)\rho + w_0\lambda$ and that this weight occurs with multiplicity 1. □

We conclude by some results whose quantized version play a role in the construction of invariants of 3-manifolds by Reshetikhin and Turaev [RT]. The key is

Theorem 12.8 (([An4, Theorem 3.4]))   *Let* $M \in \mathfrak{C}_t$, *and suppose* $f \in \mathrm{End}_G(M)$. *If* M *has no summands of the form* $D(\lambda)$ *where* $\lambda \in C$, *then* $\mathrm{Tr}(f) = 0$.

(In the quantum case, we (of course) have to use the quantum trace Tr_q. It is defined by $\mathrm{Tr}_q(f) = \mathrm{Tr}(K_{2\rho}f)$.)

Proof: By Theorem 12.4 we immediately reduce to the case $M = D(\lambda)$ with $\lambda \in X(T)^+\backslash C$. Moreover, since $D(\lambda)$ is indecomposable, any $f \in \mathrm{End}_G(D(\lambda))$ may be written $f = a \cdot \mathrm{Id} + f'$ for some $a \in k$ and some nilpotent $f' \in \mathrm{End}_G(D(\lambda))$. Since $\mathrm{Tr}(f') = 0$, we have reduced the theorem to the following statement:

$$\text{If } \lambda \in X(T)^+\backslash C, \text{ then } p|\dim D(\lambda). \tag{12.1}$$

By Weyl's dimension formula, [Hu1, Corollary 24.3], we have

$$\dim H^0(\mu) = \prod_{\alpha \in R_+} \frac{\langle \mu + \rho, \alpha^\vee \rangle}{\langle \rho, \alpha^\vee \rangle}.$$

In particular, we see that if μ is p-singular (i.e., $p|\langle \mu + \rho, \alpha^\vee \rangle$ for some $\alpha \in R_+$), then $p|\dim H^0(\mu)$ (we assume $\langle \rho, \alpha^\vee \rangle < p$ for all α). This observation gives (12.1) for all p-singular λ, because by the linkage principle $[D(\lambda) : H^0(\mu)] = 0$ unless $\mu \in W_p.\lambda$.

In general, $D(\lambda)$ is a summand of $D(\mu) \otimes V$ for some p-singular μ and some module V. Hence, we are done via the following lemma. $\square$

Lemma 12.9 *For any two finite dimensional G-modules M and N with M indecomposable, we have*

$$p|\dim M \Rightarrow p|\dim Q \qquad \text{for all } G\text{-summands } Q \text{ of } M \otimes N.$$

Proof: Exercise (Hint: look at $\mathrm{Tr}(\phi)$ for $\phi \in \mathrm{End}_G(M \otimes N)$). $\square$

Theorem 12.8 allows us to define a new tensor product on $\mathfrak{C}_t$. First, let for $M \in \mathfrak{C}_t$ the integers $(a_\lambda(M))_{\lambda \in X(T)^+}$ be determined by

$$M = \bigoplus_{\lambda \in X(T)^+} D(\lambda)^{a_\lambda(M)}.$$

Then $a_\lambda(M) = 0$ for all but finitely many $\lambda \in X(T)^+$.

Definition 12.10   Let $M_1, M_2 \in \mathfrak{C}_t$. The reduced tensor product is defined by

$$M_1 \underline{\otimes} M_2 = \bigoplus_{\lambda \in C} D(\lambda)^{a_\lambda(M_1 \otimes M_2)}.$$

Proposition 12.11 $\underline{\otimes}$ *is associative.*
Proof: Use Theorem 12.8. $\square$

For further results in this direction, see, e.g., [AP].

References

[An1]   H. H. Andersen, *The first cohomology group of a line bundle on G/B*, Invent. Math. **51** (1979), 287–296.

[An2]   _____, *The strong linkage principle*, J. reine angew. Math. **315** (1980), 53–59.

[An3]   _____, *The Frobenius morphism on the cohomology of homogeneous vector bundles on G/B*, Ann. of Math. (2) **112** (1980), 113–121.

[An4]   _____, *Tensor products of quantized tilting modules*, Comm. Math. Phys. **149** (1992), 149–159.

[AJS]   H. H. Andersen, J. C. Jantzen, and W. Soergel, *Representations of quantum groups at a p-th root of unity and of semisimple groups in characteristic p: independence of p*, Asterisque **220** (1994).

[AP]   H. H. Andersen and J. Paradowski, *Fusion categories arising from semisimple Lie algebras*, Comm. Math. Phys. **169** (1995), 563–588.

[APW]   H. H. Andersen, P. Polo, and K. Wen, *Representations of quantum algebras*, Invent. Math. **104** (1991), 1–59.

[Bo]   R. Bott, *Homogeneous vector bundles*, Ann. of Math. (2), **66** (1957), 203–248.

[Cu]   C. W. Curtis, *Representations of Lie algebras of classical type with applications to linear groups*, J. Math. Mech. **9** (1960), 307–326.

[De]   M. Demazure, *A very simple proof of Bott's theorem*, Invent. Math. **33** (1976), 271–272.

[Do]   S. Donkin, *Rational representations of algebraic groups*, Lecture Notes in Mathematics **1140**, Springer-Verlag, 1985.

[Ha]   R. Hartshorne, *Algebraic Geometry*, Graduate Texts in Mathematics **52**, Springer-Verlag, 1977.

[Hu1]   J. E. Humphreys, *Introduction to Lie Algebras and Representation Theory*, Graduate Texts in Mathematics **9**, Springer-Verlag, 1972.

[Hu2]   ———, *Linear Algebraic Groups*, Graduate Texts in Mathematics **21**, Springer-Verlag, 1975.

[Ja]   J. C. Jantzen, *Representations of Algebraic Groups*, Academic Press, 1987.

[KT]   M. Kashiwara and T. Tanisaki, *Characters of the negative level highest weight modules for affine Lie algebras*,Int. Math. Res. Not. 1994, 151–160.

[KL]   D. Kazhdan and G. Lusztig, *Tensor structures arising from affine Lie algebras*, J. Amer. Math. Soc. **6** (1993), 905–947, 949–1011; **7** (1994), 335–381, 383–453.

[Ke1]   G. Kempf, *Linear systems on homogeneous spaces*, Ann. of Math. (2), **103** (1976), 557–591.

[Ke2]   ———, *Representations of algebraic groups in prime characteristics*, Ann. Sci. Ecole Norm. Sup. **14** (1981), 61–74.

[Lu1]   G. Lusztig, *Finite dimensional Hopf algebras arising from quantized universal enveloping algebras*, J. Amer. Math. Soc. **3** (1990), 257–296.

[Lu2]   ———, *Introduction to quantum groups*, Progress in Mathematics, Birkhäuser, 1993.

[Ma]   O. Mathieu, *Filtrations of G-modules*, Ann. Sci. Ecole Norm. Sup. **23** (1990), 625–644.

[Pa]   J. Paradowski, *Filtrations of modules over quantum algebras*, Proc. Symp. Pure Math. **56** (1994), Part 2, 93–108.

[Ri]   C. M. Ringel, *The category of modules with good filtrations over a quasi-hereditary algebra has almost split sequences*, Math. Z. **208** (1991), 209–223.

[RT]   N. Reshetikhin and V. Turaev, *Invariants of 3-manifolds via link polynomials and quantum groups*, Invent. Math. **103**, 547–597.

[Sw]   M. Sweedler, *Hopf Algebras*, Benjamin, 1969.

[Wa]   J.-P. Wang, *Sheaf cohomology of G/B and tensor products of Weyl modules*, J. Alg. **66** (1982), 162–185.

Chapter 2

Orbital Varieties, Goldie Rank Polynomials and Unitary Highest Weight Modules

ANTHONY JOSEPH

Weizmann Institute of Science

CONTENTS

Lecture 1   Introductory lecture

These lectures combine a number of related topics inspired by the orbit method and a desire to complete our knowledge of the primitive spectrum of the enveloping algebra of a semisimple Lie algebra.

Let $\mathfrak{g}$ be a complex finite dimensional Lie algebra with enveloping algebra $U(\mathfrak{g})$. Its dual $\mathfrak{g}^*$ admits the action of an irreducible algebraic group G, and the orbit method suggests the existence of a bijection between $\mathfrak{g}^*/G$ and $\operatorname{Prim} U(\mathfrak{g})$. This works rather well for $\mathfrak{g}$ solvable [D], and it is now even known that the method of induction provides a homeomorphism

[Ma] of these topological spaces. For $\mathfrak{g}$ semisimple, however, this point of view is not as good. Basically, the greater noncommutativity of $\mathfrak{g}$ forces discrepancies between these objects, which are a priori unrelated.

If we consider that our basic aim is to determine $\operatorname{Prim} U(\mathfrak{g})$ along with some interesting (for example, unitary) modules, then it will be enough that the orbit method provides a (key) part of $\operatorname{Prim} U(\mathfrak{g})$. For example, the zero orbit in $\mathfrak{g}^*$ should correspond to the primitive ideals of finite codimension, which for $\mathfrak{g}$ semisimple is an infinite set containing nevertheless just one completely prime member. One could therefore just replace $\operatorname{Prim} U(\mathfrak{g})$ by $\operatorname{Prim}_c U(\mathfrak{g}) := \{J \in \operatorname{Prim} U(\mathfrak{g}) \mid U(\mathfrak{g})/J \text{ is a domain}\}$; however, it would be a mistake to ignore the structure imposed by the presence of the remaining ideals. This is indicated by the following discussion. Here and from now on, it will be assumed that $\mathfrak{g}$ is semisimple. Identify $\mathfrak{g}^*$ with $\mathfrak{g}$ through the Killing form. Call $x \in \mathfrak{g}^*$ nilpotent if $\operatorname{ad}_{\mathfrak{g}} x$ is a nilpotent derivation. This property is preserved under the action of G, and so it is natural to define a nilpotent orbit in $\mathfrak{g}^*$ to be one consisting of just nilpotent elements.

Theorem 1.1 ([BBr2, J5, Vo3])   *For each $J \in \operatorname{Prim} U(\mathfrak{g})$, the associated variety $V(\operatorname{gr} J)$ of J is the closure of a nilpotent orbit.*

Thus, for each nilpotent orbit $\mathcal{O}$ in $\mathfrak{g}^*$ it makes sense to consider the subset $\mathbf{X}(\mathcal{O})$ of $\mathbf{X} := \operatorname{Prim} U(\mathfrak{g})$ of primitive ideals whose associated variety is $\bar{\mathcal{O}}$. Now for $J \in \mathbf{X}$ one may form $\operatorname{Fract}(U(\mathfrak{g})/J)$, which is a matrix algebra over a skew field. The rank of this matrix algebra is denoted $\operatorname{rk}(U(\mathfrak{g})/J)$ and called the Goldie rank of $U(\mathfrak{g})/J$. Rank 1 corresponds to $U(\mathfrak{g})/J$ being a domain or, equivalently, to P being completely prime.

Let $\mathfrak{h}$ denote a Cartan subalgebra of $\mathfrak{g}$. For each $\lambda \in \mathfrak{h}^*$ one may associate a simple highest weight module $L(\lambda)$ of highest weight $\lambda - \rho$ (where ρ is the sum of the fundamental weights), and hence a primitive ideal $J(\lambda) := \operatorname{Ann}_{U(\mathfrak{g})} L(\lambda)$. After Duflo [Du], the map $\lambda \mapsto J(\lambda)$ of $\mathfrak{h}^*$ into $\mathbf{X}$ is *surjective*. In other words, the elements of $\mathbf{X}$ (and of $\mathbf{X}(\mathcal{O})$) are parametrized by $\mathfrak{h}^*$. One may ask, how does the Goldie rank depend on this parametrization? The answer is that this dependence is essentially a polynomial one; that is, there is a finite set of polynomials the values of which determine these Goldie ranks. Moreover, for each $\mathbf{X}(\mathcal{O})$ it is appropriate to select a subset of these polynomials. Remarkably, this subset forms a basis for an irreducible representation of the Weyl group W for the pair $(\mathfrak{g}, \mathfrak{h})$, and this representation can be identified with the Springer representation attached to $\mathcal{O}$ defined by quite different (geometrical) considerations. It would clearly be almost a ridiculous coincidence if just one of these polynomials took the value 1 and moreover just once in the appropriate parameter set.

These considerations will be made more precise in Lecture 3. For the moment, let us just stress a few salient points.

1) Up to scale factors, the Goldie rank polynomials are known and indeed given by a simple formula involving the Kazhdan–Lusztig polynomials (or rather their values at $q = 1$).

2) The scale factors can be largely determined just by knowing that $\mathbf{X}(\mathcal{O}) \cap \operatorname{Prim}_c U(\mathfrak{g}) \neq \emptyset$ for each $\mathcal{O}$. Hence an interest in the foregoing question. So far, one can really only check this case by case. In other words, we have no satisfactory general theory for constructing $\operatorname{Prim}_c U(\mathfrak{g})$.

3) The Goldie rank polynomials are intimately related to the geometry of the nilpotent orbits via what was at first just a strange coincidence with the Springer theory. However, one can now directly attach to a nilpotent orbit polynomials that are not quite the same, but that span the same space. These are the characteristic polynomials of the orbital varieties attached to a given orbit. It turns out that these are given by exactly the same formulae, but involving geometric analogs of the Kazhdan–Lusztig polynomials that can be and are different. This gives an entirely new twist to the orbit method and insight into geometry from representation theory rather than vice versa. For example, one can conjecture that the inclusion relation of orbital variety closure is determined by these geometric Kazhdan–Lusztig polynomials (the values of these are as yet unknown) in a manner analogous to the known ordering [J1, Vo1] in $\operatorname{Prim} U(\mathfrak{g})$.

4) The relationship between Goldie rank and characteristic polynomials allows one to conclude that the latter (and hence the former) can be expressed as a positive linear combination of products of (distinct) positive roots. Consequently, Goldie rank 1 occurs only at small parameters. This is also true for annihilators of unitary modules, though the reason is not so clear. Yet D. A. Vogan [Vo2] has noted that if the simple quotient $U(\mathfrak{g})/J$ is a unitary module for $\mathfrak{g} \times \mathfrak{g}$ (in the appropriate sense), then it is necessarily a domain. This establishes at least

one link between unitarity and small parameters. In the case when $L(\lambda)$ is unitary, it need not be that $\operatorname{Ann} L(\lambda) \in \operatorname{Prim}_c U(\mathfrak{g})$; yet there is a natural bound (Theorem 8.2 and remarks following) on the Goldie rank, which fails completely in the nonunitary case.

So far we have gained at least one motivation for studying orbital varieties. Let us give their precise definition. Fix a triangular decomposition $\mathfrak{g} = \mathfrak{n}^- \oplus \mathfrak{h} \oplus \mathfrak{n}^+$, where $\mathfrak{n}^+$ (resp., $\mathfrak{n}^-$) denotes the span of the positive (resp., negative) root vectors. Set $\mathfrak{b}^\pm = \mathfrak{h} \oplus \mathfrak{n}^\pm$. Then, for each nilpotent orbit $\mathcal{O}$, the intersection $\mathcal{O} \cap \mathfrak{n}^+$ is a locally closed subset of $\mathfrak{n}^+$ and, therefore, may be expressed as a union of its irreducible components, which by definition are just the orbital varieties attached to $\mathcal{O}$. As we shall see in the next lecture, the Spaltenstein–Steinberg theory allows one to describe them more precisely as follows. Let B be a Borel subgroup corresponding to $\mathfrak{b}^+$. Then, for each $w \in W$, there is a unique orbital variety $V(w)$ the closure of which coincides with the B saturation set of $\mathfrak{n}^+ \cap w(\mathfrak{n}^+)$; that is, $\overline{V(w)} = \overline{\{b(\mathfrak{n}^+ \cap w(\mathfrak{n}^+))b^{-1} : b \in B\}}$. Moreover, the map $w \mapsto V(w)$ of W to the set $\mathcal{V}$ of orbital varieties is surjective. Finally, there is a unique nilpotent orbit $\mathcal{O}(w)$ the closure of which coincides with the G saturation set of $\mathfrak{n}^+ \cap w(\mathfrak{n}^+)$, and the map $w \mapsto \mathcal{O}(w)$ of W to the set $\mathcal{N}$ of nilpotent orbits is surjective. Thus, one has surjections $W \twoheadrightarrow \mathcal{V} \twoheadrightarrow \mathcal{N}$, and the orbital varieties attached to a given orbit are just fibers of the second map. The fibers of the first map are called the geometric cells of W. They differ ever so slightly from the left cells of W defined by the theory of primitive ideals.

One may note that an orbital variety is Lagrangian and therefore a natural subject for quantization, a process that may be interpreted as follows. Can one give the algebra of regular functions $A[V]$ on the closure V of an orbital variety a $U(\mathfrak{g})$-module structure that at least preserves (up to a shift) its $\mathfrak{h}$-module structure? A more demanding version of this question is whether one can find a highest weight module $N(\lambda)$ such that $\operatorname{gr} \operatorname{Ann}_{U(\mathfrak{n}^-)} e_\lambda$ coincides with the ideal of definition of V. This is quite possibly so, and also one would expect $N(\lambda)$ and in particular λ to be completely determined (modulo some obvious reservations). A different but related approach is to study the algebra $D[V]$ of differential operators on V. Obviously, A is a faithful D-module, and so one can ask if there is a natural map of $U(\mathfrak{g})$ into D making A a $U(\mathfrak{g})$-module of not worse than finite length? Since V can be singular, it is not immediate (and not known) if $A[V]$ is simple as a $D[V]$-module.

These questions have particularly nice answers when V lies in the nilradical $\mathfrak{m}^+$ of a parabolic subalgebra $\mathfrak{p}$ of $\mathfrak{g}$ with $\mathfrak{m}^+$ commutative. Then $A[V]$ has even the structure of a unitary highest weight module. Moreover, except for easily described special cases, $D[V]$ coincides with $U(\mathfrak{g})/\operatorname{Ann} A[V]$. Conversely, every unitary highest weight module L is associated to a parabolic subalgebra with a commutative nilradical $\mathfrak{m}^+$, and remarkably $\operatorname{Ann}_{U(\mathfrak{m}^-)} L$ is a prime ideal and hence the ideal of definition of the closure of an orbital variety contained in $\mathfrak{m}^+$. The proof of these results will be indicated in Lectures 3–6. Some further perspectives including the use of quantum groups and of differential operators are given in Lectures 7 and 8.

Lecture 2   The geometry of orbital varieties

A key observation of the orbit method of Kirillov–Kostant–Souriau is that a G-orbit in $\mathfrak{g}^*$ admits the structure of a G-equivariant symplectic variety. Let us brush aside the geometric sophistication for the moment and consider what this means in just algebraic terms. Recall that $S(\mathfrak{g})$ identifies with the algebra of polynomial (and hence regular) functions on $\mathfrak{g}^*$. It is a commutative algebra and admits a Poisson bracket $\{\ ,\ \}$, which we recall for any algebra A is a bilinear antisymmetric map $A \times A \longrightarrow A$ satisfying the Jacobi identity such that for each $a \in A$ the map $b \mapsto \{a, b\}$ is a derivation of A. Clearly, $\{\ ,\ \}$ is defined by its value on generators, and for $S(\mathfrak{g})$ one just takes the Lie product on $\mathfrak{g}$. A basic question is to determine the prime Poisson ideals of A, namely those prime ideals I satisfying $\{A, I\} \subset I$. Notice one can take $\{\ ,\ \}$ to be the commutator bracket, and this is one reason that their study for $S(\mathfrak{g})$ and for $U(\mathfrak{g})$ are supposed to be related. On the other hand, $S(\mathfrak{g})$ being commutative gives rise to ideals that are not obviously present in $U(\mathfrak{g})$, that is, the ideal I of definition of any closed subvariety V of $\mathfrak{g}^*$. If the variety in question is irreducible and G-invariant, then I is prime and Poisson. An important special case is when V is the closure of a G-orbit $\mathcal{O}$. Then, the Poisson bracket gives a nondegenerate antisymmetric bilinear form on the cotangent space $T^*_{x,V}$ to any point $x \in \mathcal{O}$. Indeed, for all $a \in A/I$, $x \in \mathcal{O}$, define $d_x a = a - a(x) \bmod \mathfrak{m}_x^2$, where $\mathfrak{m}_x$ is the ideal of definition of $\{x\}$. Then $d_x a \in T^*_{x,V}$, and $d_x a \times d_x b \mapsto \{a, b\}(x)$ is the required form.

The construction of prime (or primitive) ideals in $U(\mathfrak{g})$ follows a quite different path. For the latter, one constructs a simple $U(\mathfrak{g})$-module M. In terms of growth rate, M should be one-half the size of $U(\mathfrak{g})/\operatorname{Ann} M$. Thus, a geometric equivalent should be a subvariety of the corresponding orbit, in the sense of Theorem 1.1 having half the dimension. It turns out that there is a natural candidate, namely the variety of zeros of $\operatorname{gr}_{\operatorname{Ann} U(\mathfrak{g})} m$, which is independent of the choice of $0 \neq m \in M$. (See exercises at the end of this lecture.) It is called the associated variety $V(M)$ of M and is trivially contained in $V(\operatorname{Ann} M)$. Moreover, by a deep result of O. Gabber [G], it is an involutive subvariety of $V(\operatorname{Ann} M)$, and this remembers to a large extent the $U(\mathfrak{g})$ module structure of M. Here, being involutive can be expressed by saying that the ideal $I(M)$ of definition of $V(M)$, namely $\sqrt{\operatorname{gr}_{\operatorname{Ann} U(\mathfrak{g})} m}$, satisfies $\{I(M), I(M)\} \subset I(M)$. This is equivalent to $(d_x a, d_x b) := \{a, b\}(x) = 0$ for all $a, b \in I(M)$, $x \in V(M)$. Thus, the subspace $\{d_x a : a \in I(M)\}$ of $T^*_{x, V(\operatorname{Ann} M)}$ is isotropic and so can have at most half the dimension. Its orthogonal in $T_{x, V(\operatorname{Ann} M)}$ identifies with $T_{x, V(M)}$, and this forces $\dim V(M) \geq \frac{1}{2} \dim V(\operatorname{Ann} M)$. So far no appeal to the proposed simplicity of M has been made, and this should provide the reverse inequality.

One calls a subvariety of an orbit $\mathcal{O}$ (or more generally, of a symplectic variety) Lagrangian if it is involutive and of exactly half the dimension. The preceding discussion suggests that one should be able to attach to a Lagrangian subvariety $\mathcal{L}$ a simple module M satisfying $\operatorname{Ann} M \in \mathbf{X}(\mathcal{O})$ and with associated variety $\mathcal{L}$. However, it is too much to expect a bijective map, as involutivity cannot be expected to capture the full meaning of M being a $U(\mathfrak{g})$-module, nor need a simple module necessarily have the correct dimension. On a finer level, modules may differ appreciably in size and yet still have the same associated variety because the latter only sees the radical of $\operatorname{gr}_{\operatorname{Ann} U(\mathfrak{g})} m$.

Recall $V(w)$ of Lecture 1. The involutivity of $V(w)$ has an easy proof based on the observation that the orthogonal of $\mathfrak{n}^+ \cap w(\mathfrak{n}^+)$ in $\mathfrak{n}^-$ is a subalgebra. That $V(w)$ is Lagrangian results from the Spaltenstein–Steinberg equality (see Theorem 2.1), which asserts that $\dim(\mathcal{O} \cap \mathfrak{n}^+) = \frac{1}{2} \dim \mathcal{O}$ for any nilpotent orbit $\mathcal{O}$.

A deeper analysis of orbital varieties results [St] from considering the Steinberg variety $\mathcal{S}$. This is defined as follows. Let $\mathcal{U}$ denote the set of all unipotent elements of G, and $\mathcal{B}$ the set of all Borel subgroups of G.

By definition, $\mathcal{S} := \{(u, B_1, B_2) \in \mathcal{U} \times \mathcal{B} \times \mathcal{B} \mid u \in B_1 \cap B_2\}$. For each conjugacy class $\mathcal{C} \subset \mathcal{U}$, let $\mathcal{S}(\mathcal{C})$ denote the inverse image of $\mathcal{C}$ in $\mathcal{S}$ under projection onto the first component. Obviously, $\mathcal{S}$ is a disjoint union of the $\mathcal{S}(\mathcal{C})$, and since the number of conjugacy classes is finite, this is a finite union.

Under conjugation, $\mathcal{B}$ identifies with G/B for some fixed Borel subalgebra B of G. Set $gBg^{-1} = {}^gB$ for all $g \in G$. Bruhat decomposition implies that under diagonal action, the G orbits Z_w in $\mathcal{B} \times \mathcal{B}$ have as representatives the $(B, {}^wB) : w \in W$. Set $\mathcal{S}_w(\mathcal{C}) = \{(u, B_1, B_2) \in \mathcal{S}(\mathcal{C}) \mid (B_1, B_2) \in Z_w\}$. These form a finite disjoint union of $\mathcal{S}(\mathcal{C})$. Set $U = B \cap \mathcal{U}$ and $U_w = U \cap {}^wU$. The fibers of the projection $\mathcal{S}_w(\mathcal{C}) \longrightarrow \mathcal{B} \times \mathcal{B}$ are all isomorphic to $\mathcal{C} \cap U_w$. Since Z_w identifies with $G/(B \cap {}^wB)$, it follows that $\dim \mathcal{S}_w(\mathcal{C}) = \dim(G/B \cap {}^wB) + \dim(\mathcal{C} \cap U_w) \leq \dim G - \operatorname{rk} G$, with equality if and only if $\mathcal{C} \cap U_w$ is dense in U_w, equivalently if $\overline{(GU_w)} = \bar{\mathcal{C}}$. We denote the subset of $w \in W$ satisfying this last condition by $\operatorname{St}^{-1}(\mathcal{C})$. In this case, $\mathcal{C} \cap U_w$ is irreducible. Now $\mathcal{S}_w(\mathcal{C})$ can be identified with the set of G translates of the fiber $\mathcal{C} \cap U_w$ over $(B, {}^wB)$. Since G is irreducible, it follows that so is $\mathcal{S}_w(\mathcal{C})$. Its closure is hence an irreducible component of $\mathcal{S}(\mathcal{C})$ when $w \in \operatorname{St}^{-1}(\mathcal{C})$.

At least for characteristic zero, the Bala–Carter–Dynkin theory gives a specific procedure for determining for each conjugacy class $\mathcal{C}$ an element $w \in W$ such that $\mathcal{C} \cap U_w$ is dense in U_w. The idea is as follows. Consider $e \in \mathfrak{g}$ nilpotent. By the Jacobson–Morosov theorem, there exists an $\mathfrak{sl}(2)$ triple (e, h, f) containing e. Let

$$\mathfrak{g} = \bigoplus_{i \in \mathbb{Z}} \mathfrak{g}_i$$

be the decomposition of $\mathfrak{g}$ into its $\mathfrak{h}$ eigenspaces with $\mathfrak{g}_i$ corresponding to h-eigenvalue i. Now $[e, \mathfrak{g}_0] = \mathfrak{g}_2$ from $\mathfrak{sl}(2)$ theory, and so $\dim \mathfrak{g}_2 < \dim \mathfrak{g}_0$ implies $\mathfrak{g}_0^e \neq 0$. Moreover, $\mathfrak{g}_0^e = \mathfrak{g}_0^f$, and so this subalgebra is reductive. Let $\mathfrak{h}_0 \subset \mathfrak{g}_0^e$ be a Cartan subalgebra. Then $\mathfrak{g}^{\mathfrak{h}_0}$ is a proper Levi factor of $\mathfrak{g}$ containing e. This reduces the problem to a strictly lower rank case.

If $\dim \mathfrak{g}_2 = \dim \mathfrak{g}_0$, then $\mathfrak{g}_1 = 0$ by the Bala–Carter–Dynkin theorem. This was at first checked case by case, but then J. C. Jantzen found the following "one-line" proof. Consider the Dynkin parabolic $\mathfrak{p} = \bigoplus_{i \geq 0} \mathfrak{g}_i$, whose nilradical is $\mathfrak{m} = \bigoplus_{i \geq 1} \mathfrak{g}_i$. By Richardson's theorem (which is also proved using Bruhat decomposition), the corresponding parabolic group

P has a dense orbit in $\mathfrak{m}$. In Lie algebra terms, this means that there exists $x \in \mathfrak{m}$ such that $[\mathfrak{p}, x] = \mathfrak{m}$. Decomposing x into h eigenvectors x_i, one obtains $[\mathfrak{g}_0, x_1 + x_2] + [\mathfrak{g}_1, x_1] = \mathfrak{g}_1 \oplus \mathfrak{g}_2$. The first term on the left hand side has dimension $\leq \dim \mathfrak{g}_0 = \dim \mathfrak{g}_2$, whereas the second term has dimension $< \dim \mathfrak{g}_1$ unless $\mathfrak{g}_1 = 0$. Hence the assertion.

On the other hand, setting $\mathfrak{m}' = \bigoplus_{i \geq 2} \mathfrak{g}_i$ one checks again from $\mathfrak{sl}(2)$ theory that $\dim Ge = \dim \mathfrak{g} - \dim \mathfrak{g}^e = \dim \mathfrak{m} + \dim \mathfrak{m}'$. Yet $Pe \subset \mathfrak{m}'$ and $\dim Pe \geq \dim Ge - \dim \mathfrak{m} = \dim \mathfrak{m}'$, and so Pe is dense in $\mathfrak{m}'$. In the present case $\mathfrak{g}_1 = 0$, so that Pe is dense in $\mathfrak{m}$. Finally, for an appropriate choice of $w \in W$, one has Lie $U_w = \mathfrak{m}$.

There is a second natural way to obtain irreducible components of $\mathcal{S}(\mathcal{C})$. Fix $u \in \mathcal{C}$, and set $\mathcal{B}_u = \{B \in \mathcal{B} \mid u \in B\}$. An argument of N. Spaltenstein [S] based on Bruhat decomposition (which we use later in a related context) shows how to move from component to component of $\mathcal{B}_u$. From this, one concludes that $\mathcal{B}_u$ is equidimensional. Now let X be a Stab$_G \, u$ orbit of an irreducible component of $\mathcal{B}_u \times \mathcal{B}_u$, hence in particular a finite union of pairs of irreducible components of $\mathcal{B}_u$. Let Y be the inverse image of X under the continuous map $(gug^{-1}, B_1, B_2) \mapsto (g^{-1}B_1, g^{-1}B_2)$ of $\mathcal{S}(\mathcal{C})$ onto $\mathcal{B}_u \times \mathcal{B}_u$. Then Y is closed, irreducible, and has dimension $\dim \mathcal{C} + \dim X = \dim \mathcal{C} + 2 \dim \mathcal{B}_u$. Since $\mathcal{S}(\mathcal{C})$ is a finite irredundant union of such Y, the latter run over its irreducible components.

On the other hand, we have seen that at least one such component has dimension equal to $\dim G - \operatorname{rk} G$. Consequently, every component has this dimension and furthermore,

$$2 \dim \mathcal{B}_u + \dim \mathcal{C} = \dim G - \operatorname{rk} G. \tag{2.1}$$

A further comparison of these two descriptions of irreducible components shows that the Stab$_G \, u$ orbits of an irreducible component of $\mathcal{B}_u$ are the closures of the $C_w := \{{}^g B \mid \forall \, g \in G \text{ such that } {}^g(B \cap {}^w B) \ni u\} : w \in$ St$^{-1}(\mathcal{C})$.

Finally, let $V_1, V_2, \ldots, V_n$ be the irreducible components of $\mathcal{C} \cap U = \mathcal{C} \cap B$, and $C_1, C_2, \ldots, C_m$ the irreducible components of $\mathcal{B}_u$. Consider the surjective morphisms $\pi_1(g) = g^{-1}ug$ of G onto $\mathcal{C}$ and $\pi_2(g) = {}^g B$ of G onto $\mathcal{B}$. Identifying $\mathcal{B}$ with G/B identifies $\pi_2^{-1}(C_i)$ with $C_i B$. Since C_i and B are irreducible, so is $\pi_2^{-1}(C_i)$, and so these sets form the irreducible components of $\pi_2^{-1}(\mathcal{B}_u) = \pi_1^{-1}(\mathcal{C} \cap U)$. Thus, the $V_i : i = 1, 2, \ldots, n$ are among the

$\pi_1\pi_2^{-1}(C_j) : j = 1, 2, \ldots, m$. Moreover, $\pi_1\pi_2^{-1}(C_i)$ is independent of the choice of C_i in its Stab_G u-orbit. Thus, the V_i are among the closures of the $\pi_1(\pi_2^{-1}(C_w)) = {}^B(U_w \cap \mathcal{C})$.

Notice that from the definition of π_1, π_2 one has

$$\dim C_j + \dim B = \dim \pi_1\pi_2^{-1}(C_j) + \dim \text{Stab}_G u. \qquad (2.2)$$

Combined with (2.1) and the equidimensionality of $\mathcal{B}_u$, this implies that $U \cap \mathcal{C}$ is equidimensional and of dimension $\dim(U \cap \mathcal{C}) = \dim \mathcal{B}_u + \dim B - \dim \text{Stab}_G u = \frac{1}{2}\dim\mathcal{C}$.

The preceding result may be summarized by the following.

Theorem 2.1 *Let $\mathcal{O}$ be a nilpotent orbit of a complex semisimple Lie algebra $\mathfrak{g}$ with triangular decomposition $\mathfrak{g} = \mathfrak{n}^+ \cap \mathfrak{h} \cap \mathfrak{n}^-$. Then the irreducible components of $\mathfrak{n}^+ \cap \mathcal{O}$ are Lagrangian subvarieties of $\mathcal{O}$ and take the form $V(w) := \overline{{}^B(\mathfrak{n}^+ \cap w(\mathfrak{n}^+))} \cap \mathcal{O} : w \in \text{St}^{-1}(\mathcal{O})$.*

Note that $V(w) \not\subset {}^B(\mathfrak{n}^+ \cap w(\mathfrak{n}^+))$. This point must be carefully taken into account in what follows.

Spaltenstein's operation on the components of $\mathcal{B}_u$ caries over to the components of $\mathfrak{n}^+ \cap \mathcal{O}$ and takes the following form. Let π be the set of simple roots defined with respect to the foregoing triangular decomposition.

Let V be a component of $\mathfrak{n}^+ \cap \mathcal{O}$, and take $\alpha \in \pi$. Let $\mathfrak{p}_\alpha$ denote the parabolic subalgebra $\mathfrak{b}^+ \oplus \mathfrak{g}_{-\alpha}$, and $\mathfrak{m}_\alpha$ its nilradical. Let P_α be the connected subgroup of G with Lie algebra $\mathfrak{p}_\alpha$. The τ-invariant of V is defined to be

$$\tau(V) := \{\alpha \in \pi \mid V \subset \mathfrak{m}_\alpha\}.$$

Since $\mathfrak{m}_\alpha$ is P_α-stable, it follows that $\alpha \in \tau(V(w))$ if and only if $\mathfrak{n}^+ \cap w(\mathfrak{n}^+) \subset \mathfrak{m}_\alpha$, which in turn is equivalent to $w^{-1}\alpha \in -\mathbb{N}\pi$ or that $s_\alpha w < w$ for the Bruhat order. In particular, V is P_α stable if $\alpha \in \tau(V)$.

Suppose $\alpha \notin \tau(V)$. Let us show that

$$\mathfrak{m}_\alpha \cap {}^{P_\alpha}(V(w)) \subset V(w) \cup V(s_\alpha w) \cup \left(\bigcup_{z \in w \mid s_\alpha z < z} V(z) \right). \qquad (2.3)$$

Set $V_1(w) = {}^B(\mathfrak{n}^+ \cap w(\mathfrak{n}^+))$. By Bruhat decomposition and the hypothesis, one has $P_\alpha wB = BwB \cup Bs_\alpha wB$. Then, since $\mathfrak{m}_\alpha$ is P_α stable,

$$\mathfrak{m}_\alpha \cap {}^{P_\alpha}(V_1(w)) = {}^{P_\alpha}(\mathfrak{m}_\alpha \cap w(\mathfrak{n}^+))$$
$$= m_\alpha \cap ({}^B w(\mathfrak{n}^+) \cup {}^B(s_\alpha w)(\mathfrak{n}^+))$$
$$= (\mathfrak{m}_\alpha \cap V_1(w)) \cup V_1(s_\alpha w).$$

Set $V_2(w) = V_1(w) \cap \mathcal{O}$. Then $\mathfrak{m}_\alpha \cap {}^{P_\alpha}(V_2(w)) = (\mathfrak{m}_\alpha \cap V_2(w)) \cup V_2(s_\alpha w)$. Now consider $V(w) \setminus V_2(w)$. This has codimension ≥ 1 in $V(w)$ and is B-stable. Hence $\dim^{P_\alpha}(V(w) \setminus V_2(w)) \leq \dim V(w)$. Since $\mathcal{O} \cap \mathfrak{n}^+$ is equidimensional, it follows that $\mathfrak{m}_\alpha \cap {}^{P_\alpha}(V(w) \setminus V_2(w))$ is precisely those components of $\mathcal{O} \cap \mathfrak{n}^+$ that lie in $\mathfrak{m}_\alpha$ and whose union contains $V(w) \setminus V_2(w)$. Together with the first observation, this proves (2.3).

Notice it is not guaranteed that $V(s_\alpha w)$ is actually a component of $\mathfrak{n}^+ \cap \mathcal{O}$, as it may be of codimension ≥ 1. The contribution in the far right hand side of (2.3) may be quite sizable.

Starting from a given orbital variety V, the operation defined by (2.3) iterated sufficiently often eventually gives all the orbital varieties lying in the unique dense orbit $\mathcal{O}$ contained in GV. This was originally proved by Spaltenstein [S] using an argument of Steinberg involving Bruhat decomposition. It implies that $\mathcal{O} \cap \mathfrak{n}^+$ is equidimensional. A similar argument can be used to show that $\bar{\mathcal{O}} \cap \mathfrak{n}^+$ is equidimensional. Using the characteristic polynomials introduced in Lecture 3, we can follow this operation more closely. By this means, it is possible to generate interesting examples of what can occur on the right hand side of (2.3). However, a complete description is still an important open problem.

Exercises and results on associated varieties

Let $\{\mathcal{F}^m(U(\mathfrak{g}))\}_{m \in \mathbb{N}}$ be the canonical filtration [D, 2.3] of $U(\mathfrak{g})$. By the Poincaré–Birkhoff–Witt theorem, $\mathrm{gr}_\mathcal{F} U(\mathfrak{g})$ identifies with $S(\mathfrak{g})$. Let M be a finitely generated $U(\mathfrak{g})$-module with generators $m_1, m_2, \ldots, m_s$. Then M can be made a filtered module for $(U(\mathfrak{g}), \mathcal{F})$ through

$$\mathcal{F}^m(M) := \sum_{i=1}^{s} \mathcal{F}^m(U(\mathfrak{g}))m_i.$$

1) Show that

$$V(\mathrm{Ann}_{S(\mathfrak{g})} \, \mathrm{gr}_\mathcal{F} M) = V(\mathrm{gr}(\bigcap_{i=1}^{s} \mathrm{Ann}_{U(\mathfrak{g})} \, m_i)), \qquad (2.4)$$

and that both sides of this expression are independent of the choice of generators. (It is important in Lecture 3 to know that these assertions also hold in the scheme theoretic sense.)

2) One writes $V(M)$ for the variety in (2.4), and, by a slight abuse of
notation, $V(\mathrm{Ann}\,M)$ for $V(U(\mathfrak{g})/\mathrm{Ann}\,M)$. Show that

$$V(\mathrm{Ann}\,M) = V(\mathrm{gr}\,\mathrm{Ann}_{U(\mathfrak{g})}\,M) \supset \overline{GV(M)}. \qquad (2.5)$$

One knows that equality holds in (2.5) under certain circumstances.
Suppose M is homogeneous for Gelfand–Kirillov dimension, for example, if M is simple. Then a result of O. Gabber [LS, Appendix] asserts
that $V(M)$ is equidimensional. For further results on associated varieties, see [BE].

3) Show that $\mathrm{gr}_{\mathcal{F}}\,\mathrm{Ann}\,M$ prime implies that $\mathrm{Ann}\,M$ is completely prime
[J7, 8.1].

4) Assume $\mathfrak{g}$ semisimple. Given $J \in \mathrm{Prim}\,U(\mathfrak{g})$, show that $V(\mathrm{gr}_{\mathcal{F}}\,J)$ is
contained in the cone of nilpotent elements of $\mathfrak{g}^*$.

Lecture 3   Goldie rank and characteristic polynomials

Let us describe more precisely how $\mathrm{rk}\,U(\mathfrak{g})/J(\lambda)$ depends on $\lambda \in \mathfrak{h}^*$.
First some notation is needed. Let R (resp., R^+, π) denote the set of
nonzero (resp., positive, simple) roots relative to our triangular decompo-
sition of $\mathfrak{g}$. For each $\alpha \in R$, set $\alpha^\vee = 2\alpha/(\alpha, \alpha)$. The lattice of integral
weights is defined to be $P(R) := \{\lambda \in \mathfrak{h}^* \mid (\alpha^\vee, \lambda) \in \mathbb{Z}, \ \forall \ \alpha \in R\}$.
The Weyl group W leaves $P(R)$ invariant, and this action admits the
subset $P(R)^+ := \{\lambda \in P(R) \mid (\alpha, \lambda) \geq 0, \ \forall \ \alpha \in R^+\}$ of dominant
weights as a fundamental domain. Finally, a weight λ is called regular
if $(\alpha, \lambda) \neq 0, \ \forall \ \alpha \in R$. A regular dominant weight is just one which
belongs to the subset $P(R)^{++} := P(R)^+ + \rho$.

The fibers of the map $w \mapsto J(w\lambda)$ of W into $\mathrm{Prim}\,U(\mathfrak{g})$ are known
after W. Borho and J. C. Jantzen [BJ] to be independent of the choice of
$\lambda \in P(R)^{++}$ (with some degenerations for $\lambda \in P(R)^+$). They are called
the *left cells* of W. Their conjectured [J1] description in terms of the
Jordan–Hölder multiplicities $[M(y\lambda) : L(w\lambda)]$ was settled by D. A. Vogan
[Vo1], a result that was one of the motivations for the Kazhdan–Lusztig
polynomials [KL1] now known to determine these multiplicities. This result
leads naturally to a conjectural description [J4, 9.9] of the geometric cells of
W (Lecture 1), but one that is as yet unresolved. (There is unfortunately a
gap in the proof in [T, 2.11].) One calls the set of ideals $\{J(\lambda) : \lambda \in P(R)\}$

the integral fiber (of the Duflo map). Except in $\mathfrak{sl}(n)$, not all nilpotent orbits occur as associated varieties (cf. Theorem 1.1) of ideals in the integral fiber. Those that do coincide with the special orbits of G. Lusztig, which were defined in a rather different manner [Lu]. Notice that this already means that left cells and geometric cells must differ outside $\mathfrak{sl}(n)$.

If we let $a(w, y)$ denote the coefficients of the matrix inverse to $[M(y\lambda) : L(w\lambda)]_{y,w \in W}$ (known by [Jan1] to be independent of $\lambda \in P(R)^{++}$), then the formal character of $L(w\lambda)$ takes the form

$$ch \ L(w\lambda) = \sum_{y \in W} \frac{a(w, y)e^{y\lambda - \rho}}{\Pi_{\alpha \in R^+}(1 - e^{-\alpha})}. \tag{3.1}$$

It turns out that the required Goldie ranks are obtained by studying the asymptotics of the right hand side of (3.1). Indeed, fix $\nu \in P(R)^{++}$. Then it makes sense to consider $e^{(w\lambda, \nu)}$ as a genuine exponential and to expand the function

$$\sum_{y \in W} a(w, y)e^{(y\lambda, \nu)}$$

in powers of $(y\lambda, \nu)$. The first nonvanishing term in the expansion has the remarkable property that it factors as a product of two polynomials; namely, one has

$$\sum_{y \in W} a(w, y)(y\lambda, \nu)^m = p_w(\lambda)q_w(\nu). \tag{3.2}$$

This fact (observed by a student, D. King, of D. A. Vogan) resulted from the interpretation of the first factor $p_w(\lambda)$ as the required Goldie rank, namely, one has [J3, Theorem 5.1]

Theorem 3.1   *For all* $\lambda \in P(R)^+$, $w \in W$,

$$\text{rk}(U(\mathfrak{g})/J(w\lambda)) = p_w(\lambda).$$

Notice one can express this result by saying that

$$p_w = \sum_{y \in W} a(w, y)y^{-1}\rho^m \tag{3.3}$$

up to some scalar. These scalars are not yet completely known [J6]. However, since p_w takes integer values on $P(R)^+$, it must take integer values
on the whole of $P(R)$. Then the scalar would be determined by knowing
that it takes at least once the value 1. I conjectured this to be always true
[J7, 8.4(i)]. Parts (ii), (iii) of the cited conjecture have counterexamples
[B, T].

The factorization (3.2) has some further remarkable consequences,
which are quite easy to prove [J3, Section 2].

A) $\mathbb{Q}Wp_w$ is a simple W-module. Moreover, this simple module
occurs with multiplicity one in the space of homogeneous polynomials on $\mathfrak{h}^*$
of degree m and not in lower degree. To prove this, one should observe that
replacing ρ by $z\rho : z \in W$ in the expression (3.3) for p_w only changes the
latter by a scalar. This result generalizes an observation of I. G. Macdonald
[M]. We call such modules univalent. Not all W-modules have this property.

B) p_w is a W-harmonic function on $\mathfrak{h}^*$, that is, it satisfies the relation

$$\sum_{w \in W} p(w\mu + \nu) = |W|p(\nu), \qquad \text{for all} \ \ \mu, \nu \in \mathfrak{h}^*.$$

For this, one should recall the vanishing of the sum $\sum_{y \in W} a(w, y)(y\lambda, \nu)^n$
for $n < m$.

The second polynomial $q_w(\nu)$ occurring in the factorization (3.2) has
also an important interpretation and leads to a determination of the associated variety of $L(w\lambda)$. Formalizing this leads to the notion of the
characteristic polynomial of an orbital variety, and this has some important geometric consequences. This will be examined. First note, however,
after [J2, 5.4] one has $a(w, y) = a(w^{-1}, y^{-1})$ and so

C) $q_{w^{-1}} = p_w$ up to a scalar.

Let V be a closed irreducible $\mathfrak{h}$-stable subvariety of $\mathfrak{n}^+$. View its ideal
of definition $I(V)$ as an $\mathfrak{h}$-invariant ideal of $S(\mathfrak{n}^-)$. From the previous
discussion, it is natural to study the asymptotics of the formal character
of $S(\mathfrak{n}^-)/I(V)$. Indeed, fix $\nu \in P(R)^{++}$ and consider

$$\sum_{(\xi, \nu) \leq n} \dim(S(\mathfrak{n}^-)/I(V))_{-\xi}$$

as a function of n. From Hilbert–Samuel theory, it is a polynomial in n the
degree of which is the dimension of V. Moreover, the leading coefficient

$r_V(\xi)$ is a function of ξ, and the simple calculation following will show that this takes the form

$$r_V(\xi) = \frac{p_V(\xi)}{\Pi_{\alpha \in R^+}(\alpha, \xi)}$$

for some polynomial p_V on $\mathfrak{h}^*$, called the characteristic polynomial of V. If V is the associated variety of $L(w\lambda)$ and the latter is irreducible as in, say, $\mathfrak{sl}(n)$ by [Me1], then comparison with the previous analysis shows that

$$p_V = p_{w^{-1}}$$

up to a nonzero scalar. More generally, $p_{w^{-1}}$ is a sum with strictly positive rational coefficients of the characteristic polynomials of the components of the associated variety of $L(w\lambda)$.

The general form that p_V can take is easily deduced using hyperplane intersection [J8, Section 8]. Let $\mathfrak{n}^+ = \mathfrak{n}_1 \supsetneq \mathfrak{n}_2 \supsetneq \cdots = 0$ be any $\mathfrak{h}$-stable flag of $\mathfrak{n}^+$. It will be sufficient to suppose that $V \subset \mathfrak{n}_i$ and to relate r_V to $r_{V \cap \mathfrak{n}_{i+1}}$. Here, $V \cap \mathfrak{n}_{i+1}$ is interpreted scheme theoretically; that is, $V \cap \mathfrak{n}_{i+1}$ is a linear combination of the components V_j of $V \cap \mathfrak{n}_{i+1}$, each possibly occurring several times; then,

$$r_{V \cap \mathfrak{n}_{i+1}} = \sum_j r_{V_j}.$$

It remains to prove the following lemma. Let α_i be the weight of a vector in $\mathfrak{n}_i/\mathfrak{n}_{i+1}$.

Lemma 3.2 i) $r_{V \cap \mathfrak{n}_{i+1}} = r_V$, *if* $V \subset \mathfrak{n}_{i+1}$.

ii) $r_V = \frac{1}{\alpha_i} r_{V \cap \mathfrak{n}_{i+1}}$, *otherwise*.

Proof: Only (ii) is nontrivial. Write $\alpha = \alpha_i$, $\mathfrak{n} = \mathfrak{n}_i$, $\mathfrak{m} = \mathfrak{n}_{i+1}$, $I = I(V)$, and

$$R_V(\nu) = \sum_{\xi \in \mathbb{N}\pi} \dim(S(\mathfrak{n}^*)/I)_{-\xi} e^{(\xi, \nu)}$$

for all $\nu \in P(R)^{++}$. Set $J = I + S(\mathfrak{n}^*)x_{-\alpha}$. For each $\xi \in \mathbb{N}\pi$, choose a sub-space $V_{-\xi}$ of $(S(\mathfrak{n}^*)/I)_{-\xi}$ that under the canonical projection $S(\mathfrak{n}^*)/I \longrightarrow S(\mathfrak{n}^*)/J$ maps bijectively to $(S(\mathfrak{n}^*)/J)_{-\xi}$. Set $V = \bigoplus V_{-\xi}$, which as an $\mathfrak{h}$-module is isomorphic to $S(\mathfrak{m}^*)/I \cap S(\mathfrak{m}^*)$.

The hypothesis of (ii) implies that $x_{-\alpha}$ is a nonzero divisor in $S(\mathfrak{n}^*)/I$, and so the sum $\sum_k V x^k_{-\alpha}$ is direct and equals $S(\mathfrak{n}^*)/I$. One concludes that

$$R_V(\nu) = \sum_{k \in \mathbb{N}} \sum_{\xi \in \mathbb{N}\pi} (\dim \ V_{-\xi}) e^{(\xi - k\alpha, \nu)}$$

$$= \left(\frac{1}{1 - x^{-(\alpha, \nu)}} \right) R_{V \cap \mathfrak{m}}(\nu),$$

and hence the assertion of (ii) follows by a simple asymptotic analysis [J4, 2.3]. $\square$

This simple argument gives the following significant positivity result [J8, 8.6], not apparent from our previous explicit formula.

Corollary 3.3 *The Goldie rank polynomial p_w is a linear combination with positive rational coefficients of products of distinct positive roots.*

This helps [J8, 8.6] to locate completely prime primitive ideals. For example, one can show that $U(\mathfrak{g})/\operatorname{Ann} L(w\lambda) : \lambda \in P(R)^{++}$ is a domain if and only if $L(w\lambda)$ is induced from a one-dimensional representation of a parabolic subalgebra. Unfortunately, the restriction on λ is rather strong in this context. (See [J9, 5.3] for a slight improvement.)

Examples 1) Take $\mathfrak{g} = \mathfrak{sl}(4)$ and V the orbital variety closure with ideal of definition $\langle e_{21}, e_{43}, e_{31}e_{42} - e_{32}e_{41} \rangle$. Then $p_V = \alpha_1\alpha_3(\alpha_1 + 2\alpha_2 + \alpha_3)$.

 2) Take $\mathfrak{g} = \mathfrak{sl}(6)$ and V the orbital variety closure with ideal of definition $\langle e_{21}, e_{43}, e_{54}, e_{53}, e_{31}e_{42} - e_{32}e_{41}, \ e_{41}e_{52} - e_{42}e_{51}, \ e_{31}e_{52} - e_{32}e_{51} \rangle$. Then $p_V = \alpha_1\alpha_3\alpha_4(\alpha_3 + \alpha_4)[(\alpha_1 + \alpha_2)(\alpha_1 + 2\alpha_2 + 2\alpha_3 + \alpha_4) + (\alpha_2 + \alpha_3)(\alpha_2 + \alpha_3 + \alpha_4)]$.

 3) Take $\mathfrak{g} = \mathfrak{so}(7)$ and $\pi = \{\alpha_1, \alpha_2, \alpha_3\}$ with α_3 short. Then $\pi' = \{\alpha_2, \alpha_3\}$ is a simple system for a Lie algebra $\mathfrak{g}'$ of type $\mathfrak{so}(5)$. This admits a unique orbital variety of dimension 2, the closure V' of which has characteristic polynomial $p_{V'} = \alpha_3(\alpha_2 + \alpha_3)$. Set $V = \mathfrak{n}^+ + V'$, which is the closure of an orbital variety of $\mathfrak{g}$. Then $p_V = \alpha_3(\alpha_2 + \alpha_3)$. Neither $p_{V'}$ nor p_V can be proportional to Goldie rank polynomials attached to ideals in the integral fiber. However, in the $\mathfrak{so}(7)$ example one has the additional surprise that GV is the associated variety of a primitive ideal in the integral fiber. This in turn implies that the associated varieties of the corresponding highest weight modules cannot all

be irreducible and illustrates the discussion in (3) following Theorem 1.1. This example was developed from a note of T. Tanisaki, and we refer the reader to [J9, 8.6] for further details. Determining all such occurrences is still very much an open problem.

In general, p_V need not be W-harmonic, nor need it generate a simple W-module. To prove that these properties do hold when V is an orbital variety closure motivates a similar but independent presentation of these characteristic polynomials analogous to that already established for the Goldie rank polynomials. This is provided by [J8] using the geometric considerations sketched briefly in what follows.

Fix $\lambda \in \mathfrak{h}^*$, $h \in \mathfrak{h}$. One knows that expressions of the form

$$\sum_{y \in W} \frac{A(w,y)e^{(y\lambda)(h)}}{\prod_{\alpha \in R^+} \alpha(h)} \quad : A(w,y) \in \mathbb{Z}$$

can be obtained as integrals. Then the key factorization property should be obtained by reexpressing this integral in product form by some change of variables. Now, the orbital variety $V(w)$ can be viewed as the image under the moment map of the conormal

$$Y(w) := \{bwB, b(\mathfrak{n}^+ \cap w(\mathfrak{n}^+))b^{-1} : b \in B\}$$

in $T^*(G/B)$. Consequently, one should integrate over $Y(w)$.

To determine the relevant integrals, one appeals to the remarkable theory of M. F. Atiyah and R. Bott as developed more recently by N. Berline and M. Vergne [BGV] and W. Rossmann [R]. Essentially, the integrand is chosen so that Stokes theorem can be applied. The integral over $V(w)$ becomes $p_{V(w)}$ by [J8, 5.7]. Explicitly, this means that, up to a nonzero constant,

$$p_{V(w)} = \sum_{y \in W} A(w,y)y\rho^m, \tag{3.4}$$

where m is the least integer ≥ 0 for which such an expression is nonvanishing. This result was first obtained by W. Rossmann using Hotta's result [Ho] and answered a conjecture of [J4].

A key point in the required factorization is that the integral over a fiber of the moment map is *independent* of which fiber is chosen. This is because the integral represents the class of the fiber in homology [J8, 3.6], which through the G action is independent of the fiber chosen. Finally, it is possible to express the $A(w, y)$ in terms of Euler numbers [J8, 4.8] and in particular to show that $A(w, y) = A(w^{-1}, y^{-1})$. Then by the analog of (C) following Theorem 3.1, the $p_{V(w)}$ must form a basis for (Springer's) W-module $M_{\mathcal{O}(w)}$ attached to $\mathcal{O}(w)$ lying in the homology space of a fixed point set $\mathcal{B}_u : u \in \mathcal{O}(w)$. This *recovers* the aforementioned result of R. Hotta [Ho].

Set $A(w) = \sum_{y \in W} A(w, y) y$. As in (A) following Theorem 3.1, the factorization of the first nonvanishing term $(A(w)\lambda, \nu)^m$ implies that $M_{\mathcal{O}(w)}$ is invariant. In particular, an isomorphism $M_{\mathcal{O}(w)} \xrightarrow{\sim} M_{\mathcal{O}(w')}$ of W-modules must be an equality. Moreover, according to Springer's well-known result, one has

Theorem 3.4 *The $M_{\mathcal{O}} : \mathcal{O} \in \mathcal{N}$ are pairwise nonisomorphic.*

As first noted by Kazhdan–Lusztig [KL2], following a suggestion of Springer, this may be obtained from a study of the Steinberg variety (see also [T] and references therein). It gives the following important and a priori surprising result.

Corollary 3.5 *The $p_{\bar{V}} : V \in \mathcal{V}$ are linearly independent.*

Exercise Take $u \in V(w)$. Show that the fiber over u under the moment map $Y(w) \longrightarrow V(w)$ is just a $\mathrm{Stab}_G\, u$-orbit of an irreducible component of $\mathcal{B}_u$. (See Lecture 2.)

Except possibly in type $\mathfrak{sl}(n)$, the $A(w, y)$ and $a(w, y)$ do not quite coincide [J8, 4.9]. Thus, we obtain the following rather subtle version of the orbit method correspondence. Orbital varieties and primitive ideals satisfy exactly analogous theories, but the data that must be used to calculate their properties (viz. left cells or geometric cells, Goldie rank, or characteristic polynomials) differ slightly. Again the $A(w, y)$ contain data (for example, the characteristic polynomials) concerning *all* the nilpotent orbits, whereas $a(w, y)$ concern only those occurring in the integral fiber (Lusztig's special orbits). For example, take $\mathfrak{g} = \mathfrak{so}(5)$. Then $\mathfrak{g}$ admits four nilpotent orbits of dimensions 8, 6, 4, 0. Exactly the four-dimensional one does not occur in the integral fiber. The unique orbital variety attached to it has the characteristic polynomial given in example (3). Again using $\ell(\cdot)$ for reduced length and $\geq$ for Bruhat order, one has $a(w, y) = (-1)^{\ell(w)+\ell(y)}$ if $y \geq w$ and is zero otherwise. One concludes from Theorem 3.1 that the Goldie rank polynomials are proportional to 1, α_2, α_3, $\alpha_2\alpha_3(\alpha_2 + \alpha_3)(\alpha_2 + 2\alpha_3)$. In particular, the polynomial of example (3) does not appear. Yet it must appear in the right hand side of (3.4). Consequently, $A(w, y)$ must differ from $a(w, y)$ in this case. Further reasons that $A(w, y)$ and $a(w, y)$ must differ are given in [J8, 4.9].

Let us go into how the integrand is chosen in a little more detail. Let M be a C^∞ manifold with symplectic form σ, and x a vector field on M. If f is a smooth function on M, the compatibility condition we need is that $df + c(x)\sigma = 0$, where $c(x)$ denotes contraction by x. Then $d_x := d + c(x)$ is just the "covariant differential." One has $d_x(\sigma + f) = 0$, and so $d_x(\sigma + f)^n = 0$ for all integer $n \geq 0$. The integrand is chosen to be $\exp(\sigma + f) = \exp\sigma \, \exp f$, with the convention that only the term σ^n is retained if the integrand is over a variety Y of real dimension $2n$. Under suitable conditions (in particular that the set Ω of zeros of x is finite), the Atiyah–Bott–Berline–Vergne–Rossmann theory asserts that

$$\int_Y e^f \sigma^n = \sum_{\omega \in \Omega} e^{f(\omega)} n(\omega)$$

for certain real numbers $n(\omega)$.

In the present case we identify G/B with the compact real variety K/T, where K is a maximal compact subgroup (see Lecture 4) of G viewed as a real Lie group and T a maximal torus for K. Set $\mathfrak{t} = \mathrm{Lie}\,T$. Then $T^*(G/B)$ identifies as a real manifold with $K \times_T \mathfrak{t}^\perp$. With respect to the elements $x_\alpha : \alpha \in R$ of a Chevalley basis for $\mathfrak{k}_{\mathbb{C}}$, one has Cartisian coordinates $e_\alpha := i(x_\alpha + x_{-\alpha})$, $f_a := x_\alpha - x_{-\alpha} : \alpha \in R^+$ on $\mathfrak{t}^\perp$. Fix $h \in \mathfrak{h}$, $i\lambda \in \mathfrak{t}^*$ regular. Then one takes f to be the function on $K \times_T \mathfrak{t}^\perp$ defined by

$$f(k, a) = \langle h, k\lambda \rangle - \frac{1}{2} \sum_{\alpha \in R^+} (e_\alpha^2 + f_\alpha^2),$$

where $(e_\alpha, f_\alpha) : \alpha \in R^+$ denotes the coordinates of $ka : k \in K$ under the projection onto $\mathfrak{t}^\perp$ defined by the decomposition $\mathfrak{k} = \mathfrak{t}^\perp \oplus \mathfrak{t}$. This function satisfies the preceding (integrability) condition for $\sigma := \sigma_\lambda + \tau_h$, where σ_λ is the 2-form on K/T defined through $\sigma_\lambda(x, y)(kT) = \langle [x, y], k\lambda \rangle$, $\forall\, x, y \in \mathfrak{k}$, and where

$$\tau_h = \sum_{\alpha \in R^+} \frac{1}{i\alpha(h)} \, (de_\alpha \wedge df_\alpha).$$

Examples and exercises

Let π' be a subset of π, and set $R'^+ = \mathbb{N}\pi' \cap R^+$. One easily checks that there exists an orbital variety the closure V of which has characteristic polynomial $p_{\pi'} := \prod_{\alpha \in R'^+} \alpha$. According to the preceding theory, the W-module generated by p_V has a basis formed by the characteristic polynomials of the orbital varieties attached to the unique dense orbit in GV. Moreover, these polynomials are given by (3.4), though here the $A(w, y)$ are unfortunately not yet known. However, in low rank cases one can obtain these polynomials just by following the rules imposed by Spaltenstein's operation described in Lecture 2 in the following manner.

Let V be an orbital variety, and recall (Lecture 2) the definition of the τ-invriant $\tau(V)$.

1) Set $\pi' = \tau(V)$. Show that $p_{\pi'}$ divides $p_{\bar{V}}$.
2) Suppose $\alpha \notin \pi'$. Show that $s_\alpha p_{\bar{V}} - p_{\bar{V}}$ is a linear combination with nonnegative rational coefficients of characteristic polynomials of orbital varieties.
3) Assume there exists $\alpha \in \pi \setminus \pi'$, $\beta \in \pi'$ such that $(\alpha, \alpha) = (\beta, \beta) = -2(a, \beta)$. Show that there exists a unique irreducible component V' in $\mathfrak{m}_\alpha \cap {}^{P_\alpha}V$ such that $\alpha \in \tau(V')$, $\beta \notin \tau(V)$.

   For more details see [J4, Sect. 9].
4) Calculate $p_{\bar{V}}$ for each of the 26 orbital varieties V occurring in $\mathfrak{sl}(5)$ using the information provided by (1)–(3).

Lecture 4   Unitarizable highest weight modules

Let $\mathfrak{g}$ be a complex semisimple Lie algebra. The notion of a complex unitarizable $\mathfrak{g}$-module depends on the choice of a real form $\mathfrak{g}_0$ of $\mathfrak{g}$. Let $j: \mathfrak{g} \longrightarrow \mathfrak{g}$ denote complex conjugation with respect to $\mathfrak{g}_0$.

Let $\{x_\alpha : \alpha \in R;\ h_\beta := [x_\beta, x_{-\beta}] : \beta \in \pi\}$ be a Chevalley basis for $\mathfrak{g}$. This basis generates a real Lie subalgebra of $\mathfrak{g}$, and we let j_c denote complex conjugation with respect to this basis. The (compact) Chevalley antiautomorphism σ_c associated to this basis is defined by $\sigma_c(x_\alpha) = x_{-\alpha}$. Define an involution θ through $\theta(x) = -j_c\sigma_c(x)$, $\forall\ x \in \mathfrak{g}$. One may check that the Killing form is negative definite with respect to the fixed points of θ, and this defines the compact real form of $\mathfrak{g}$.

By, say [H, Chapter 2, Theorem 7.1], one may choose the Chevalley basis so that $\theta j = j\theta$. Consequently, $\mathfrak{g}_0$ is θ-stable. θ is called the Cartan involution of $\mathfrak{g}_0$ and gives $\mathfrak{g}_0$ a $\mathbb{Z}_2$ grading. Indeed, setting

$$\mathfrak{k}_0 = \{x \in \mathfrak{g}_0 \mid \theta(x) = x\}, \qquad \mathfrak{p}_0 = \{x \in \mathfrak{g}_0 \mid \theta(x) = -x\},$$

one has

$$\mathfrak{g}_0 = \mathfrak{k}_0 \oplus \mathfrak{p}_0, \qquad [\mathfrak{k}_0, \mathfrak{k}_0] \subset \mathfrak{k}_0, \qquad [\mathfrak{k}_0, \mathfrak{p}_0] \subset \mathfrak{p}_0, \qquad [\mathfrak{p}_0, \mathfrak{p}_0] \subset \mathfrak{k}_0.$$

The adjoint action of $\mathfrak{g}_0$ on itself defines a Lie algebra homomorphism $x \mapsto \mathrm{ad}_{\mathfrak{g}_0} x$ of $\mathfrak{g}_0$ into $\mathfrak{gl}(n, \mathbb{R})$: $n = \dim \mathfrak{g}_0$, which is injective since $\mathfrak{g}$ is semisimple. The compact real form can be viewed as $\mathfrak{k}_0 \oplus i\mathfrak{p}_0$. With respect to the (anti)orthonormal basis it defines for the Killing form, one may identify θ with $x \mapsto -{}^t x$, where t denotes transpose of a matrix. Let G be the connected (real) subgroup of $\mathrm{GL}(n, \mathbb{R})$ of elements of the form $\exp x : x \in \mathfrak{g}_0$. Then $\mathrm{Lie}\, G = \mathfrak{g}_0$. Let K be the subgroup of G of all orthogonal matrices (that is, ${}^t g = g^{-1}$). Then K is a compact subgroup of G, in fact, maximal compact. Moreover, $\mathrm{Lie}\, K = \mathfrak{k}_0$.

After Harish-Chandra (see D. A. Vogan's lectures in Chapter 5), classifying irreducible unitary representations of G is equivalent to classifying simple admissible $(\mathfrak{g}, K)$-modules admitting a positive definite invariant Hermitian form $\langle\ ,\ \rangle$. If K is connected, then a $(\mathfrak{g}, K)$-module M is simply a $\mathfrak{g}$-module, admissibility means that M is a direct sum of simple $\mathfrak{k}$-modules each occurring with finite multiplicity, and invariance of the form means that

$$0 = \langle xm, n \rangle + \langle m, xn \rangle, \qquad \forall\, m, n \in M,\ x \in \mathfrak{g}_0.$$

We shall call such modules *unitarizable*.

Let $\mathfrak{g} = \mathfrak{m}^+ \oplus \mathfrak{k} \oplus \mathfrak{m}^-$ be a decomposition of $\mathfrak{g}$ into Lie subalgebras such that $\mathfrak{p} := \mathfrak{m}^+ \oplus \mathfrak{k}$ is a parabolic subalgebra with nilradical $\mathfrak{m}^+$ and Levi factor $\mathfrak{k}$ leaving $\mathfrak{m}^+$ stable. One may assume that $\mathfrak{k} \supset \mathfrak{h}$, and then any such decomposition corresponds to writing $\pi = \pi_1 \amalg \pi_2$ taking the root vectors in $\mathfrak{k}$ (resp., $\mathfrak{m}^+$, $\mathfrak{m}^-$) to have roots in the subset $R \cap \mathbb{Z}\pi_1$ (resp., $R^+ \setminus (R \cap \mathbb{N}\pi_1)$, $R^- \setminus (R \cap -\mathbb{N}\pi_1)$). In particular, $\sigma_c(\mathfrak{k}) = \mathfrak{k}$ and $\sigma_c(\mathfrak{m}^+) = \mathfrak{m}^-$.

If $\mathfrak{m}^+$ (and hence $\mathfrak{m}^-$) is commutative, one may define an involution θ on $\mathfrak{g}$ taking the value 1 on $\mathfrak{k}$ and -1 on $\mathfrak{m}^- \oplus \mathfrak{m}^+$. Suppose this holds, and set $\sigma = \theta \sigma_c$. Then $j(x) = -j_c \sigma(x)$ defines a new complex structure on $\mathfrak{g}$. Set $\mathfrak{g}_0 = \{x \in \mathfrak{g} \mid j(x) = x\}$. Explicitly, $\mathfrak{g}_0$ has an $\mathbb{R}$-basis formed from $\{ih_\beta : \beta \in \pi, \; i(x_\alpha + x_{-\alpha}), \; (x_\alpha - x_{-\alpha}) : \alpha \in R_c^+ := R^+ \cap \mathbb{N}\pi_1\}$ and from $\{x_\alpha + x_{-\alpha}, i(x_\alpha - x_{-\alpha}) : \alpha \in R_n^+ := R^+ \setminus (R^+ \cap \mathbb{N}\pi_1)\}$. Moreover, the former (resp., latter) gave an $\mathbb{R}$-basis for $\mathfrak{k}_0 = \mathfrak{g}_0 \cap \mathfrak{k}$ (resp., $\mathfrak{p}_0$) taking $\theta \mid_{\mathfrak{g}_0}$ to be the Cartan involution.

The preceding condition on $\mathfrak{m}^+$ implies that $R_n^+ \cap \pi$ is a singleton, called the noncompact simple root, which must furthermore occur with multiplicity one in any root β of $\mathfrak{m}^+$.

Let $Q(\lambda)$ be a highest weight module with highest weight λ, that is, a quotient of a Verma module $M(\lambda)$. (Note our change of convention compared with Lecture 3.)

By the Poincaré–Birkhoff–Witt theorem, the triangular decomposition of $\mathfrak{g}$ gives a direct sum decomposition $U(\mathfrak{g}) = U(\mathfrak{h}) \oplus (\mathfrak{n}^- U(\mathfrak{g}) + U(\mathfrak{g})\mathfrak{n}^+)$. Let p be the projection of $U(\mathfrak{g})$ onto $U(\mathfrak{h})$ defined by this decomposition. Define a Hermitian form on $U(\mathfrak{g})$ through $\langle a, b \rangle \mapsto p((j_c\sigma)(a)b)(\lambda)$. One easily checks that its kernel contains $\mathrm{Ann}_{U(\mathfrak{g})} \, e_\lambda$, where e_λ denotes a highest weight vector for $M(\lambda)$. Writing $m = ae_\lambda$, $n = be_\lambda$, one thus obtains a Hermitian form on $M(\lambda)$ satisfying $\langle m, xn \rangle = \langle j_c\sigma(x)m, n \rangle$ for all $m, n \in M(\lambda)$, $x \in \mathfrak{g}$. One easily checks that its kernel is precisely the unique maximal submodule of $M(\lambda)$, and so the form passes to $Q(\lambda)$ and in particular to $L(\lambda)$. Notice that $j_c\sigma = -1$ on $\mathfrak{g}_0$, and hence $\langle \, , \, \rangle$ is the unique (up to scalars) form on $Q(\lambda)$ satisfying the required invariance condition for unitarity. In particular, $Q(\lambda)$ is simple if this form is positive definite. Again, $L(\lambda)$ being admissible implies that its highest weight vector generates a finite dimensional $\mathfrak{p}$-module V. Then $L(\lambda)$ is an image of the induced module $U(\mathfrak{g}) \otimes_{U(\mathfrak{p})} V$. Thus, $L(\lambda)$ is admissible and hence is unitarizable if and only if the form is positive definite. All unitarizable highest weight modules for noncompact connected real Lie groups arise in this fashion.

Let $\alpha \in \pi_n := R_n^+ \cap \pi$ denote the noncompact simple root and ω the fundamental weight corresponding to α. Set $\pi_c = \pi \setminus \pi_n$. In $P(R)$, $P(R)^+$ let a subscript c denote the condition that the corresponding weight is orthogonal to α. Given $\tau \in P(R)_c^+$, let $V(\tau)$ denote the simple finite dimensional $\mathfrak{k}$-module with highest weight τ. For all $u \in \mathbb{C}$, set $\lambda_u^\tau := \tau + u\omega$. The simple highest weight $\mathfrak{k}$-module it defines can be identified as $V(\tau) \otimes \mathbb{C}_{u\omega}$, which we view as a $\mathfrak{p}$-module by letting $\mathfrak{m}^+$ act by zero. The induced module $N(\lambda_u^\tau) := U(\mathfrak{g}) \otimes_{U(\mathfrak{p})} (V(\tau) \otimes \mathbb{C}_{u\omega})$ is isomorphic to $S(\mathfrak{m}^-) \otimes V(\tau) \otimes \mathbb{C}_{u\omega}$ as a $\mathfrak{p}$-module. For its simple quotient to be unitarizable, one requires $u \in \mathbb{R}$. It remains to describe for which pairs $(\tau, u) \in P(R)_c^+ \times \mathbb{R}$ this quotient $L(\lambda_u^\tau)$ is unitarizable. In this, it is enough to assume $\mathfrak{g}$ simple.

One defines a totally ordered sequence $\beta_1, \beta_2, \ldots, \beta_t$ of (strongly orthogonal) noncompact positive roots as follows. Let β_1 be the highest root. Delete the simple roots in the Dynkin diagram of $\mathfrak{g}$ not orthogonal to β, and take the unique connected component containing α. Let β_2 be the unique highest root of this component, and continue until α itself is deleted.

One may show that the partial sums

$$\mu_i := \sum_{j=1}^{i} \beta_j : i = 1, 2, \ldots, t$$

belong to $P(R)^+$. Moreover, up to scalars there exists for each i a unique nonzero $v_i \in S(\mathfrak{m})^{\mathfrak{n}^-}$ of weight $-\mu_i$. Let V_i denote the simple $\mathfrak{k}$-submodule of $S(\mathfrak{m})$ generated by v_i under adjoint action.

Fix $\tau \in P(R)_c^+$. According to a result of K. R. Parthasarathy, R. Ranga Rao, and V. S. Varadarajan [PRV], there exists a unique component P_i of $V(\tau) \otimes V_i$ having extreme weight $\tau - \mu_i$. Call this the PRV component. Finally, define τ to be of level s if $(\beta_i - \beta_j, \tau) = 0$, $\forall\, i, j = 1, 2, \ldots, s$, and s is maximal $\leq t$ with this property.

The classification of unitarizable highest weight modules now goes as follows [EJ]. Take $\tau \in P(R)_c^+$ and suppose it is of level s. Take $i \in \{1, 2, \ldots, s\}$. Then there is a unique value u_i^τ of u such that $S(\mathfrak{m})P_i \otimes \mathbb{C}_{u\omega}$ is a $U(\mathfrak{g})$-submodule of $N(\lambda_u^\tau)$. The $u_i^\tau : i = 1, 2, \ldots, s$ are decreasing, and for $u < u_s^\tau$ the induced module $N(\lambda_u^\tau)$ is unitarizable. The remaining unitarizable modules are its quotients by the previous submodules. One may also calculate the u_i^τ. Notably, the differences $u_i^\tau - u_{i-1}^\tau : 2 \leq i \leq s$ are independent of τ and of i. One calls u_s^τ the first reduction point, and u_1^τ the last place of unitarity.

One may show, [J10, 2.3] for example, that the $S(\mathfrak{m})V_i : i = 1, 2, \ldots, t$ run through the ideals of definition of the closures of the orbital varieties lying in $\mathfrak{m}^+$. Thus, when $\tau = 0$, the corresponding unitarizable highest weight modules identify with the spaces of regular functions on these varieties. Their highest weights are the $u_i^0 \omega : i = 1, 2, \ldots, t$.

Example   Take $\mathfrak{g} = \mathfrak{sl}(n)$, and choose $m \in \{1, 2, \ldots, n-1\}$. One may take $\alpha = \alpha_m$ to be the noncompact simple root. Then, $t = \min\{m, n - m\}$.

Moreover, $v_i : i = 1, 2, \cdots, t$ is the lower left hand corner $i \times i$ minor, and V_i is the subspace spanned by all $i \times i$ minors lying in the bottom left hand $m \times (n - m)$ corner. In particular, if $n = 4$, $m = 2$, one recovers examples (1) and (2) of Lecture 3.

In the last example, take $\tau = \omega_1$. Then $V(\tau)$ is a two-dimensional $\mathfrak{k}$-module (here, $\mathfrak{k} \simeq \mathfrak{gl}(2) \times \mathfrak{gl}(2)$). One may choose a basis $\{x, y\}$ for $V(\tau)$ and a basis $\{a, b, c, d\}$ for $\mathfrak{m}^-$ such that $v_2 = ad - bc$ and P_1 has basis $ay - bx$, $cy - dx$. Then,

$$(ad - bc)x = c(ay - bx) - a(cy - dx) \qquad (4.1a)$$

and

$$(ad - bc)y = d(ay - bx) - b(cy - dx), \qquad (4.1b)$$

which shows that $v_2 V(\tau)$ is contained in $\mathfrak{m}P_1$. Consequently, we have $\text{Ann}_{U(\mathfrak{m}^-)} L(\lambda_1^\omega) = U(\mathfrak{m}^-)v_2$, which identifies with a *prime* ideal of $S(\mathfrak{m}^-)$. Of course, $(4.1a, b)$ is just Cramer's rule of linear algebra, which here translates to a statement about $\text{Ann}_{U(\mathfrak{m}^-)} L(\lambda)$ being prime [J10, 5.16] whenever $L(\lambda)$ is unitarizable! An interesting example also occurs in E_7, and the corresponding polynomial $v_3 \in V_3$ is of degree 3 in 27 variables [J10, 8.1]. It was studied in an old paper of A. Freudenthal [F] concerning geometry over an octionic field.

Lecture 5   The Parthasarathy equality

Recall the setup and notation of Lecture 4. Take $\lambda = \tau + u\omega$ with $\tau \in P(R)_c^+$, $u \in \mathbb{C}$, and let $N(\lambda)$ be the induced module $N(\lambda) = U(\mathfrak{g}) \otimes_{U(\mathfrak{p})} V(\lambda)$. Let e be a $\mathfrak{k}$ highest weight vector of highest weight μ. Let $(\, , \,)$ denote the Cartan inner product on $\mathfrak{h}^*$, and set $\|\nu\|^2 = (\nu, \nu)$. The key to studying when $L(\lambda)$ is unitarizable obtains from the following result due to K. R. Parthasarathy [P].

Lemma 5.1

$$(\|\lambda + \rho\|^2 - \|\mu + \rho\|^2)\langle e, e \rangle = -2 \sum_{\alpha \in R_n^+} \langle x_\alpha e, \ x_\alpha e \rangle.$$

Proof: Let I (resp., J) denote the Casimir invariant for $\mathfrak{g}$ (resp., $\mathfrak{k}$). Then, up to appropriate choices, $Ie = \|\lambda + \rho\|^2 e$, $Je = \|\mu + \rho\|^2 e$. On the other hand, for these choices

$$I - J = 2 \sum_{\alpha \in R_n^+} x_{-\alpha} x_\alpha.$$

Recalling that $\sigma(x_{-\alpha}) = -x_\alpha$ for all $\alpha \in R_n^+$, the assertion results. $\square$

Remark Of course, for the "compact" Chevalley antiautomorphism one has $\sigma(x_{-\alpha}) = x_\alpha$, and the resulting conclusion would lead to quite different results. A similar result holds for any real form of $\mathfrak{g}$, and fairly easy calculations similar to those following can be used to show that only the real forms we are considering lead to unitary highest weight modules.

It is clear that the assertion of the lemma applies to any quotient of $N(\lambda)$, in particular to $L(\lambda)$. This gives the following important

Corollary 5.2 *The Hermitian form on $L(\lambda)$ is positive definite if and only if $\|\mu + \rho\| > \|\lambda + \rho\|$ for every $\mathfrak{k}$ highest weight $\mu \neq \lambda$ occurring in $L(\lambda)$.*

Proof: Let $\mathcal{F}^m(L(\lambda))$ denote the image of $\mathcal{F}^m(U(\mathfrak{m}^-)) \otimes V(\lambda)$ in $L(\lambda)$. Obviously, $x_\alpha \mathcal{F}^m(L(\lambda)) \subset \mathcal{F}^{m-1}(L(\lambda))$ for all $\alpha \in R_n^+$. Since K is compact and $\mathfrak{k}_0 = \mathrm{Lie}\ K$, an invariant form on any simple finite dimensional $\mathfrak{k} = (\mathfrak{k}_0)_{\mathbb{C}}$-module is either positive or negative definite. In particular $\langle\ ,\ \rangle$ can be assumed to be positive definite on $\mathcal{F}^0(L(\lambda)) = V(\lambda)$. Take $m \in \mathbb{N}$, and assume $\langle\ ,\ \rangle$ is positive definite on $\mathcal{F}^m(L(\lambda))$. Assume $e \in \mathcal{F}^{m+1}(L(\lambda))$. The sum occurring in the right hand side of the assertion of the lemma is ≥ 0, with equality if and only if $x_\alpha e = 0$, $\forall\ \alpha \in R_n^+$, which forces $\lambda = \mu$. Hence the required assertion follows. $\square$

For each dominant weight ν, let $V(\nu)$ denote the simple finite dimensional $\mathfrak{k}$-module with highest weight ν. One calls ν the $\mathfrak{k}$-type of $V(\nu)$. Let ν_1, ν_2 be dominant weights, and let $V(\nu)$ be a composition factor of $V(\nu_1) \otimes V(\nu_2)$, which one may recall is completely reducible. One calls ν a minimal $\mathfrak{k}$-type if $\|\nu + \rho\|$ takes its minimal value among those $\mathfrak{k}$-types occurring in $V(\nu_1) \otimes V(\nu_2)$. The PRV component of $V(\nu_1) \otimes V(\nu_2)$ is by definition the composition factor with extreme weight $\nu_1 + w_0 \nu_2$.

Lemma 5.3 *The PRV component of $V(\nu_1) \otimes V(\nu_2)$ is its unique minimal $\mathfrak{k}$-type.*

Proof: Let ν be a $\mathfrak{k}$-type of $V(\nu_1) \otimes V(\nu_2)$. By complete irreducibility, one has a surjective map $V(\nu_1) \otimes V(\nu_2) \longrightarrow V(\nu)$. Now $e_{\nu_1} \otimes e_{w_0\nu_2}$ is a *cyclic* vector for the preceding tensor product and hence has a nonzero image in $V(\nu)$. Let ν_0 be the highest weight of the PRV component. This takes the form $y(\nu_1 + w_0\nu_2)$ for some $y \in W$. Then

i) $\|\nu\| \geq \|\nu_1 + w_0\nu_2\| = \|\nu_0\|$.

ii) $(\nu - \nu_0, \rho) \geq 0$, with equality if and only if $\nu = \nu_0$.

Hence $\|\nu + \rho\| \geq \|\nu_0 + \rho\|$, with equality if and only if $\nu = \nu_0$. This gives the lemma. $\square$

Exercises Let $\lambda_1, \lambda_2, \lambda_3$ be dominant weights.

1) One knows (but this is quite difficult) that

$$\mathrm{Ann}_{U(\mathfrak{n}^+)}\, e_{-\lambda_1} = \sum_{\alpha \in \pi} U(\mathfrak{n}^+) x_\alpha^{(\alpha^\vee, \lambda_1)+1}.$$

2) Recall that $e_{\lambda_3} \otimes e_{-\lambda_1}$ is a cyclic vector for $V(\lambda_3) \otimes V(-\lambda_1)$. (Here, $V(-\lambda_1)$ denotes the simple module with *lowest* weight $-\lambda_1$. It is isomorphic to $V(\lambda_1)^*$.) Let $V(\lambda_2)_{\lambda_3-\lambda_1}^{\lambda_1}$ denote the subspace of vectors in $V(\lambda_2)$ of weight $\lambda_3 - \lambda_1$ and annihilated by $\mathrm{Ann}_{U(\mathfrak{n}^+)}\, e_{-\lambda_1}$. Construct an isomorphism

$$\mathrm{Hom}_{\mathfrak{g}}(V(\lambda_3) \otimes V(-\lambda_1), \ V(\lambda_2)) \xrightarrow{\sim} V(\lambda_2)_{\lambda_3-\lambda_1}^{\lambda_1}.$$

3) Deduce from Frobenius reciprocity that

$$\dim \mathrm{Hom}_{\mathfrak{g}}(V(\lambda_1) \otimes V(\lambda_2), V(\lambda_3)) = \dim V(\lambda_2)_{\lambda_3-\lambda_1}^{\lambda_1}.$$

4) Define ν_1, ν_2, ν_0 as in Lemma 5.3. Using (1), verify that

$$V(\nu_0)_{\nu_1+w_0\nu_2}^{-w_0\nu_2} = V(\nu_0)_{\nu_1+w_0\nu_2}.$$

5) Deduce from (1)–(4) that $\dim \mathrm{Hom}_{\mathfrak{g}}(V(\nu_1) \otimes V(\nu_2), V(\nu_0)) = 1$.

This proves the existence and uniqueness of the PRV component. (See [W2, Section 9] for further details.)

Recall the notation of Lecture 4. Suppose $\tau \in P(R)_c^+$ has level s. Take $j \in \{1, 2, \ldots, s\}$, and let ν_j denote the highest weight of the PRV component P_j of $V_j \otimes V(\tau)$. Set

$$u_j^\tau = \frac{\|\nu_j + \rho\|^2 - \|\tau + \rho\|^2}{2j(\alpha, \omega)}.$$

Take $\lambda = \tau + u_j^\tau \omega$.

Theorem 5.4 $\overline{N(\lambda)} := U(\mathfrak{m})(P_j \otimes \mathbb{C}_{u_j^\tau \omega}) : j = 1, 2, \ldots, s$ *is a* $U(\mathfrak{g})$-*submodule of* $N(\lambda)$.

We give the proof for $j = 1$, which is a little easier. By definition, the simple $\mathfrak{k}$-module P_1 has highest weight $\mu = \nu_1 + u_1^\tau \omega$. Then

$$\|\mu + \rho\|^2 - \|\lambda + \rho\|^2 = \|\nu_1 + \rho\|^2 - \|\tau + \rho\|^2 - 2(\tau - \nu_1, \omega)u_1^\tau$$

$$= 2u_1^\tau(\tau - \nu_1 - \alpha, \omega).$$

Yet α occurs with multiplicity 1 in any root $-\beta$ of $\mathfrak{m}^-$, and so $(\tau - \nu_1 - \alpha, \omega) = 0$. Lemma 5.1 shows that $\sum_{\alpha \in R_n^+} \langle x_\alpha e_\mu, x_\alpha e_\mu \rangle = 0$. Yet $x_\alpha e_\mu \in V(\lambda)$, on which the form is positive definite. Thus, $x_\alpha e_\mu = 0$, $\forall \alpha \in R_n^+$, as required. $\square$

Remarks For the general case, one notes that $(\tau - \nu_k - k\alpha, \omega) = 0$. Moreover, the $u_k^\tau : k = 1, 2, \ldots, s$ are strictly decreasing and

$$[\mathfrak{m}^+, V_k] \subset \mathfrak{k}V_{k-1} + V_{k-1} \tag{5.1}$$

by [EJ, 2.5, 3.6].

One proves inductively that $\langle \, , \, \rangle$ is positive definite on $V_k \otimes V_\tau \otimes \mathbb{C}u_j^\tau \omega$ for $k < j$ and then the required assertion of the theorem by the procedure used for the case $j = 1$.

When $\tau = 0$, one has

$$u_j^0 = \frac{\|u_j + \rho\|^2 - \|\rho\|^2}{2j(\alpha, \omega)} = \frac{\|\mu_j\|^2 + 2(\rho, \mu_j)}{2j(\alpha, \omega)}.$$

Now the β_i are always long roots and pairwise orthogonal. Thus, $\|\mu_j\|^2 = j(\beta_1, \beta_1)$. Consequently,

$$u_{j-1}^0 - u_j^0 = \frac{1}{j(j-1)(\alpha, \omega)} \sum_{i=1}^{j}(\beta_i - \beta_j, \rho) > 0,$$

and so the fact that the u_j^0 are strictly decreasing is quite easy. Thus, apart from verifying (5.1), we have also proved the theorem for $\tau = 0$. This proves that the space of regular functions on the closure of an orbital variety contained in $\mathfrak{m}^+$ admits the structure of a $U(\mathfrak{g})$-module of highest weight, namely $N(\lambda)/\overline{N(\lambda)}$ where $\lambda = u_i^0 \omega : i = 1, 2, \ldots, t$. Further estimates of this type show that this quotient is unitarizable and hence simple [EJ, 8.2].

Lecture 6   The last place of unitarity

The description of the unitary module for $\tau = 0$, namely the $L(\lambda_i^0)$:
$i = 1, 2, \ldots, t$, was given by N. Wallach [W1]. The set $\lambda_1^0 = 0 > \lambda_2^0 > \cdots >$
λ_t^0 is often called the Wallach set. Their relation with orbital varieties
(using a slightly different language) is due to Harris and Jakobsen [HJ]. It
was realized (particularly by M. Kashiwara and M. Vergne [KV]), that the
tensor product of unitary modules is again unitary. In particular, given
that the $L(\lambda_1^\tau)$ are unitary, then so are the

$$L(\lambda_1^\tau) \otimes L(\lambda_i^0) : i = 1, 2, \ldots, t,$$

and moreover, for $i \leq s$ (the level of τ) one recovers the $L(\lambda_i^\tau)$: $i =$
$1, 2, \ldots, s$ in this fashion as the orthogonal direct summand with highest
weight $\lambda_1^\tau + \lambda_i^0$. (The $L(\lambda_i^\tau)$: $i > s$ are also unitary but correspond to
members of the continuous family of unitary modules occurring before the
first reduction point is reached.) Moreover, one can show that all unitary
modules are obtained in this fashion (excluding the continuous families of
induced ones). This can be regarded as explaining the independence of
$u_i^\tau - u_j^\tau$ on τ and on i, j, since $\lambda_i^\tau = \lambda_1^\tau + \lambda_i^0$. Note again that $\lambda_1^0 = 0$
and corresponds to the trivial module, whereas tensoring $L(\lambda_2^0)$ with itself
i times gives $L(\lambda_{i+1}^0)$. One may call $L(\lambda_2^0)$ the generalized metaplectic
representation.

Notice we have now solved the purely algebraic problem of quantizing
all orbital varieties lying in any nilradical (of a parabolic subalgebra) that
is commutative. In this, the positive definiteness of the form leads to a
significant simplification. However following [EJ, Section 2] and as noted in
Lecture 7, one also obtains this result by more naive algebraic calculations.

It thus essentially remains to establish unitarity at the last place λ_1^τ,
and this was a point of view emphasized by H. P. Jakobsen [Ja]. However,
his analysis was case by case and rather complicated. Here, we give an
easy intrinsic proof. Simultaneously with to the Jakobsen classification,
one should also mention, the unitary highest weight modules were classified
by T. J. Enright, R. Howe, and N. R. Wallach [EHW]. The present analysis
comes from joint work with T. J. Enright [EJ].

Set $\lambda = \lambda_1^\tau$. Theorem 5.4 established the existence of the quotient
$N(\lambda)/\overline{N(\lambda)}$. Now we prove its unitarity. Set

$$R_< = \{\nu \in R_n^+ \mid (\gamma^\vee, \lambda + \rho) < 1\}$$

and $\Omega = \bigcup_{i \in \mathbb{N}} \Omega_i$, where $\Omega_i = \{\tau - \gamma_1 - \gamma_2 - \cdots - \gamma_i \mid \gamma_i \in R_<\}$. (Here, the γ_i need not be distinct.) Let Ω^+ denote the set of $\mathfrak{k}$-dominant elements in Ω; that is, $\Omega^+ = \{\delta \in \Omega \mid (\alpha^\vee, \delta) \geq 0, \ \forall \ \alpha \in \pi_c\}$. Our aim is to prove the following [EJ, 7.9].

Proposition 6.1   *Every $\mathfrak{k}$-type of $N(\lambda)/\overline{N(\lambda)}$ belongs to $\Omega^+ + u_1^\tau \omega$.*

Let us first show how this establishes unitarity at the last place.

Theorem 6.2   *Take $\tau \in P(R)_c^+$, and set $\lambda = \tau + u_1^\tau \omega$. Then $N(\lambda)/\overline{N(\lambda)}$ is unitarizable (and simple).*

Proof: If $N(\lambda)/\overline{N(\lambda)}$ were not simple, then some highest weight ξ of a submodule would have to satisfy $\|\xi + \rho\| = \|\lambda + \rho\|$. Further, taking account of Corollary 5.2, it is enough to show that $\|\nu + \rho\| > \|\lambda + \rho\|$ for every $\mathfrak{k}$-dominant weight $\nu \neq \lambda$ of $N(\lambda)/\overline{N(\lambda)}$.

By the proposition, we can write $\nu = \tau - \delta + u_1^\tau \omega$ with $\delta \in \mathbb{N}R_< \setminus \{0\}$. Then the required inequality will result from $2(\delta, \lambda + \rho) < (\delta, \delta)$. Now,

$$\delta = \sum_{\gamma_i \in R_<} \gamma_i$$

and $(\gamma_i, \gamma_j) \geq 0$, because in R_n^+ the sum of two roots is not a root. Hence,

$$(\delta, \delta) \geq \sum_{\gamma_i \in R_<} (\gamma_i, \gamma_i) \geq 2(\delta, \lambda + \rho),$$

by definition of $R_<$. Moreover, the second inequality is strict, as $\delta \neq 0$. $\square$

We now sketch a proof of the proposition. Set $R_> = R_n \setminus R_<$, and let $\mathfrak{m}_<$ (resp., $\mathfrak{m}_>$) denote the subspace of $\mathfrak{m} := \mathfrak{m}^-$ spanned by the root vectors $x_{-\gamma} : \gamma \in R_<$ (resp., $R_>$). An easy verification gives

Lemma 6.3   i) $[\mathfrak{n}_c^+, \mathfrak{m}_<] \subset \mathfrak{m}_<$.
   ii) $[\mathfrak{n}_c^-, \mathfrak{m}_>] \subset \mathfrak{m}_>$.

Let $M_c(\tau)$ denote the Verma module for $\mathfrak{k}$ with highest weight τ. Consider $M := \mathfrak{m} \otimes M_c(\tau)$ as $\mathfrak{k}$-module for the diagonal action. It has a Verma flag [D, 7.6.14] with quotients isomorphic to the $M_c(\tau - \gamma) : \gamma \in R_n^+$. Moreover, writing $M_\gamma = U(\mathfrak{n}_c^-)(x_{-\gamma} \otimes e_\tau)$ and

$$M_< = \bigoplus_{\gamma \in R_<} M_\gamma, \qquad M_> = \bigoplus_{\gamma \in R_>} M_\gamma,$$

then by Lemma 6.3(i), $M_<$ identifies with a submodule of M and $M_>$ can be viewed as the quotient $M/M_<$; that is, we have an exact sequence

$$0 \longrightarrow M_< \longrightarrow M \longrightarrow M_> \longrightarrow 0 \tag{6.1}$$

of $U(\mathfrak{k})$-modules. One checks that $\|\tau - \gamma + \rho_c\| \neq \|\tau - \gamma' + \rho_c\|$ for all $\gamma \in R_<$ and $\gamma' \in R_>$. Using the center of $U(\mathfrak{k})$, it follows that (6.1) splits. Let $M'_>$ denote the image of $M_>$ in M.

Lemma 6.4 P_1 *is the image of* $M'_>$ *under the map*

$$\mathfrak{m} \otimes M_c(\tau) \longrightarrow \mathfrak{m} \otimes V(\tau).$$

Proof: Observe that M_γ identifies in the Verma flag of M with $M_c(\tau - \gamma)$. Now,

$$\|\lambda + \rho - \gamma\|^2 = \|\lambda + \rho\|^2 + (\gamma, \gamma) - 2(\gamma, \lambda + \rho) \leq \|\lambda + \rho\|^2$$

if and only if $\gamma \in R_>$. Then the assertion follows from Theorem 5.4. $\square$

It is not hard to deduce Proposition 6.1 from Lemma 6.4, though in working out the details it is convenient to work with the tensor algebra of $\mathfrak{m}$ rather than the symmetric algebra, as the appropriate quotients are easier to identify.

This completes the proof of unitarity in the last place and with it the classification of unitarizable highest weight modules. Of course, far more detailed results are available [EJ, E]. For example, one has [EJ, 4.2] an explicit general formula for $u_i^\tau : i = 1, 2, \ldots, s$, which for $i = 1$ gives

$$u_1^\tau = 1 + \frac{1}{2}|S_{1,\tau}| + 2(\rho_c, \beta^\vee) - (\tau + \rho, \beta^\vee),$$

where β is the unique highest root and

$$S_{1,\tau} = \{\gamma \in R_c^+ \mid (\tau, \gamma^\vee) = 1, (\beta, \gamma^\vee) = 2\}.$$

(This is empty in the simply laced case.) That $\overline{N(\lambda)}$ is in fact the maximal submodule of $N(\lambda)$ for an arbitrary unitary parameter λ results from [E] or [DES].

Lecture 7   Quantization of orbital varieties

Let $V \subset \mathfrak{n}^+$ be the closure of an orbital variety and $I(V) \subset S(\mathfrak{n}^-)$ its ideal of definition. As noted in Lecture 1, one may interpret the quantization of V as the construction of a highest weight module $N(\lambda) : \lambda \in \mathfrak{h}^*$ with highest weight vector e_λ of weight λ such that

$$\operatorname{gr} \operatorname{Ann}_{U(\mathfrak{n}^-)} e_\lambda = I(V). \tag{7.1}$$

A particularly intriguing aspect of (7.1) is the determination and meaning of $\lambda + \rho$, which in general can *neither be integral nor regular*. Physicists may recognize λ as the analog of the zero point energy that arises in the quantization of a Hamiltonian system.

One may remark that (7.1) implies [J7, Section 8.2] that $\operatorname{Ann} N(\lambda)$ is completely prime. Thus, a positive answer to quantization would construct a large part if not all of $\operatorname{Prim}_c U(\mathfrak{g})$.

As noted in Lecture 6, the preceding problem was completely solved in the case when $N(\lambda)$ can be a unitarizable module. Then, the possible values of λ define the Wallach set.

For $\mathfrak{g} = \mathfrak{sl}(n) : n \leq 6$, this problem was solved by E. Benlolo in her thesis. To appreciate the amount of work involved, we remark that for $\mathfrak{sl}(6)$ there are 76 orbital varieties that have to be individually quantized. An important result of hers [B] is that there are two of these orbital varieties in $\mathfrak{sl}(6)$ that *cannot* be made to satisfy (7.1) if $N(\lambda)$ is required to be simple, but which paradoxically do satisfy (7.1) for some *nonsimple* $N(\lambda)$. This introduces some obvious difficulties, in particular the significance of λ, which can now also serve for some other orbital variety. However, one can "canonically" replace λ by the highest weight of the necessarily unique simple submodule of $N(\lambda)$ whose associated variety is V.

A sophisticated approach to (7.1) would no doubt cunningly exploit the fact that V is Lagrangian and is even (see Lecture 8) the image under the moment map of a conormal to a B-orbit in G/B. At present, we do not even understand the possible nonintegrality of λ in geometric terms. Thus, one can at present only propose the naive approach indicated in the following.

The solution to (7.1) naturally decomposes into three stages.

1) Find a system of generators for $I(V)$.

2) Find a left ideal $J(V)$ of $U(\mathfrak{n}^-)$ such that $\operatorname{gr} J(V) = I(V)$.

3) Give $U(\mathfrak{n}^-)/J(V)$ a left $U(\mathfrak{g})$-module structure.

The work of Benlolo indicates that (1) has the following solution when $\mathfrak{g} = \mathfrak{sl}(n)$. Let M be the $n \times n$ matrix with 0 above the diagonal, 1 on the diagonal, and for $i > j$ having $e_{ij} \in \mathfrak{n}^- \subset S(\mathfrak{n}^-)$ as its ijth entry. Then $I(V)$ should take the form $\operatorname{gr}\langle a_1, a_2, \cdots, a_n \rangle$, where the a_i are among the minors of M. Notice that then $I(V)$ contains $\langle \operatorname{gr} a_1, \operatorname{gr} a_2, \cdots, \operatorname{gr} a_n \rangle$, and we remark that there was one case found by Benlolo in $\mathfrak{sl}(6)$ where this inclusion was strict.

In the preceding, we can also interpret the e_{ij} as elements of $U(\mathfrak{n}^-)$. This makes it possible to write down elements $b_1, b_2, \ldots, b_n \in U(\mathfrak{n}^-)$ as corresponding "minors," though there is an obvious and significant ambiguity in doing so. These elements should generate $J(V)$ as a left ideal. Whereas one may anticipate no difficulty in checking that $\operatorname{gr} J(V) \supset I(V)$, equality should only be a deeper consequence of the involutivity of V.

Finally, for (3) it is sufficient to linearly order the b_i (possibly using degree) and to solve the system of equations

$$ x_\alpha b_i e_\lambda = 0 \quad \bmod \quad \sum_{j=1}^{i-1} U(\mathfrak{n}^-) b_j e_\lambda : \qquad \alpha \in \pi, \; i = 1, 2, \ldots, n, $$

for λ. Heuristically, this should be possible because V is B-invariant. It leads to a condition on $(\lambda, \alpha^\vee)$ that if not empty should be (miraculously) independent of i. This point of view does not imply that $\lambda + \rho$ be either integral or regular, nor that $U(\mathfrak{n}^-)/J(V)$ be a simple $U(\mathfrak{g})$-module. In the unitary case, one finds that $(\lambda, \alpha^\vee) = 0$ if $\alpha \in \pi_c$, whereas for the noncompact root one has $-(\lambda, \alpha^\vee) = m - 1$, where m is the size of the minor that (Lecture 4) is the unique highest weight vector of the $\mathfrak{k}$-module generating $J(V)$.

It turns out that the use of quantum groups facilitates steps (1) and (2). First, it gives a natural interpretation of the mysterious M matrix. Second, no ambiguities arise in step (2). Finally, the formalism naturally generalizes to arbitrary $\mathfrak{g}$ semisimple. All this is now detailed, the notation being essentially that of [J11].

Let $U_q(\mathfrak{g})$ denote the Drinfeld–Jimbo quantization of the enveloping algebra $U(\mathfrak{g})$ of a complex semisimple Lie algebra $\mathfrak{g}$. It is a Hopf algebra over $\mathbb{C}(q)$ that has essentially the same finite dimensional modules as $U(\mathfrak{g})$.

Set $A = \mathbb{C}[q, q^{-1}]$, and identify $\mathbb{C}$ with $A/\langle q - 1 \rangle$. Let $U_A(\mathfrak{g})$ be the A-subalgebra of $U_q(\mathfrak{g})$ generated by the obvious "canonical generators" of $U_q(\mathfrak{g})$. Up to a minor triviality, $U_A(\mathfrak{g}) \otimes_A \mathbb{C}$ identifies [J11, 4.1.15] with $U(\mathfrak{g})$.

One may adopt the construction of the algebra $R[G]$ of regular functions on G to construct the Hopf dual of $U_q(\mathfrak{g})$ as follows. For each $\omega \in P(R)^+$, let $V(\omega)$ be the simple finite dimensional $U_q(\mathfrak{g})$-module with highest weight ω. For all $\xi \in V(\omega)^*$, $v \in V(\omega)$, let $c^\omega_{\xi,v}$ be the element $a \mapsto \xi(av)$ of $U_q(\mathfrak{g})^*$. Then the $c^\omega_{\xi,v}$ generate a Hopf subalgebra $R_q[G]$ of $U_q(\mathfrak{g})^*$. We write $c^\omega_{\xi,v} = c^\omega_{\mu,v}$ when ξ (resp., v) has weight μ (resp., ν). Define an order relation on $P(R)$ through $\mu \geq \nu$ if $\mu - \nu \in \mathbb{N}\pi$. Set $R_A[G] = \{c \in R_q[G] \mid c(U_A(\mathfrak{g})) \subset A\}$. One easily checks that $R_q[G] \otimes_A \mathbb{C} \simeq R[G]$.

Corresponding to the triangular decomposition $\mathfrak{g} = \mathfrak{n}^+ \oplus \mathfrak{h} \oplus \mathfrak{n}^-$, one has subalgebras $U_q(\mathfrak{n}^\pm)$ of $U_q(\mathfrak{g})$. Let τ be an isomorphism of the additive abelian group $P(R)$ to a multiplicative abelian group T. In what follows, we must assume that $U_q(\mathfrak{g})$ is its so-called simply connected version with torus T, that is, $U_q(\mathfrak{g}) = U_q(\mathfrak{n}^-)TU_q(\mathfrak{n}^+)$. The subalgebras $U_q(\mathfrak{n}^\pm) := U_q(\mathfrak{n}^\pm)T$ are Hopf subalgebras. Writing $U_A(\mathfrak{n}^\pm) = U_A(\mathfrak{g}) \cap U_q(\mathfrak{n}^\pm)$, one has $U_A(\mathfrak{n}^\pm) \otimes_A \mathbb{C} = U(\mathfrak{n}^\pm)$ by [J11, 3.4.5].

The advantage of quantum groups arises from the Drinfeld isomorphism, first given in the context of topological Hopf algebras but reinterpreted in a purely algebraic context in [J11, 9.2.12] as follows. Consider the map $R_q[G] \to U_q(\mathfrak{b}^\pm)^*$ defined by restriction, the kernel of which we denote by $J^\pm$. To describe its image, we remark (see, for example, [J11, 9.2.10]) that there exists a nondegenerate skew-Hopf pairing $\varphi : U_q(\mathfrak{b}^-) \times U_q(\mathfrak{b}^+) \to \mathbb{C}(q)$. This defines in particular an injection $\Phi' : b \mapsto (a \mapsto \varphi(a, b))$ of $U_q(\mathfrak{b}^-)$ into $U_q(\mathfrak{b}^+)^*$, which is a homomorphism of algebras and an antihomomorphism of coalgebras. The result in [J11, 9.2.11] asserts that $Im\,\Phi' = Im(R_q[G] \to U_q(\mathfrak{b}^+)^*)$. This and a corresponding result with $\pm$ interchanged eventually gives

Theorem 7.1   *There is an isomorphism* $\Phi^\pm : R_q[G]/J^\pm \xrightarrow{\sim} U_q(\mathfrak{b}^\pm)$ *of Hopf algebras.*

In what follows, it is convenient to consider the algebra isomorphism $\Phi' : U_q(\mathfrak{b}^-) \xrightarrow{\sim} R_q[G]/J^+$, since this more closely corresponds to the vector space isomorphism $\mathfrak{b}^- \xrightarrow{\sim} (\mathfrak{b}^+)^*$ obtained from the Killing form and which allows one to consider $S(\mathfrak{n}^-)$ as the algebra of regular functions on $\mathfrak{n}^+$ and as the graded dual of $U(\mathfrak{n}^+)$ viewed as a graded algebra through weight space decomposition.

Set $\mathfrak{h}_{\mathbb{Q}}^* = \{\lambda \in \mathfrak{h}^* \mid (\lambda, \alpha^\vee) \in \mathbb{Q}, \ \forall \alpha \in \pi\}$. By adjoining nth roots of q to $\mathbb{C}(q)$, one may view $\lambda \in \mathfrak{h}_{\mathbb{Q}}^*$ as the character on T given by $\tau(\mu) \mapsto q^{(\mu,\lambda)}$. From this, one may define a highest weight module $N_q(\lambda)$ of $U_q(\mathfrak{g})$. Such a module is given by a left ideal L of $U_q(\mathfrak{b}^-)$. Through Φ', one may identify L as a left ideal of $R_q[G]/J^+$. Moreover, the latter identifies with the algebra of matrix coefficients $c_{\mu,\nu}^\omega$ in which the entries strictly above the diagonal (i.e., when $\mu \not\geq \nu$) are set equal to zero. Again, the torus element $c_{\mu,\mu}^\omega$ must act by a scalar on the highest weight vector e_λ of $N_q(\lambda)$ (specifically, the scalar $q^{-(\mu,\lambda)}$ via [J11, 9.2.11(ii)]). Thus, L contains the left ideal L_λ of $R_q[G]/J^+$ generated by the $c_{\mu,\mu}^\omega - q^{-(\mu,\lambda)}1 : \omega \in P(R)^+$, $\mu \in P(R)$. Consequently, the diagonal matrix coefficients specialize to 1 in the $q \to 1$ limit. Finally, when $G = \mathrm{SL}(n)$, the algebra $R_q[G]$ is generated by the matrix coefficients $c_{\mu,\nu}^\omega$, where ω corresponds to the n-dimensional defining representation. These matrix coefficients may be written as $c_{ij} : i, j = 1, 2, \dots, n$ and then, for an appropriate ordering, J^+ is generated by the $c_{ij} : j > i$. For the k-fundamental weight ω_k (corresponding to the kth exterior power of ω), the matrix coefficients $c_{x\omega_k, y\omega_k}^{\omega_k} : x, y \in W$ are the quantum $k \times k$ minors in the c_{ij}, and of course these specialize to ordinary minors in the $q \to 1$ limit. From all this, we can understand that our previous conjectured description of $I(V)$ has a more intrinsic formulation.

Our program for quantizing an orbital variety closure V can be expressed as determining some $\lambda \in \mathfrak{h}_{\mathbb{Q}}^*$ and a left ideal $J_q(V, \lambda)$ of $U_q(\mathfrak{b}^-)$ containing $J_\lambda := \sum_{\mu \in P(R)} U_q(\mathfrak{b}^-)(\tau(\mu) - q^{(\lambda,\mu)}1)$ such that

1) The $U_q(\mathfrak{b}^-)$-module structure on $N_q(\lambda) := U_q(\mathfrak{b}^-)/J_q(V, \lambda)$ extends to a $U_q(\mathfrak{g})$-module structure.

2) $I(V) = \mathrm{gr}(J_A'(V) \otimes_A \mathbb{C})$, where $J_A'(V) = \Phi'(J_q(V, \lambda)) \cap \mathrm{Im}\, R_A[G]$.

For (2), we note that $\Phi'(U_q(\mathfrak{n}^-)) \cap \mathrm{Im}\, R_A[G] = \{c \in \Phi'(U_q(\mathfrak{n}^-)) \mid c(U_A(\mathfrak{n}^+)) \subset A\}$ and so coincides with the A-subalgebra S_A^- defined analogously to S_A^+ of [J11, 3.4.5]. Thus, $S_A^- \otimes_A \mathbb{C}$ coincides with $S(\mathfrak{n}^-)$ viewed as the graded dual of $U(\mathfrak{n}^+)$ (see the paragraph following Theorem 7.1). Since $\Phi'(J_\lambda) = L_\lambda$, it follows that the $(c_{\mu,\mu}^\omega - 1)$ belong to $J_A'(V) \otimes_A \mathbb{C}$, which may therefore be viewed as an ideal of $S(\mathfrak{n}^-)$, which is $\mathfrak{h}$-stable but not necessarily graded for gradation by *degree*. Defining formal character ch as in [J11, 3.4.7], one obtains from (2) that

$$\operatorname{ch} N_q(\lambda) = \operatorname{ch}(S(\mathfrak{n}^-)/I(V)). \tag{7.2}$$

On the other hand, we can define a second specialization $J(V)$ of $J_q(V,\lambda)$ by setting $J(V) = J_A(V) \otimes_A \mathbb{C}$, where $J_A(V) = J_q(V,\lambda) \cap U_A(\mathfrak{n}^-)T$. It identifies with a left ideal of $U(\mathfrak{n}^-)$. Moreover, $N(\lambda) := U(\mathfrak{n}^-)/J(V)$ has by (1) the structure of $U(\mathfrak{g})$-module of highest weight λ, and

$$\operatorname{ch} N_q(\lambda) = \operatorname{ch} N(\lambda). \tag{7.3}$$

The obvious diagram involving the preceding two specializations fails to be commutative. Yet (7.2) and (7.3) imply that $p_{V(N(\lambda))} = p_V$. Now, $p_{V(N(\lambda))}$ is a sum of the characteristic polynomials of the orbital varieties occurring in $V(N(\lambda))$, and so by their linear independence (Corollary 3.5) we deduce that $V(N(\lambda)) = V$. Thus, $\operatorname{gr} J(V) \subset I(V)$, and equality holds by (7.2) and (7.3) again.

Although (1) and (2) can be expected to essentially uniquely determine $J_q(V,\lambda)$ if it exists, we would clearly like to have a formula for it analogous to that for $I(V)$. Toward this aim we note that Φ' gives $U_q(\mathfrak{b}^-)$ a $U_q(\mathfrak{b}^+)$ bimodule structure! Then to satisfy (1) we need only require that $\Phi'(J_q(V,\lambda))$ be invariant with respect to the adjoint action of $U_q(\mathfrak{b}^+)$ (defined by $(a,b) \mapsto a_1 b S(a_2)$ for all $a \in U_q(\mathfrak{b}^+)$, $b \in R_q[G]/J^+$, where $\Delta(a) = a_1 \otimes a_2$ in the sum convention of [J11, 1.1.8] and S is the antipode). This is the quantum analog of the orbital variety closure V being B-stable. Finally, $\Phi'(J_q(V,\lambda))$ is a left ideal of $R_q[G]/J^+ \hookrightarrow U_q(\mathfrak{b}^+)^*$, and so it is natural to think of it as an ideal of zeros of some subset. We already require it to contain L_λ and to be $\operatorname{ad} U_q(\mathfrak{b}^+)$-stable. If $V = \overline{V(w)}$, one should therefore further require it to vanish on some quantum analog of $\mathfrak{n}^+ \cap w(\mathfrak{n}^+)$. This should be a subspace of $U_q(\mathfrak{n}^+)$ with basis formed by choosing for each $\gamma \in R^+ \cap w^{-1}R^-$ a weight vector a_γ of weight γ. Such vectors can be obtained using the Lusztig–Soibelman automorphisms [J11, 10.2], but this procedure is not canonical. An alternative would be to make the choice specified by the conclusion of [J13, 4.5]. Finally, λ must be chosen to ensure that $J_q(V,\lambda)$ is big enough to satisfy (2). Although this point of view may seem a little far-fetched, it is nevertheless true that the generic prime (two-sided) ideals of $R_q[G]$ can be viewed as ideals of zeros [J14, Theorem 1].

Exercises

1)   Assume $N(\lambda)$ satisfies (7.1). Show that $N(\lambda)$ has a simple socle, say $L(\lambda')$, and that $V(N(\lambda)/L(\lambda'))\subsetneq V(N(\lambda))$.

2)   Take $\mathfrak{g}$ of type G_2 with $\pi = \{\alpha, \beta\}$, where α is the short root. Let V be the closure of the unique 3-dimensional orbital variety. Show that $I(V)$ is generated by $x_{-\alpha}$ and by three further quadratic generators that span the unique 3-dimensional representation of $M \times M$, where $M = \mathbb{C}x_{-\beta} \oplus \mathbb{C}x_{-(\alpha+\beta)} \oplus \mathbb{C}x_{-(2\alpha+\beta)} \oplus \mathbb{C}x_{-(3\alpha+\beta)}$ viewed as an $\mathfrak{h} \oplus \mathbb{C}x_{-\alpha} \oplus \mathbb{C}x_{\alpha}$-module. Let $a_{-2\beta-2\alpha}$ denote its highest weight vector. Solve the equation $x_\beta a_{-2\beta-2\alpha} e_\lambda = 0$ for $(\lambda, \beta^\vee)$ and hence determine the solution to (7.1) in this case. Show that $N(\lambda) = L(\lambda)$. This constructs a completely prime ideal, namely $\operatorname{Ann} L(\lambda)$, the associated variety of which is the unique 6-dimensional orbit. In this example, one fails [Z] to have an embedding $U(\mathfrak{g})/\operatorname{Ann} N(\lambda) \hookrightarrow D(V(N(\lambda)))$ unless one localizes at the Ore set generated by the image of the highest weight vector.

Lecture 8   Further results and perspectives

Take $\lambda = \lambda_i^0 : i = 1, 2, \ldots, t$, that is, in the Wallach set. One has [LS, Chapter IV for $\mathfrak{g}$ classical; J10, 4.2 in general].

Theorem 8.1 $\operatorname{Ann} L(\lambda)$ *is a maximal ideal.*

This is false for arbitrary unitary parameters. It can happen that $\operatorname{Ann} L(\lambda)$ is strictly contained in the augmentation ideal. E. Gvirsk (unpublished) classified all such occurrences for $\mathfrak{g} = \mathfrak{sl}(n)$. One has a simple sufficiency [J10, 6.8] criterion for $\operatorname{Ann} L(\lambda)$ not to be maximal, but this does not quite seem to be a necessary one.

Take λ in the Wallach set. That $\operatorname{Ann} L(\lambda)$ is completely prime is an easy consequence of the fact that $\operatorname{Ann}_{U(\mathfrak{m}-)} e_\lambda$ is prime. Again, this fails for arbitrary unitary parameters. However, one does have the following curious fact [J10, 5.16].

Theorem 8.2   *Assume λ a unitary parameter, say $\lambda = \tau + u_i^\tau \omega$. Then $Q(\lambda) := \operatorname{Ann}_{U(\mathfrak{m}-)} L(\lambda)$ is a prime ideal.*

As a fairly easy consequence, one can show [J10, 6.5] that the rank $\operatorname{rk}(U(\mathfrak{g})/\operatorname{Ann} L(\lambda))$ divides $\dim V(\tau)$, but it need not be 1 nor $\dim V(\tau)$.

It can be 1 even if $\dim V(\tau) > 1$ by [J10, 8.9]. A table describing $Q(\lambda)$ as a function of λ is given in [J10, 7.13]. We remark that the proof of Theorem 8.2 is a rather delicate matter. The result is quite false for nonunitary parameters and especially those which are unitary for the compact real form. This seems to exclude any easy shortcuts.

One has a remarkable surjectivity theorem [J7]. Call a simple highest weight module L quasi-induced if it is a submodule of some $E \otimes M$ with E of finite dimension and M a simple highest weight module induced from a proper parabolic subalgebra $\mathfrak{p}$ of $\mathfrak{g}$. One may easily show that L, M have the same associated varieties, and the latter is in some obvious sense nontrivially induced. Thus, it is easy to find L that are not quasi-induced. For example, $L(\lambda)$ for λ in the Wallach set is quasi-induced if and only if its associated variety is the nilradical of a parabolic subalgebra, and in $\mathfrak{sl}(n)$, for instance, this only happens when the noncompact simple root is either α_1 or α_{n-1}. Let $F(L)$ (resp., $A(L)$) denote the set of $\mathfrak{g}$ (resp., $\mathfrak{n}^-$) locally finite elements of $\operatorname{End} L$.

Theorem 8.3 *Assume L is not quasi-induced. Then $A(L) = F(L)$.*

If $\operatorname{Ann} L$ is maximal, then $F(L)$ can only differ from $U(\mathfrak{g})/\operatorname{Ann} L$ if their rings of fractions differ. Using this, one may check that $F(L(\lambda)) = U(\mathfrak{g})/\operatorname{Ann} L(\lambda)$ for λ in the Wallach set. On the other hand, in this case $A(L(\lambda))$ is just the algebra of differential operators on $\mathcal{V}(\operatorname{Ann} \operatorname{gr}_{\mathcal{F}} L(\lambda))$. This gives the following result [J10, 4.5], due to T. Levasseur and J. T. Stafford for $\mathfrak{g}$ classical [LS].

Corollary 8.4 *Except if $\mathfrak{g} = \mathfrak{sl}(n)$ and $\alpha = \alpha_1$ or α_{n-1}, one has*

$$U(\mathfrak{g})/L(\lambda) \overset{\sim}{\longrightarrow} D[\mathcal{V}(\operatorname{Ann} \operatorname{gr}_{\mathcal{F}} L(\Lambda))]$$

for all λ in the Wallach set.

A key step in the proof of Theorem 8.3 comes from the study of $\mathcal{O}$-algebras, that is, $U(\mathfrak{g})$-algebras that as modules belong to the $\mathcal{O}$ category. There are surprisingly few such objects, especially if one demands that the algebra in question be semiprime [J12, 5.6]. In the completely prime case, this can be generalized to the case when $\mathfrak{g}$ is Kac–Moody as in what follows.

Let $\mathfrak{m}$ be a Lie algebra and A an associative algebra. We say that an algebra homomorphism $\psi \colon U(\mathfrak{m}) \longrightarrow \operatorname{End} A$ is a Leibniz action of $U(\mathfrak{m})$ on

A if $\psi(x) \in \mathrm{Der}\,A$ for all $x \in \mathfrak{m}$. By definition, this makes A a $U(\mathfrak{m})$-algebra. We say that the action is nondegenerate if $A^{\mathfrak{m}}$ reduces to scalars. Take $\mathfrak{g}$ a symmetrizable Kac–Moody Lie algebra with triangular decomposition $\mathfrak{g} = \mathfrak{n}^+ \oplus \mathfrak{h} \oplus \mathfrak{n}^-$. Let σ_c be the Chevalley antiautomorphism of $\mathfrak{g}$ (defined on generators through $\sigma_c(x_{\pm\alpha}) = x_{\mp\alpha}$, $\forall\, \alpha \in \pi$). Given $M \in \mathrm{Ob}\,\mathcal{O}$, consider M^* as a left $\mathfrak{g}$-module through σ_c and set $\delta M \in \{m \in M^* \mid \dim U(\mathfrak{h})m < \infty\}$. Then δM is a $U(\mathfrak{g})$-submodule and belongs to $\mathcal{O}$.

Lemma 8.5 *Let ψ, ψ' be $\mathfrak{h}$ equivariant nondegenerate Leibniz actions of $U(\mathfrak{n}^+)$ on $S(\mathfrak{n}^-)$. Then there is an algebra isomorphism intertwining ψ, ψ' and commuting with $\mathfrak{h}$.*

Proof: Let $\varepsilon\colon S(\mathfrak{n}^-) \to \mathbb{C}$ be the augmentation, and define a pairing $U(\mathfrak{n}^+) \times S(\mathfrak{n}^-) \to \mathbb{C}$ by $\langle x, a \rangle = \varepsilon(\psi(x)a)$. Then $\langle xy, a \rangle = \varepsilon(\psi(xy)a) = \langle x, \psi(y)a \rangle$ for all $x, y \in U(\mathfrak{n}^+)$, $a \in S(\mathfrak{n}^-)$. Clearly, $\langle 1, 1 \rangle = 1$ and $\langle\,,\,\rangle$ respects weight space decomposition. Let $\beta \in \mathbb{N}\pi$ be minimal such that some nonzero $a_{-\beta} \in S(\mathfrak{n}^-)$ of weight $-\beta$ lies in $\ker\langle\,,\,\rangle$. Then $\beta > 0$, and the foregoing formula shows that $a_{-\beta}$ is invariant hence zero. Since $\dim U(\mathfrak{n}^+)_\beta = \dim S(\mathfrak{n}^-)_{-\beta} < \infty$, $\forall\, \beta \in \mathbb{N}\pi$, one concludes that $\langle\,,\,\rangle$ is nondegenerate.

Similar assertions hold for the pairing defined by $\langle x, a \rangle' = \varepsilon(\psi'(x)a)$. We may therefore define an $\mathfrak{h}$-linear isomorphism φ of $S(\mathfrak{n}^-)$ by $\langle x, \varphi(a) \rangle = \langle x, a \rangle'$. Then,

$$\langle x, \psi(y)(\varphi(a)) \rangle = \langle xy, \varphi(a) \rangle = \langle xy, a \rangle'$$
$$= \langle x, \psi'(y)a \rangle' = \langle x, \varphi(\psi'(y)a) \rangle,$$

which shows that φ intertwines the actions.

Clearly, $\varphi(1) = 1$. To establish $\varphi(ab) - \varphi(a)\varphi(b) = 0$, it is enough to show that the left hand side is annihilated by $\psi(x) : x \in \mathfrak{n}^+$ when x, a, b are all weight vectors. This results from $\psi(x)$, $\psi'(x)$ being derivations and from φ intertwining ψ, ψ' using an obvious induction on weight. $\square$

Lemma 8.5 is needed to justify the claim just preceding $(*)$ of [J13, 4.4]. Such an action obtains from, say, the conclusion of [J11, 3.4.6], which holds without any assumptions on the Cartan matrix (in the sense of [J11, 3.1.2]). However, we shall also want $S(\mathfrak{n}^-)$ to have a $U(\mathfrak{g})$-algebra structure. For this recall that, after Kac, $\mathfrak{g}$ admits a nondegenerate invariant bilinear form K that identifies $(\mathfrak{b}^-)^*$ with $\mathfrak{b}^+$ (this only needs the Cartan matrix to be symmetrizable [J11, 4.2.2]). The Verma module $M(0)$ of highest weight 0 is just $U(\mathfrak{g})/U(\mathfrak{g})\mathfrak{b}^+ \simeq U(\mathfrak{n}^-)$. Since $U(\mathfrak{g})\mathfrak{b}^+$ is a coideal and a left ideal, $(U(\mathfrak{g})/U(\mathfrak{g})\mathfrak{b})^*$ is a subalgebra of $U(\mathfrak{g})^*$ admitting an action of $\mathfrak{g}$ on the right through the coproduct [J11, 1.1.6] and hence by derivations. This gives $M(0)^*$ a $U(\mathfrak{g})$-algebra structure. The submodule $\delta M(0)$ is also a subalgebra and identifies with $\sigma_c(S(\mathfrak{n}^+)) = S(\mathfrak{n}^-)$ as a $U(\mathfrak{h})$-algebra. Finally, $S(\mathfrak{n}^-)^{\mathfrak{n}^+} = H^0(\mathfrak{n}^+, \delta M(0)) = H_0(\mathfrak{n}^-, M(0)) = \mathbb{C}$ by [J11, 4.2.9].

This result can be improved as follows. By transposing the adjoint action of $\mathfrak{b}^+$ on itself under K, one obtains a *linear action* of $\mathfrak{b}^+$ on $\mathfrak{b}^-$ and hence an $\mathfrak{h}$-equivariant Leibniz action of $U(\mathfrak{n}^+)$ on $S(\mathfrak{b}^-)$. For each positive root β, let $\{a_\beta^i\}$ be a basis for $\mathfrak{n}_\beta^+$ and $\{b_{-\beta}^j\}$ a basis for $\mathfrak{n}_{-\beta}^-$. One checks that $a_\beta^i(b_{-\beta}^j) = \beta \delta_{ij}$. Let I be a maximal ideal of $S(\mathfrak{h})$ not containing any positive root. Then the Leibniz action of $U(\mathfrak{n}^+)$ on $S(\mathfrak{b}^-)$ factors to an $\mathfrak{h}$-equivariant Leibniz action of $U(\mathfrak{n}^+)$ on $S(\mathfrak{n}^-) : \xleftarrow{\sim} S(\mathfrak{b}^-)/I\, S(\mathfrak{n}^-)$, in which the pairing $\mathfrak{n}^- \times \mathfrak{n}^+ \to \mathbb{C}$, defined by restricting $\langle\ ,\ \rangle$ of the lemma, is nondegenerate. From this, one easily checks that $S(\mathfrak{n}^-)^{\mathfrak{n}^+}$ reduces to scalars. By the lemma, any such action is equivalent to the foregoing (linear) one. In particular, it follows that $\delta M(0)$ admits a subspace V isomorphic to $\mathfrak{n}^-$ as an $\mathfrak{h}$-module satisfying $U(\mathfrak{b}^+)V \subset V \oplus \mathbb{C}$. Moreover, given *any* compatible nontrivial $U(\mathfrak{b}^+)$-algebra structure on $\delta M(0)$ (that is, compatible with its $U(\mathfrak{g})$-module structure and for which $(\delta M(0))^{\mathfrak{n}^+} = \mathbb{C}$ as an *algebra*), it follows from [J11, 3.4.4(i)] that $\delta M(0)$ is isomorphic to $S(V)$ as a $U(\mathfrak{b}^+)$-algebra.

The foregoing may be generalized as follows. Given $\pi' \subset \pi$, let $\mathfrak{p}_{\pi'} \supset \mathfrak{b}^+$ be the parabolic subalgebra of $\mathfrak{g}$ whose Levi factor $\mathfrak{r}_{\pi'}$ contains $\mathfrak{h}$ and the root subspaces corresponding to $R' := \mathbb{Z}\pi' \cap R$. Let $\mathfrak{m}_{\pi'}^+$ be the nilradical of $\mathfrak{p}_{\pi'}$, and set $\mathfrak{m}_{\pi'}^- = \sigma_c(\mathfrak{m}_{\pi'}^+)$. Set $M_{\pi'}(0) := M(0)/\sum_{\alpha \in \pi'} M(-\alpha)$, which is just the generalized Verma module obtained by inducing from the trivial one-dimensional $\mathfrak{p}_{\pi'}$-module. Then $\delta M_{\pi'}(0)$ is isomorphic to $S(\mathfrak{m}_{\pi'}^-)$ as an $\mathfrak{h}$-module and hence is actually a $U(\mathfrak{g})$-subalgebra of $\delta M(0)$, which we can view as $S(\mathfrak{m}_{\pi'}^-)$. Again, $(\delta M_{\pi'}(0))^{\mathfrak{m}_{\pi'}^+}$ identifies with $(M_{\pi'}(0)/\mathfrak{m}_{\pi}^- M_{\pi'}(0))^*$ and hence reduces to scalars. On the other hand, taking I to contain just the positive roots in R'^+ gives an $\mathfrak{h}$-equivariant Leibniz action of $U(\mathfrak{m}_{\pi'}^+)$ on $S(\mathfrak{m}_{\pi'}^-)$ with a nondegenerate pairing $\mathfrak{m}_{\pi'}^- \times \mathfrak{m}_{\pi'}^+ \to \mathbb{C}$. Then, by the analog of Lemma 8.5 with $\mathfrak{n}^\pm$, $\mathfrak{h}$ replaced by $\mathfrak{m}_{\pi'}^\pm, \mathfrak{r}_{\pi'}$, we conclude that $\delta M_{\pi'}(0)$ admits a subspace V isomorphic to $\mathfrak{m}_{\pi'}^-$ as an $\mathfrak{r}_{\pi'}$-module and satisfying $U(\mathfrak{p}_{\pi'})V \subset V \oplus \mathbb{C}$. Then, for *any* compatible nontrivial $U(\mathfrak{p}_{\pi'})$-

algebra structure on $\delta M_{\pi'}(0)$, we conclude from [J11, 3.4.4(i)] that $\delta M_{\pi'}(0)$ is isomorphic to $S(V)$ as a $U(\mathfrak{p}_{\pi'})$-algebra.

Let us now show that one may obtain a nondegenerate pairing $\mathfrak{m}_{\pi'}^- \times \mathfrak{m}_{\pi'}^+ \to \mathbb{C}$ from the adjoint action of $\mathfrak{m}_{\pi'}^-$ on $\mathfrak{g}$ (rather than the coadjoint action we have considered). This requires that the Cartan matrix be integrable in the sense of [J12, 3.1.2] and not just symmetric.

Set $\tilde{\mathfrak{m}}_{\pi'}^- = \mathrm{ad}\ U(\mathfrak{n}^-)(\bigoplus_{\alpha \in \pi \setminus \pi'} \mathbb{C} x_{-\alpha}) \subset \mathfrak{m}_{\pi'}^-$.

Lemma 8.6   *One has $\tilde{\mathfrak{m}}_{\pi'}^- = \mathfrak{m}_{\pi'}^-$.*

Proof: If not, there exists $\beta \in R^+ \setminus R'^+$ minimal and a nonzero vector $b_\beta \in \mathfrak{m}_{\pi'}^+$ of weight β such that $K(b_\beta, \tilde{\mathfrak{m}}_{\pi'}^-) = 0$. Certainly, $\beta \notin \pi$. Assume $\alpha \in \pi$ is such that $\beta - \alpha \in R^+ \setminus R'^+$. Then, the invariance and nondegeneracy of K combined with the choice of β imply that $(\mathrm{ad}\ x_{-\alpha})b_\beta = 0$. Setting $b_{-\beta} = \sigma_c(b_\beta)$ gives $(\mathrm{ad}\ x_\alpha)b_{-\beta} = 0$. Let e_0 be the canonical generator of $M_{\pi'}(0)$ of weight 0. The preceding property of $b_{-\beta}$ implies that $x_\alpha(b_{-\beta}e_0) = 0$ for all $\alpha \in \pi$. Then $(\rho - \beta, \rho - \beta) = (\rho, \rho)$ by [J11, 4.2.9], which is impossible by, say, [J11, A.1.3]. $\square$

For each $\alpha \in \pi \setminus \pi'$, one may choose $p_{-\alpha} \in \delta M_{\pi'}(0)$ of weight $-\alpha$ such that $x_\alpha p_{-\alpha} = K(x_{-\alpha}, x_\alpha)$. Take $\beta \in R^+ \setminus R'^+$, $\alpha \in \pi \setminus \pi'$. One checks that

$$K((\mathrm{ad}\ a_{-(\beta-\alpha)})x_{-\alpha}, b_\beta) = b_\beta a_{-(\beta-\alpha)}p_{-\alpha} \qquad (8.1)$$

for all $a_{-(\beta-\alpha)} \in \mathfrak{n}^-$ of weight $-(\beta - \alpha)$ and $b_\beta \in \mathfrak{m}_{\pi'}^+$ of weight β. Set $(\mathrm{ad}\ a_{-(\beta-\alpha)})x_{-\alpha} = \kappa(a_{-(\beta-\alpha)}p_{-\alpha})$.

Lemma 8.7   *κ extends to a linear surjection of $\sum_{\alpha \in \pi \setminus \pi'} U(\mathfrak{n}^-)p_{-\alpha}$ onto $\mathfrak{m}_{\pi'}^-$ satisfying $K(\kappa(p_{-\beta}), b_\beta) = b_\beta p_{-\beta}$.*

Proof: Suppose that $\sum_{\alpha \in \pi \setminus \pi'} a_{-(\beta-\alpha)}p_{-\alpha} = 0$, and set

$$a_{-\beta} = \sum_{\alpha \in \pi \setminus \pi'} (\mathrm{ad}\ a_{-(\beta-\alpha)})x_{-\alpha}.$$

The formula (8.1) shows that $K(a_{-\beta}, b_\beta) = 0$ for all $b_\beta \in \mathfrak{m}_{\pi'}^+$ of weight β. Then $a_{-\beta} = 0$ by the nondegeneracy of K. Hence κ is a well defined map, which by Lemma 8.6 is surjective. By construction, it satisfies the required identity. $\square$

Let V' be an $\mathfrak{h}$-stable complement of $\ker \kappa$. The identity of Lemma 8.7 and the nondegeneracy of K imply that we have an injection $S(V') \hookrightarrow$

$\delta M_{\pi'}(0)$ of $\mathfrak{h}$-modules. Yet as $\mathfrak{h}$-modules both sides are isomorphic to $S(\mathfrak{m}_{\pi'}^-)$, and so $S(V') \xrightarrow{\sim} \delta M_{\pi'}(0)$.

Remark   It does not seem obvious that we can choose $V = V'$.

Proposition 8.8   *Let A be an $\mathcal{O}$-algebra such that $A^{\mathfrak{n}^+} = \mathbb{C}$ as an algebra. Then there exists $\pi' \subset \pi$ such that $A \simeq \delta M_{\pi'}(0)$ as a $U(\mathfrak{g})$-algebra.*

Proof: One defines $\pi' := \{\alpha \in \pi \mid A_{-\alpha} = 0\}$. Since $H_0(\mathfrak{n}^-, \delta A) = \mathbb{C}$ by the hypothesis, it follows that δA is a quotient of $M_{\pi'}(0)$. Then $A \hookrightarrow \delta M_{\pi'}(0)$. Yet A contains V' in the conclusion of Lemma 8.7, and so $A \xrightarrow{\sim} \delta M_{\pi'}(0)$. It remains to check that $\delta M_{\pi'}(0)$ can admit only one (non-trivial) $U(\mathfrak{g})$-algebra structure. In this it is enough to show that the $U(\mathfrak{p}_{\pi'})$-algebra isomorphism $\varphi \colon S(V) \xrightarrow{\sim} \delta M_{\pi'}(0)$ satisfies $x_{-\alpha}\varphi(b_{-\beta}) = \varphi(x_{-\alpha}b_{-\beta})$ for all $\alpha \in \pi$ and all $b_{-\beta} \in \delta M_{\pi'}(0)$ of weight $-\beta$. This follows by the obvious induction on weights applying $x_\gamma : \gamma \in \pi$ to both sides and using that $(\delta M_{\pi'}(0))^{\mathfrak{n}^+}$ reduces to scalars. $\square$

Remark.   This gives an alternative proof of [J12, 3.6].

Exercise.   Let ψ be a nondegenerate $\mathfrak{h}$-equivariant Leibniz action of $U(\mathfrak{n}^-)$ on $S(\mathfrak{n}^-)$. Let ψ' be an $\mathfrak{h}$-equivariant action of $U(\mathfrak{n}^+)$ on $S(\mathfrak{n}^-)$ commuting with ψ. Show that ψ' is Leibniz.

One may place the question of quantizing orbital varieties in a wider context as follows. Let $\mathcal{D}$ denote the (possibly twisted) sheaf of differential operators on the flag variety G/B. The Beilinson–Bernstein theory [BB] relates $U(\mathfrak{g})$-modules with an appropriate central character to coherent $\mathcal{D}$-modules on the flag variety G/B. This allows one to cut down to the family of $U(\mathfrak{g})$-modules corresponding to holonomic $\mathcal{D}$-modules, which seem to be the most interesting. Now the associated variety $\mathcal{V}(\mathcal{M})$ in $T^*(G/B)$ of a coherent $\mathcal{D}$-module $\mathcal{M}$ can be defined in an analogous way to the associated variety $\mathcal{V}(M) \subset \mathfrak{g}^*$ of a $U(\mathfrak{g})$-module M, using the filtration on differential operators defined by degree. As before, $\mathcal{V}(\mathcal{M})$ is homogeneous and involutive. By definition, $\mathcal{M}$ is holonomic if $\dim \mathcal{V}(\mathcal{M}) = \frac{1}{2}\dim T^*(G/B) = \dim G/B$. Then, $\mathcal{V}(\mathcal{M})$ must be the closure of the conormal $C(Y)$ of some smooth subvariety $Y \subset G/B$. By definition, this is the subvariety of $T^*(G/B)$ with base Y such that the fiber over each $y \in Y$ is just the cotangent space $T^*_{y,G/B}$. Obviously, $C(Y)$ has dimension n and is involutive and homogeneous. Let $\pi \colon T^*(G/B) \longrightarrow \mathfrak{g}^*$ be the moment map. Expressing $T^*(G/B)$ as $G \times_B \mathfrak{n}^*$, this is just the map $(g,x) \mapsto gx$ defined by the (coadjoint) action of G on $\mathfrak{g}^*$. Suppose M is the $U(\mathfrak{g})$-module corresponding to $\mathcal{M}$ under the Beilinson–Bernstein equivalence. By a result of W. Borho and J.-L. Brylinski [BBr, Theorem 1.9c], one has $\pi(\mathcal{V}(\mathcal{M})) = \mathcal{V}(M)$.

Holonomic $\mathcal{D}$-modules arise in the study of $(\mathfrak{g}, K)$-modules. Here, K may be (the complexification of) a maximal compact subgroup of (a real form of) G or, more generally, a closed subgroup of G for which $K \setminus G/B$ is finite (for example, B). The classification of simple $(\mathfrak{g}, K)$-modules in terms of holonomic $\mathcal{D}$-modules was indicated by A. A. Beilinson and J. Bernstein [BB]. For such a module, the variety Y is a union of K orbits in G/B.

The foregoing discussion (see also [Vo3]) motivates the following general question. Let K be as before. Let Y be a K-orbit in G/B, and set $V = \pi(\overline{C(Y)})$. Can one give $R[V]$ a natural structure of an admissible $(\mathfrak{g}, K)$-module of finite length? This generalizes our previous question when V is orbital. One may anticipate that this will not be too easy to solve.

Comments and further reading

The main content of Lectures 1–6 and 8 was derived from the author's papers given in the bibliography. Lecture 7 was partly based on [B] and some new results on quantum groups [J11]. A further study of characteristic polynomials and cells is made in [BBM], [Lu], and [T]. For an introduction to enveloping algebras, see [D] and [Jan2]. For the orbit method, one may consult the introduction to [BBG], and for a further point of view on the geometric approach to representation theory, the book of N. Chriss and V. Ginzburg [CG]. The integration method used in Lecture 3 is described in [BGV] and has been used notably in [R] and [V] to obtain results of a similar nature. For recent results on completely prime ideals, see [MG]. For the combinatorics of orbital varieties in $\mathfrak{sl}(n)$, see [Me2]. Finally, the primitive ideal theory of quantum groups is developed in [J11].

Acknowledgments
I should like to express my appreciation to Professors Bent Ørsted and Henrik Schlichtkrull for organizing a delightful meeting in Sandbjerg

in which I had the opportunity to give these lectures. I should also like to thank Dr. Anna Melnikov for useful comments on the manuscript.

References

[BB]   A. Beilinson and J. Bernstein, *Localisation de* $\mathfrak{g}$ *modules*, C.R. Acad. Sci. Paris (A) **292** (1981), 15–18.

[B]   E. Benlolo, *Sur la quantification de certaines variétés orbitales*, Bull. Sci. Math. **3** (1994), 225–241.

[BGV]   N. Berline, E. Getzler, and M. Vergne, *Heat Kernels and the Dirac Operator*, Springer-Verlag, Berlin, 1991.

[BE]   J.-E. Björk and E. K. Ekström, *Filtered Auslander–Gorenstein rings*, pp. 425–448 in *Operator Algebras, Unitary Representations, Enveloping Algebras, and Invariant Theory*, Progress in Mathematics **92**, Birkhäuser, Boston, 1990.

[BBr1] W. Borho and J.-L. Brylinski, *Differential operators on homogeneous spaces*. II, Bull. Soc. Math. France **117** (1989), 167–210.

[BBr2] ———, *Differential operators on homogeneous spaces*. III, Invent. Math. **80** (1985), 1–68.

[BBM]   W. Borho, J.-L. Brylinski, and R. MacPherson, *Nilpotent Orbits, Primitive Ideals and Characteristic Classes*, Progress in Mathematics **78**, Birkhäuser, Boston, 1989.

[BJ]   W. Borho and J. C. Jantzen, *Über primitive Ideale in der Einhüllenden einer halbeinfachen Lie-Algebra*, Invent. Math. **39** (1977), 1–53.

[BBG]   J. L. Brylinski, R. K. Brylinski, V. Guillemin and V. G. Kac (eds.), *Lie Theory and Geometry: In Honor of Bertram Kostant*, Progress in Mathematics **123**, Birkhäuser, Boston, 1994.

[CG]   N. Chriss and V. Ginzburg, *Representation Theory and Complex Geometry*, Birkhäuser, Boston, 1994.

[DES]   M. G. Davidson, T. J. Enright and R. J. Stanke, *Differential Operators and Highest Weight Representations*, Mem. Amer. Math. Soc. **455** (1991).

[D]   J. Dixmier, *Algèbres enveloppantes*, Gauthier-Villars, Paris, 1974.

[Du] M. Duflo, *Sur la classification des ideaux primitifs dans l'algèbre enveloppante d'une algèbre de Lie semi-simple*, Ann. of Math. **105** (1977), 107–120.

[E] T. J. Enright, *Analogues of Kostant's $\mathfrak{u}$ cohomology formulas for unitary highest weight modules*, J. reine angew. Math. **392** (1988), 27–36.

[EHW] T. J. Enright, R. Howe and N.R. Wallach, *A classification of unitary highest weight modules*, pp. 97–143 in *Representation Theory of Reductive Groups*, Progress in Mathematics **40**, Birkhäuser, Boston, 1983.

[EJ] T. J. Enright and A. Joseph, *An intrinsic classification of unitary highest weight modules*, Math. Ann. **288** (1990), 571–594.

[F] A. Freudenthal, *Zur eben Oktavengeometrie*, Indag. Math. **15** (1953), 195–200.

[G] O. Gabber, *The integrability of the characteristic variety*, Amer. J. Math. **103** (1981), 445–468.

[HJ] M. Harris and H. P. Jakobsen, *Singular holomorphic representations and singular modular forms*, Math. Ann. **259** (1982), 227–244.

[H] S. Helgason, *Differential Geometry and Symmetric Spaces*, Academic Press, London, 1962.

[Ho] R. Hotta, *On Joseph's construction of Weyl group representations*, Tohoku Math. J. **36** (1984), 49–74.

[Ja] H. P. Jakobsen, *Hermitian symmetric spaces amd their unitary highest weight modules*, J. Funct. Anal. **52** (1983), 385–412.

[Jan1] J. C. Jantzen, *Moduln mit einem höchsten Gewicht*, Lecture Notes in Mathematics **750**, Springer-Verlag, Berlin, 1979.

[Jan2] ———, *Einhüllende Algebren halbeinfachen Lie-Algebren*, Springer-Verlag, Berlin, 1983.

[J1] ———, *W module structure in the primitive spectrum of the enveloping algebra of a semisimple Lie algebra*, pp. 116–136 in *Non-Commutative Harmonic Analysis*, Lecture Notes in Mathematics **728**, Springer-Verlag, Berlin, 1979.

[J2] ———, *Dixmier's problem for Verma and principal series submodules*, J. Lond. Math. Soc. **20** (1979), 193–204.

[J3]   _____, *Goldie rank in the enveloping algebra of a semisimple Lie algebra. II*, J. Alg. **65** (1980), 284–306.

[J4]   _____, *On the variety of a highest weight module*, J. Alg. **88** (1984), 238–278.

[J5]   _____, *On the associated variety of a primitive ideal*, J. Alg. **93** (1985), 509–523.

[J6]   _____, *A sum rule for scale factors in the Goldie rank polynomials*, J. Alg. **118** (1988), 276–311, Addendum 312–321.

[J7]   _____, *A surjectivity theorem for rigid highest weight modules*, Invent. Math. **92** (1988), 567–596.

[J8]   _____, *On the characteristic polynomials for orbital varieties*, Ann. Ec. Norm. Sup. **22** (1989), 569–603.

[J9]   _____, *The surjectivity theorem, characteristic polynomials and induced ideals*, pp. 85–98 in *The Orbit Method in Representation Theory*, Progress in Mathematics **82**, Birkhäuser, Boston, 1990.

[J10]   _____, *Annihilators and associated varieties of unitary highest weight modules*, Ann. Ec. Norm. Sup. **25** (1992), 1–45.

[J11]   _____, *Quantum Groups and their Primitive Ideals*, Springer-Verlag, Berlin, 1995.

[J12]   _____, *Rings which are modules in the Bernstein–Gelfand–Gelfand $\mathcal{O}$ category*, J. Alg. **113** (1988), 110–126.

[J13]   _____, *Enveloping algebras: Problems old and new*, pp. 385–413 in *Lie Theory and Geometry: In Honor of Bertram Kostant*, Progress in Mathematics **123**, Birkhäuser, Boston, 1994.

[J14]   _____, *Sur les ideaux génériques de l'algèbre des fonctions sur un groupe algébrique*, C. R. Acad. Sci. Paris (I) **321** (1995), 135–140. Preprint, 1995.

[KV]   M. Kashiwara and M. Vergne, *On the Segal–Shale–Weil representations and harmonic polynomials*, Invent. Math. **44** (1978), 1–47.

[KL1]   D. A. Kazhdan and G. Lusztig, *Representations of Coxeter Groups and Hecke Algebras*, Invent. Math. **53** (1979), 165–184.

[KL2]   _____, *A topological approach to Springer's representations*, Adv. Math. **38** (1980), 222–228.

[LS]   T. Levasseur and J. T. Stafford, *Rings of Differential Operators on Classical Rings of Invariants*, Mem. Amer. Math. Soc. **412** (1989).

[Lu]   G. Lusztig, *Characters of Reductive Groups over a Finite Field*, Annals of Mathematics Studies **107**, Princeton University Press, Princeton, New Jersey, 1984.

[M]   I. G. Macdonald, *Some irreducible representations of the Weyl groups*, Bull. London Math. Soc. **4** (1972), 148–150.

[Ma]   O. Mathieu, *Bicontinuity of the Dixmier map*, J. Amer. Math. Soc. **4** (1991), 837–863.

[MG]   W. M. McGovern, *Completely Prime Maximal Ideals and Quantization*, Mem. Amer. Math. Soc. **519** (1994).

[Me1] A. Melnikov, *Irreducibility of the associated variety of simple highest weight modules in* $\mathfrak{sl}(n)$, C.R. Acad. Sci. Paris (I) **316** (1993), 53–57.

[Me2] ———, *Orbital varieties and order relations on Young tableaux*, Preprint, 1994.

[P]   K. R. Parthasarathy, *Criteria for the unitarizability of some highest weight modules*, Proc. Indian Acad. Sci. **86** (1980), 1–24.

[PRV] K. R. Parthasarathy, R. Ranga Rao, V. S. Varadarajan, *Representations of complex Lie groups and Lie algebras*, Ann. of Math. **85** (1967), 383–429.

[R]   W. Rossmann, *Nilpotent orbital integrals in a real semisimple Lie algebra and representations of Weyl groups*, pp.263–287 in *Operator Algebras, Unitary Representations, Enveloping Algebras, and Invariant Theory*, Progress in Mathematics **92**, Birkhäuser, Boston 1990.

[S]   N. Spaltenstein, *On the fixed point set of a unipotent element on the variety of Borel subgroups*, Topology **16** (1977), 203–204.

[St]   R. Steinberg, *On the desingularization of the unipotent variety*, Invent. Math. **36** (1976), 209–224.

[T]   T. Tanisaki, *Characteristic varieties of highest weight modules and primitive quotients*, Adv. Studies in Pure Math. **14** (1988), 1–30.

[V]   M. Vergne, *Polynômes de Joseph et représentations de Springer*, Ann. Ec. Norm. Sup. **23** (1990), 543–562.

98   A. Joseph

[Vo1]   D. A. Vogan, *Ordering in the primitive spectrum of a semisimple Lie algebra*, Math. Ann. **248** (1980), 195–203.

[Vo2]   ———, *The orbit method and primitive ideals for semisimple Lie algebras*, Canad. Math. Soc. Conf. Proc. **5** (1986), 281–316.

[Vo3]   ———, *Associated varieties and unipotent representations*, pp. 315–388 in *Harmonic Analysis on Reductive Groups*, Progress in Mathematics **101**, Birkhäuser, Boston, 1991.

[W1]   N. R. Wallach, *The analytic continuation of the discrete series*, I, II, Trans. Amer. Math. Soc. **251** (1979), 1–17, 19–37.

[W2]   ———, *Real Reductive Groups I*, Academic Press, New York, 1988.

[Z]   M. Zahid, *Orbite nilpotente minimale en type G_2 et operateurs différentiels*, Thèse, Partie 2, Univ. Pierre et Marie Curie, Paris, 1991.

Chapter 3

Discontinuous Groups and Clifford–Klein Forms of Pseudo-Riemannian Homogeneous Manifolds

TOSHIYUKI KOBAYASHI

University of Tokyo

CONTENTS

Introduction

Let G be a Lie group and H a subgroup. A Clifford–Klein form of the homogeneous manifold G/H is a double coset space $\Gamma \backslash G / H$, where Γ is a subgroup of G acting properly discontinuously and freely on G/H. For example, any closed Riemann surface M with genus ≥ 2 is biholomorphic to a compact Clifford–Klein form of the Poincaré plane $G/H = \mathrm{PSL}(2, \mathbb{R})/\mathrm{SO}(2)$. On the other hand, there is *no* compact Clifford–Klein form of the hyperboloid of one sheet $G/H = \mathrm{PSL}(2, \mathbb{R})/\mathrm{SO}(1, 1)$. Even more, there is no infinite discrete subgroup of G that acts properly discontinuously on G/H (the Calabi–Markus phenomenon). We discuss recent developments in the theory of discontinuous groups acting on G/H where G is a real reductive Lie group and H a noncompact reductive subgroup. Geometric ideas of various methods together with a number of examples are presented regarding the fundamental problems:

Which homogeneous manifolds G/H admit properly discontinuous actions of infinite discrete subgroups of G?

Which homogeneous manifolds admit compact Clifford–Klein forms?

1   Clifford–Klein forms

1.1   Homogeneous manifolds

First, we consider the setting:

$$G \supset H.$$

Here, G is a Lie group and H is a closed subgroup of G. Then the coset space G/H equipped with the quotient topology carries a C^∞-manifold structure, so that the natural quotient map

$$\pi : G \to G/H$$

is a C^∞-map. We say G/H is a *homogeneous manifold*.

1.2   Clifford–Klein forms

Second, we consider a more general setting:

$$\Gamma \subset G \supset H.$$

Here, G is a Lie group, H is a closed subgroup, and Γ is a discrete subgroup of G. Then one might ask

Question 1.1   *Does the double coset space $\Gamma \backslash G / H$ equipped with the quotient topology carry a C^∞-manifold structure so that the natural quotient map,*

$$\varpi : G \to \Gamma \backslash G / H,$$

is a C^∞-map ?

If *yes*, we say $\Gamma\backslash G/H$ is a *Clifford–Klein form of the homogeneous manifold G/H.*

Unfortunately, the action of a discrete subgroup Γ on G/H is not always properly discontinuous (see Definition 2.1) when H is not compact, and the quotient topology is not necessarily Hausdorff. Thus, we cannot always expect an affirmative answer to Question 1.1. This is the main difficulty of our subject. However, leaving this question aside for a moment, we first discuss the geometric aspect of Clifford–Klein forms in this section.

1.3   Clifford–Klein forms from the viewpoint of geometry

From the viewpoint of differential geometry, the important point in considering a Clifford–Klein form $\Gamma\backslash G/H$ is that we have a local diffeomorphism $p\colon G/H \to \Gamma\backslash G/H$, with the following commutative diagram:

$$
\begin{array}{ccc}
 & G & \\
\pi \swarrow & & \searrow \varpi \\
G/H & \underset{p}{\longrightarrow} & \Gamma\backslash G/H
\end{array}
$$

Therefore, any G-invariant local structure (e.g., affine, complex, symplectic, Riemannian, or pseudo-Riemannian) on a homogeneous manifold G/H induces the same kind of structure on a Clifford–Klein form $\Gamma\backslash G/H$. In other words, the double coset space $\Gamma\backslash G/H$ is a manifold enjoying the same local properties as G/H, as long as the discrete subgroup Γ allows an affirmative answer to Question 1.1.

Conversely, let us start from a differentiable manifold M endowed with some local structure $\mathcal{T}$ and then explain how a Clifford–Klein form arises. Let $\widetilde{M}$ be the universal covering manifold of M. Then, the local structure $\mathcal{T}$ is also defined on $\widetilde{M}$ through the covering map $\widetilde{M} \to M$. We set

$$
G \equiv \mathrm{Aut}(\widetilde{M}, \mathcal{T}) := \{\varphi \in \mathrm{Diffeo}(\widetilde{M}) : \varphi \text{ preserves the structure } \mathcal{T}\}.
$$

We note that G is the group of isometries if $\mathcal{T}$ is a Riemannian structure; it is the group of biholomorphic automorphisms if $\mathcal{T}$ is a complex structure.

We fix a point $o \in \widetilde{M}$ and write $\bar{o} \in M$ for its image under the covering map $\widetilde{M} \to M$. Then, the fundamental group $\pi_1(M, \bar{o})$ acts effectively on

$\widetilde{M}$ as the covering automorphism. We write Γ for the image of the injection $\pi_1(M, \bar{o}) \hookrightarrow G = \mathrm{Aut}(\widetilde{M}, \mathcal{T})$. That is, Γ is a discrete subgroup of G in the compact open topology, which is isomorphic to $\pi_1(M, \bar{o})$. Then we have a natural diffeomorphism $M \simeq \Gamma \backslash \widetilde{M}$.

Assume $G = \mathrm{Aut}(\widetilde{M}, \mathcal{T})$ is small enough to be a Lie group (see a textbook by Shoshichi Kobayashi [Ko-S] for some sufficient conditions) and is large enough to act transitively on $\widetilde{M}$. Then, $\widetilde{M}$ is represented as a homogeneous manifold $\widetilde{M} \simeq G/H$ with the isotropy subgroup H at $o \in \widetilde{M}$. Consequently, M is naturally represented as a Clifford–Klein form of G/H:

$$
\begin{array}{ccc}
\widetilde{M} & \simeq & G/H \\
\pi_1(M) \downarrow & & \downarrow \Gamma \\
M & \simeq & \Gamma \backslash G/H
\end{array}
$$

We remark that Question 1.1 has automatically an affirmative answer in this case.

We shall give a number of examples of simply connected manifolds $\widetilde{M} \simeq G/H$ equipped with some local structure in Section 1.4; and Clifford–Klein forms of these manifolds in Section 1.5.

1.4 Examples of homogeneous manifolds

First, we present some examples where $G = \mathrm{Aut}(\widetilde{M}, \mathcal{T})$ acts transitively on a simply connected manifold $\widetilde{M}$, so that $\widetilde{M}$ is represented as a homogeneous manifold of G.

Example 1.2   Let

$$
\begin{cases}
\widetilde{M} := \mathbb{R}^n, \\
\mathcal{T} := \text{ the canonical affine connection } \nabla.
\end{cases}
$$

Then $\mathrm{Aut}(\mathbb{R}^n, \nabla) \simeq \mathrm{GL}(n, \mathbb{R}) \ltimes \mathbb{R}^n$ is the affine transformation group, acting transitively on $\mathbb{R}^n$ by

$$
(a, b) \cdot x := ax + b \qquad \text{for } (a, b) \in \mathrm{GL}(n, \mathbb{R}) \ltimes \mathbb{R}^n, \quad x \in \mathbb{R}^n.
$$

The isotropy subgroup at the origin $0 \in \mathbb{R}^n$ is isomorphic to $\mathrm{GL}(n, \mathbb{R})$, and therefore the manifold $\mathbb{R}^n$ is represented as the homogeneous manifold $\mathrm{GL}(n, \mathbb{R}) \ltimes \mathbb{R}^n / \mathrm{GL}(n, \mathbb{R})$.

Example 1.3   Let

$$\begin{cases} \widetilde{M} := \mathbb{R}^n, \\ \mathcal{T} := \text{standard Riemannian structure } g. \end{cases}$$

Then $\mathrm{Aut}(\mathbb{R}^n, g) \simeq \mathrm{O}(n) \ltimes \mathbb{R}^n$ is the Euclidean motion group, acting transitively on $\mathbb{R}^n$ by

$$(a, b) \cdot x := ax + b \qquad \text{for } (a, b) \in \mathrm{O}(n) \ltimes \mathbb{R}^n, \quad x \in \mathbb{R}^n.$$

Thus, $\mathbb{R}^n$ is represented as the homogeneous manifold $\mathrm{O}(n) \ltimes \mathbb{R}^n / \mathrm{O}(n)$.

Example 1.4   Let

$$\begin{cases} \widetilde{M} := \left\{ x \in \mathbb{R}^{n+1} : x_1^2 + \cdots + x_n^2 - x_{n+1}^2 = 1 \right\} & (n \geq 3), \\ \mathcal{T} := \text{ the Lorentz metric } g \text{ induced from } dx_1^2 + \cdots + dx_n^2 - dx_{n+1}^2. \end{cases}$$

Let $\mathrm{O}(n, 1)$ be the indefinite orthogonal group preserving the quadratic form $x_1^2 + \cdots + x_n^2 - x_{n+1}^2$. Then $\mathrm{O}(n, 1)$ acts transitively and isometrically on the hyperboloid of one sheet $\widetilde{M}$, and it coincides with the group of isometries of $\widetilde{M}$. Thus, $\mathrm{Aut}(\widetilde{M}, g) = \mathrm{O}(n, 1)$ and $\widetilde{M} \simeq \mathrm{O}(n, 1)/\mathrm{O}(n - 1, 1)$.

Example 1.5   Let

$$\begin{cases} \widetilde{M} := \mathbb{C}, \\ \mathcal{T} := \text{ the standard complex structure } J. \end{cases}$$

Then the group of biholomorphic automorphisms is given by $\mathrm{Aut}(\mathbb{C}, J) \simeq \mathbb{C}^\times \ltimes \mathbb{C}$, a semidirect product of $\mathbb{C}^\times$ and $\mathbb{C}$ with $\mathbb{C}$ normal. The action of $\mathbb{C}^\times \ltimes \mathbb{C}$ on $\mathbb{C}$ is given by the complex affine transformation:

$$(a, b) \cdot z = az + b \qquad \text{for } (a, b) \in \mathbb{C}^\times \ltimes \mathbb{C}.$$

This action is obviously transitive, so that the complex plane $\mathbb{C}$ is represented as the homogeneous manifold $\mathbb{C}^\times \ltimes \mathbb{C}/\mathbb{C}^\times$.

Example 1.6   Let

$$\begin{cases} \widetilde{M} := \mathbb{C}P^1 & \text{(the complex projective space)}, \\ \mathcal{T} := \text{the standard complex structure } J. \end{cases}$$

The natural action of $\mathrm{SL}(2, \mathbb{C})$ on $\mathbb{C}^2 \setminus \{0\}$ induces a transitive action of $\mathrm{PSL}(2, \mathbb{C})$ on the projective space $\mathbb{C}P^1 = \left(\mathbb{C}^2 \setminus \{0\}\right)/\mathbb{C}^\times \simeq \mathbb{C} \cup \{\infty\}$, which is given by the linear fractional transformation:

$$\begin{pmatrix} a & b \\ c & d \end{pmatrix} \cdot z = \frac{az + b}{cz + d} \quad \text{for} \quad \begin{pmatrix} a & b \\ c & d \end{pmatrix} \in \mathrm{PSL}(2, \mathbb{C}) = \mathrm{SL}(2, \mathbb{C})/\{\pm I_2\}.$$

Then $\mathrm{PSL}(2, \mathbb{C}) \simeq \mathrm{Aut}(\mathbb{C}P^1, J)$, the group of biholomorphic automorphisms. The projective space $\mathbb{C}P^1$ is represented as a homogeneous manifold $\mathrm{PSL}(2, \mathbb{C})/B$, where B is a Borel subgroup of $\mathrm{PSL}(2, \mathbb{C})$.

Example 1.7   Let

$$\begin{cases} \widetilde{M} := \mathcal{H} \equiv \{z \in \mathbb{C} : \mathrm{Im}\, z > 0\} & \text{(the Poincaré plane)}, \\ \mathcal{T} := \text{the standard complex structure } J. \end{cases}$$

Then the group of biholomorphic automorphisms is given by $\mathrm{Aut}(\mathcal{H}, J) \simeq \mathrm{PSL}(2, \mathbb{R})$, where the action of $\mathrm{PSL}(2, \mathbb{R})$ is also defined by the linear fractional transformation:

$$\begin{pmatrix} a & b \\ c & d \end{pmatrix} \cdot z = \frac{az + b}{cz + d} \quad \text{for} \quad \begin{pmatrix} a & b \\ c & d \end{pmatrix} \in \mathrm{PSL}(2, \mathbb{R}) = \mathrm{SL}(2, \mathbb{R})/\{\pm I_2\}.$$

This action is transitive, so that the Poincaré plane $\mathcal{H}$ is represented as the homogeneous manifold $\mathrm{PSL}(2, \mathbb{R})/\mathrm{SO}(2)$.

1.5   Examples of Clifford–Klein forms

Next, we present a number of examples of Clifford–Klein forms of the homogeneous manifolds given in Examples 1.2–1.7. We also explain that these examples are closely related to the following interesting topics:

Examples 1.8 and 1.9	The Auslander Conjecture on the fundamental group of compact complete affine manifolds
Example 1.10	The Calabi–Markus phenomenon for relativistic spherical forms
Example 1.11	The uniformization theorem of Riemann surfaces due to Klein, Poincaré, and Koebe

Example 1.8 (see Example 1.2)   An *affine* manifold M is a manifold that admits a torsion free affine connection the curvature tensor of which vanishes. It is said to be *complete* if every geodesic can be defined on all time intervals. Then it is known that the universal covering of any complete affine manifold $M = M^n$ is isomorphic to $(\mathbb{R}^n, \nabla)$ as affine manifolds. Therefore, it follows from Example 1.2 that M can be represented as a Clifford–Klein form

$$M \simeq \Gamma \backslash \mathrm{GL}(n, \mathbb{R}) \ltimes \mathbb{R}^n / \mathrm{GL}(n, \mathbb{R}),$$

where Γ is a discrete subgroup of $\mathrm{GL}(n, \mathbb{R}) \ltimes \mathbb{R}^n$ that is isomorphic to the fundamental group $\pi_1(M)$.

Example 1.9 (see Example 1.3)   Retain the notation in Example 1.8. If M^n is a Riemannian manifold and if M^n is a complete affine manifold (see Example 1.8) for the Levi-Civita connection, then the universal covering of M is isometric to $\mathbb{R}^n$ endowed with the standard Riemannian metric g. Therefore it follows from Example 1.3 that M is represented as another Clifford–Klein form

$$M \simeq \Gamma \backslash \mathrm{O}(n) \ltimes \mathbb{R}^n / \mathrm{O}(n)$$

with $\Gamma \subset \mathrm{O}(n) \ltimes \mathbb{R}^n$.

Similarly, if M^n is a Lorentz manifold (namely, M carries a pseudo-Riemannian metric of type $(n-1, 1)$), and if M is a complete affine manifold for the Levi-Civita connection, then M is reduced to be a Clifford–Klein form

$$M \simeq \Gamma \backslash \mathrm{O}(n - 1, 1) \ltimes \mathbb{R}^n / \mathrm{O}(n - 1, 1)$$

with $\Gamma \subset \mathrm{O}(n - 1, 1) \ltimes \mathbb{R}^n$.

Remark (the Auslander conjecture)   Regarding to Examples 1.8 and 1.9, we mention the Auslander conjecture, which asserts that *the fundamental group π_1 of any compact complete affine manifold is virtually solvable* (see [Au], [Mi], [Ma], and references therein). In view of Example 1.8, this is equivalent to the conjecture that *a discrete group Γ is virtually solvable if $\Gamma \backslash \mathrm{GL}(n, \mathbb{R}) \ltimes \mathbb{R}^n / \mathrm{GL}(n, \mathbb{R})$ is a compact Clifford–Klein form of a homogeneous manifold* $\mathrm{GL}(n, \mathbb{R}) \ltimes \mathbb{R}^n / \mathrm{GL}(n, \mathbb{R}) \simeq \mathbb{R}^n$. Auslander's conjecture remains open except for some special cases such as $\Gamma \subset \mathrm{O}(n) \ltimes \mathbb{R}^n$ (Bieberbach's theorem, see [Ra, Corollary 8.26]) or $\Gamma \subset \mathrm{O}(n - 1, 1) \ltimes \mathbb{R}^n$ ([GK]; see also [To] for a generalization to rank one groups). The geometric meaning of $\Gamma \subset \mathrm{O}(n) \ltimes \mathbb{R}^n$ (or $\Gamma \subset \mathrm{O}(n-1,1) \ltimes \mathbb{R}^n$) is that the affine connection is the Levi-Civita connection of the standard Riemannian (or Lorentz) metric (see Example 1.9).

Example 1.10 (a relativistic spherical space form; see Example 1.4)   In the physics of relativistic cosmology, the space–time continuum is taken to be a Lorentz manifold M^4. A relativistic spherical space form is a complete Lorentz manifold M^n for $n \geq 3$ with constant sectional curvature $K = +1$. Any relativistic spherical space form M is represented as a Clifford–Klein form:

$$M \simeq \Gamma \backslash \mathrm{O}(n,1)/\mathrm{O}(n-1,1),$$

where Γ is a discrete subgroup of $\mathrm{O}(n,1)$ that is isomorphic to the fundamental group $\pi_1(M)$. We shall see that Γ must be a finite group in Section 3 (the Calabi–Markus phenomenon [CM]).

Example 1.11 (the uniformization theorem of Riemann surfaces; see Examples 1.5–1.7)   The uniformization theorem of Riemann surfaces due to Klein, Poincaré, and Koebe asserts that a simply connected Riemann surface M is biholomorphic to one of the following complex manifolds:

$$\mathbb{C}, \quad \mathbb{C}P^1, \quad \text{or } \mathcal{H}$$

We recall that the Riemann mapping theorem is the special case obtained by assuming M to be a domain of $\mathbb{C}$. We refer the reader to [Sp], [AhSa] for details.

The point here is that any of simply connected Riemann surfaces $\mathbb{C}$, $\mathbb{C}P^1$, and $\mathcal{H}$ has a transitive transformation group of biholomorphic automorphisms, as we have seen in Examples 1.5, 1.6, and 1.7, respectively. Consequently, any connected Riemann surface M is represented as a Clifford–Klein form

$$M \simeq \Gamma \backslash G/H,$$

where (G, H) is $(\mathbb{C}^\times \ltimes \mathbb{C}, \mathbb{C}^\times)$, $(\mathrm{PSL}(2,\mathbb{C}), B)$, or $(\mathrm{PSL}(2,\mathbb{R}), \mathrm{SO}(2))$, and Γ is a discrete subgroup of G that is isomorphic to $\pi_1(M)$.

In particular, suppose M is compact. The rank of the first homology group $H_1(M)$ is always even, and it is denoted by $2g$. Then, g is said to be the *genus* of the Riemann surface M, and such a Riemann surface is usually denoted by M_g. Then the universal covering of M_g is uniquely determined up to biholomorphic automorphism by the genus g. In fact, we have

$$M_g \simeq \Gamma\backslash \mathrm{PSL}(2,\mathbb{C})/B, \qquad \Gamma = \pi_1(M_g) = \{e\} \qquad (g=0),$$
$$M_g \simeq \Gamma\backslash \mathbb{C}^\times \ltimes \mathbb{C}/\mathbb{C}^\times, \qquad \Gamma = \pi_1(M_g) \simeq \mathbb{Z}^2 \qquad (g=1),$$
$$M_g \simeq \Gamma\backslash \mathrm{PSL}(2,\mathbb{R})/\mathrm{SO}(2), \qquad \Gamma = \pi_1(M_g) \qquad (g \geq 2).$$

2   Discontinuous actions (discrete and continuous version)

Let G be a Lie group, H a closed subgroup, and Γ a discrete subgroup of G. In Section 1, we have defined a *Clifford–Klein form* of a homogeneous manifold G/H as the double coset space $\Gamma\backslash G/H$ if it is Hausdorff and carries naturally a C^∞ structure (see Question 1.1). This condition is satisfied if Γ acts properly discontinuously and freely on G/H (see Definition 2.1), as we shall see in Lemma 2.2. The purpose of this section is to understand properly discontinuous actions by exhibiting a number of typical bad features of the action of noncompact groups.

2.1   Clifford–Klein form of $\mathrm{PSL}(2,\mathbb{R})/\mathrm{SO}(2)$

First, we recall Example 1.11. Let M_g be a closed Riemann surface with genus $g \geq 2$. The universal covering manifold of M_g is biholomorphic to the the Poincaré plane $\mathcal{H}$. So M_g is represented as a compact Clifford–Klein form $\Gamma\backslash \mathrm{PSL}(2,\mathbb{R})/\mathrm{SO}(2)$ of the Poincaré plane $\mathrm{PSL}(2,\mathbb{R})/\mathrm{SO}(2)$, where Γ is an infinite discrete subgroup of $\mathrm{PSL}(2,\mathbb{R})$ isomorphic to $\pi_1(M_g)$.

There are two directions of generalization of this classical example, namely,

i) generalization to higher dimensions (e.g., Theorem 5.2)
ii) generalization to noncompact isotropy subgroups

Here, we are interested in the latter direction.

2.2   Clifford–Klein form of $\mathrm{PSL}(2,\mathbb{R})/\mathrm{SO}(1,1)$

There is *no* compact Clifford–Klein form of the hyperboloid of one sheet $\mathrm{PSL}(2,\mathbb{R})/\mathrm{SO}(1,1)$. This fact can be proved in various ways, which are of importance as typical examples of later arguments. We indicate here four different proofs, which we shall elaborate in the following sections.

i) *A direct calculation* (this section). Assume that Γ is a discrete subgroup of $\mathrm{PSL}(2,\mathbb{R})$ such that $\sharp\Gamma = \infty$. Then we can show that the action of Γ on $\mathrm{PSL}(2,\mathbb{R})/\mathrm{SO}(1,1)$ is never properly discontinuous, and the corresponding quotient space $\Gamma\backslash\mathrm{PSL}(2,\mathbb{R})/\mathrm{SO}(1,1)$ is not Hausdorff. We consider first the most nontrivial case, where Γ consists of unipotent elements. For example, let us consider the case where Γ is a discrete subgroup of $\mathrm{PSL}(2,\mathbb{R})$ generated by a single unipotent element. Then, the quotient topology of $\Gamma\backslash\mathrm{PSL}(2,\mathbb{R})/\mathrm{SO}(1,1)$ is locally Hausdorff but non-Hausdorff, as we shall explain in this section (see Example 2.8 and Exercise 2.9(iii),(v)). On the other hand, if Γ does not contain a nilpotent element (still we assume $\sharp\Gamma = \infty$), then the action of Γ has an accumulating point in $\mathrm{PSL}(2,\mathbb{R})/\mathrm{SO}(1,1)$ (see Definition 2.4), so that $\Gamma\backslash\mathrm{PSL}(2,\mathbb{R})/\mathrm{SO}(1,1)$ is not Hausdorff, too (see Example 2.7(ii) and Exercise 2.9(iv),(v)).

ii) *Calabi–Markus phenomenon* (Section 3). We will give a necessary and sufficient condition in terms of the ranks of G and H for a homogeneous space G/H of reductive type to admit an infinite discontinuous group (see Theorem 3.10 for a criterion of the so-called Calabi–Markus phenomenon; see also Example 1.10). The idea of proof is to study an analogous problem in a continuous setting and to look at the infinite points by means of the Cartan decomposition.

iii) *Hirzebruch's proportionality principle* (Section 4). A Clifford–Klein form M of $\mathrm{PSL}(2,\mathbb{R})/\mathrm{SO}(1,1)$, the hyperboloid of one sheet, carries an indefinite-Riemannian metric of type $(1,1)$ induced from the Killing form. This metric gives rise to a nonvanishing vector field on M. If M were compact and orientable, this would imply the vanishing of the Euler–Poincaré class of M thanks to a theorem of Poincaré–Hopf. On the other hand, the Euler–Poincaré class of a "compact real form" $S^2 \simeq \mathrm{SU}(2)/\mathrm{SO}(2)$ does not vanish. This leads to a contradiction by a generalized Hirzebruch's proportionality principle (see Corollary 4.6). The idea here leads to a necessary condition for the existence of compact Clifford–Klein forms (see Corollary 4.12 and Example 4.13).

iv) *Semisimple orbits, invariant complex structure* (Section 5). The hyperboloid of one sheet is realized as a semisimple orbit of the adjoint representation of $\mathrm{PSL}(2,\mathbb{R})$. We shall see in Section 5 that a semisimple orbit having a compact Clifford–Klein form must be isomorphic to an elliptic orbit, so that it carries an invariant complex structure (see Corollary 5.13). But this is not the case for the hyperboloid of one sheet.

2.3   Properly discontinuous action

A distinguished feature in the setting Section 2.2 is that the isotropy subgroup $H \simeq SO(1,1)$ is noncompact. As a consequence, the action of a discrete subgroup Γ on G/H is not automatically properly discontinuous. Here, we recall the definition of properly discontinuous actions:

Suppose that a discrete group Γ acts continuously on a locally compact Hausdorff topological space X. For a subset S of X, we put

$$\Gamma_S := \{ \gamma \in \Gamma : \gamma S \cap S \neq \emptyset \}.$$

Note that if S is a singleton $\{p\}$ ($p \in X$), then Γ_S is nothing but the isotropy subgroup at p. In general, Γ_S is not a subgroup.

Definition 2.1   The action of Γ on X is said to be

i) *properly discontinuous*, if Γ_S is a finite subset for any compact subset S of X;

ii) *free*, if $\Gamma_{\{p\}}$ is trivial for any $p \in X$.

Then we have the following standard fact:

Lemma 2.2   *Suppose that a discrete group Γ acts on a [C^∞, Riemannian, complex, ...] manifold X properly discontinuously and freely [and smoothly, isometrically, holomorphically, ...]. Equipped with the quotient topology, $\Gamma \backslash X$ is then a Hausdorff topological space, on which a manifold structure is uniquely defined so that*

$$\pi : X \to \Gamma \backslash X$$

is locally homeomorphic [diffeomorphic, isometric, biholomorphic, ...].

Although this lemma is well known, we give a sketch of its proof so that readers get used to the definition of properly discontinuous actions.

Sketch of proof: 1) (Hausdorff) (We use only the assumption that the action is properly discontinuous.)   Take $x, y \in X$ so that $\pi(x) \neq \pi(y)$.

We want to find neighborhoods V, W of X such that $x \in V$, $y \in W$, and $\pi(V) \cap \pi(W) = \emptyset$. First, we take relatively compact neighborhoods V_1 and W_1 of X such that $x \in V_1$, $y \in W_1$. As the action of Γ on X is properly discontinuous, $\Gamma_{V_1 \cup W_1} = \{\gamma \in \Gamma : \gamma(V_1 \cup W_1) \cap (V_1 \cup W_1) \neq \emptyset\}$ is a finite set, say, $\{\gamma_1, \ldots, \gamma_k\}$. Second, we take neighborhoods V and W such that $x \in V \subset V_1$, $y \in W \subset W_1$, and $\gamma_j V \cap W = \emptyset$ $(j = 1, \ldots, k)$. Then, we have $\Gamma V \cap \Gamma W = \emptyset$, that is, $\pi(V) \cap \pi(W) = \emptyset$, which we wanted to prove.

2) (manifold structure) The proof is quite similar to that of (1). In fact, by using the assumption that the action is properly discontinuous and free, we can find a neighborhood $V \subset X$ at each point $x \in X$ such that

$$\{\gamma \in \Gamma : \gamma V \cap V \neq \emptyset\} = \{e\}.$$

Such an open set V is homeomorphic to $\pi(V) \subset \Gamma \backslash X$, and these sets form a basis for the open sets giving local charts of $\Gamma \backslash X$. □

Here is a necessary condition that the action of Γ on X is properly discontinuous.

Lemma 2.3 *Suppose a discrete group Γ acts properly discontinuously on a locally compact, Hausdorff topological space X. Then we have*

  i) There is no accumulating point of the action of Γ on X.
  ii) Each Γ-orbit is a closed subset of X.

Here we recall

Definition 2.4 For each element x of X, the Γ-orbit through x is a subset of X given by

$$\Gamma \cdot x := \{\gamma x : \gamma \in \Gamma\}.$$

We say y $(\in X)$ is *an accumulating point of the Γ-orbit $\Gamma \cdot x$* if

$$\sharp(U \cap \Gamma \cdot x) = \infty$$

for any neighborhood U of y in X. We say y $(\in X)$ is *an accumulating point of the action of Γ on X* if there exists $x \in X$ such that y is an accumulating point of the orbit $\Gamma \cdot x$.

Proof of Lemma 2.3:   i) We want to show that the action of Γ on X is not properly discontinuous if there exists an accumulating point of the action of Γ on X. Suppose that $y \in X$ is an accumulating point of the action of Γ on X. That is, there exists $x \in X$ such that $\sharp(U \cap \Gamma \cdot x) = \infty$ for any neighborhood U of y. We choose U to be relatively compact and define $S := U \cup \{x\}$ and $\Gamma' := \{\gamma \in \Gamma : \gamma x \in U\}$. Then we have $\sharp\Gamma' = \infty$, because $\sharp(U \cap \Gamma \cdot x) = \infty$. Then we have $\sharp\Gamma_S = \infty$, because

$$\Gamma_S = \{\gamma \in \Gamma : \gamma S \cap S \neq \emptyset\} \supset \Gamma'.$$

Hence, the action of Γ on X is not properly discontinuous.

ii) We want to show that the action of Γ on X is not properly discontinuous if there exists a nonclosed orbit $\Gamma \cdot x$. Take $y \in \overline{\Gamma \cdot x} \setminus \Gamma \cdot x$ and relatively compact, open neighborhoods U_j $(j = 1, 2, \ldots)$ of y such that $U_1 \supset U_2 \supset \cdots$ and that $\bigcap_j U_j = \{y\}$. Then, we can take $\gamma_j \in \Gamma$ such that $\gamma_j \cdot x \in U_j$ for each $j \in \mathbb{N}$, because $\Gamma \cdot x \cap U_j \neq \emptyset$. Then we have $\sharp \{\gamma_j : j = 1, 2, \ldots\} = \infty$, because $\bigcap_j U_j = \{y\}$. Finally, we put $S := U_1 \cup \{x\}$, and we have $\sharp\Gamma_S = \infty$, because

$$\Gamma_S \supset \bigcup_j \{\gamma \in \Gamma : \gamma \cdot x \in U_j\} \supset \{\gamma_j : j = 1, 2, \ldots\}.$$

Hence, the action of Γ on X is not properly discontinuous. $\square$

2.4   Discontinuous groups acting on homogeneous manifolds

Definition 2.5   A discrete subgroup Γ of G is said to be *a discontinuous group acting on* G/H if the action of Γ on G/H is properly discontinuous.

If Γ is a discontinuous group acting on G/H and if the action of Γ on G/H is free, then the double coset space $\Gamma \backslash G/H$ carries a natural C^∞-manifold structure from Lemma 2.2, so that $\Gamma \backslash G/H$ is a Clifford–Klein form of the homogeneous manifold G/H. Thus, we have an affirmative answer to Question 1.1 in this case.

Here, the point is that Γ is assumed to be a subgroup of G, so that $\Gamma \backslash G/H$ inherits any G-invariant (local) structure on the homogeneous manifold G/H. One should keep in mind the essential difference between the action on G/H of a discrete group $\Gamma \subset G$ (our case) and a discrete group $\Gamma \subset \mathrm{Diffeo}(G/H)$. For example, if $G/H = \mathrm{SL}(2, \mathbb{R})/\mathrm{SO}(1, 1)$ (a hyperboloid of one sheet), then G/H is diffeomorphic to $S^1 \times \mathbb{R}$ (a cylinder), which admits a properly discontinuous action of $\mathbb{Z}(\subset \mathrm{Diffeo}(G/H))$ along the direction of $\mathbb{R}$. But this action does not come from $\mathrm{SL}(2,\mathbb{R})$ and is not isometric with respect to a natural indefinite Riemannian metric on the hyperboloid.

2.5   A remark

In this section, proper discontinuity is essentially important, but freeness is less important. In fact, if a discrete group Γ acts on a manifold X properly discontinuously, then the isotropy group $\Gamma_{\{p\}}$ at $p \in X$ is indeed not necessarily trivial, but it is always finite. Correspondingly, $\Gamma \backslash X$ is not necessarily a smooth manifold but still has a nice structure called V-manifold in the sense of Satake [Sa] or called an *orbifold* in the sense of Thurston (see also [Car]). Moreover, if Γ acts properly discontinuously on X and if $\Gamma' \subset \Gamma$ is a torsion free subgroup (i.e., $x \in \Gamma'$, $x^n = e$ $(n \geq 1) \Rightarrow x = e$), then the action of Γ' on X is properly discontinuous and free. In view of this, the following result due to Selberg is quite useful.

Theorem 2.6 ([Sel, Lemma 8])   *A finitely generated matrix group has a torsion free subgroup of finite index.*

2.6   Examples and exercises

Example 2.7   Suppose a discrete group $\Gamma := \mathbb{Z}$ acts on a manifold $X := \mathbb{R}$ in two different manners:

i) $\Gamma \times X \to X$, $(n, x) \mapsto x + n$.
ii) $\Gamma \times X \to X$, $(n, x) \mapsto 2^n x$.

The action in (i) is properly discontinuous and free. The resulting quotient manifold $\Gamma \backslash X$ is diffeomorphic to S^1.

On the other hand, the action in (ii) is *not* properly discontinuous: in fact, $\Gamma_S = \mathbb{Z}$ is not a finite set if we take $S := [0, 1]$, the unit interval in $\mathbb{R}$. The resulting quotient space $\Gamma \backslash X$ has a non-Hausdorff topology. That is, $\Gamma \backslash X$ is homeomorphic to $S^1 \cup \{\text{point}\} \cup S^1$, which is topologized to be connected !

In Example 2.7(ii), we easily see that the action of Γ is not properly discontinuous because the origin 0 is an accumulation point of the action of Γ. The next example is more subtle, without accumulation points.

Example 2.8   Suppose a discrete group $\Gamma := \mathbb{Z}$ acts on a manifold $X := \mathbb{R}^2 \setminus \{0\}$ in the following manner:

$$\Gamma \times X \to X, \qquad (n, (x, y)) \mapsto (2^n x, 2^{-n} y).$$

Then this action is *not* properly discontinuous. In fact, we let $B_\varepsilon(x, y)$ be a ball of radius ε with the center (x, y) and put $S := \overline{B_{1/2}(1, 0)} \cup \overline{B_{1/2}(0, 1)}$. Then, it is an easy exercise to see that $\Gamma_S \equiv \{\gamma \in \mathbb{Z} : \gamma S \cap S \neq \emptyset\}$ is equal to $\mathbb{Z}$. We note that there is no accumulation point of Γ. In fact every Γ-orbit is closed in X. The resulting quotient topology of $\Gamma \backslash X$ is not Hausdorff, though it is *locally Hausdorff* in the sense that one can find a Hausdorff neighborhood of each point of $\Gamma \backslash X$. A picture of a similar topology as that of $\Gamma \backslash X$ is illustrated by the following one-dimensional example:

Exercise 2.9   In the setting as in Example 2.8, prove the following:

i) $\Gamma_S = \mathbb{Z}$ if $S = \overline{B_{1/2}(1, 0)} \cup \overline{B_{1/2}(0, 1)}$.

ii) Let $G = \mathrm{SL}(2, \mathbb{R})$ act naturally on $\mathbb{R}^2 \setminus \{0\}$ from the left. This action is transitive, and the isotropy group at $\begin{pmatrix} 1 \\ 0 \end{pmatrix}$ is

$$H = \left\{ \begin{pmatrix} 1 & x \\ 0 & 1 \end{pmatrix} : x \in \mathbb{R} \right\},$$

so that we have a diffeomorphism:

$$G/H \simeq \mathbb{R}^2 \setminus \{0\}.$$

Let

$$L := \left\{ \begin{pmatrix} y & 0 \\ 0 & y^{-1} \end{pmatrix} : y > 0 \right\}, \qquad L_{\mathbb{Z}} := \left\{ \begin{pmatrix} 2^x & 0 \\ 0 & 2^{-x} \end{pmatrix} : x \in \mathbb{Z} \right\} \simeq \mathbb{Z}.$$

Via the isomorphisms $G/H \simeq \mathbb{R}^2 \setminus \{0\}$ and $L_{\mathbb{Z}} \simeq \Gamma\,(= \mathbb{Z})$, the action of $L_{\mathbb{Z}}$ on G/H coincides with that of Γ on $\mathbb{R}^2 \setminus \{0\}$ in Example 2.8. Thus, the action of $L_{\mathbb{Z}}$ on G/H is not properly discontinuous, and the quotient topology on $L_{\mathbb{Z}} \backslash G/H$ is not Hausdorff.

iii) Let

$$H_{\mathbb{Z}} := \left\{ \begin{pmatrix} 1 & x \\ 0 & 1 \end{pmatrix} : x \in \mathbb{Z} \right\}.$$

Then the action of $H_{\mathbb{Z}}$ on G/L is not properly discontinuous, and the quotient topology on $H_{\mathbb{Z}} \backslash G/L$ is not Hausdorff.

iv) The action of $L_{\mathbb{Z}}$ on G/L has an accumulating point and is not properly discontinuous, either.

v) We define a quadratic form on $\mathbb{R}^2$ by $Q \begin{pmatrix} x \\ y \end{pmatrix} := xy$. The polarization of Q gives an indefinite metric of signature $(1,1)$. Then L is the identity component of

$$\left\{ g \in \mathrm{GL}(2,\mathbb{R}) : Q(g \begin{pmatrix} x \\ y \end{pmatrix}) = Q \begin{pmatrix} x \\ y \end{pmatrix} \quad \text{for any } x, y \in \mathbb{R} \right\} \simeq \mathrm{O}(1,1).$$

Thus, $G/L \simeq \mathrm{SL}(2,\mathbb{R})/\mathrm{SO}_0(1,1)$.

Hint: The topology of $H_{\mathbb{Z}} \backslash G/L$ has similar features as Example 2.8, whereas that of $L_{\mathbb{Z}} \backslash G/L$ is similar to Example 2.7(ii). The actions in (ii) and in (iii) are in a kind of duality, between the action of L on G/H and that of H on G/L. That is, in the setting

$$L_{\mathbb{Z}} \subset L \subset G \supset H \supset H_{\mathbb{Z}}$$

with $L/L_{\mathbb{Z}}$ and $H/H_{\mathbb{Z}}$ compact, the action of $L_{\mathbb{Z}}$ on G/H is properly discontinuous if and only if that of $H_{\mathbb{Z}}$ on G/L is properly discontinuous (see Lemma 2.16(i); see also Observation 2.12 and [Bou, Chapitre 3]).

2.7 Basic problems in a discrete setting

Here are the basic problems in the theory of discontinuous groups acting on a homogeneous manifold:

Problem 2.10 *Suppose G is a Lie group and H is a closed subgroup.*

   i) Which homogeneous manifold G/H admits an infinite discontinuous group?

   ii) Which homogeneous manifold G/H admits a compact Clifford–Klein form?

(i) is a first step in the study of discontinuous groups acting on homogeneous manifolds. The existence of a compact Clifford–Klein form in (ii) would be interesting not only from the viewpoint of geometry but also from that of harmonic analysis and representation theory (see Open problem (x) in Section 6). We will study (i) in Section 3, and (ii) in Section 5 (partly also in Sections 3 and 4).

2.8   Proper actions — as a continuous analog of properly discontinuous actions

In general, the study of a discrete group is quite difficult. Our approach is to approximate the action of discrete groups by that of connected Lie groups. For this purpose, it is crucial to find a continuous analog of a properly discontinuous action:

Definition 2.11 (see [Pa])   Suppose that a locally compact topological group L acts continuously on a locally compact topological space X. For a subset S of X, we define a subset of L by

$$L_S = \{\gamma \in L : \gamma S \cap S \neq \emptyset\} .$$

The action of L on X is said to be *proper* if and only if L_S is compact for every compact subset S of X.

Compared with the definition of proper discontinuity (Definition 2.1), compactness in Definition 2.11 has now replaced by finiteness. We note that the action of L on X is properly discontinuous if and only if the action of L on X is proper and L is discrete, because a discrete and compact set is finite.

2.9   An observation

The following elementary observation is a bridge between the action of a discrete group and that of a connected group.

Observation 2.12 ([Ko2, Lemma 2.3])   *Suppose a Lie group L acts on a locally compact space X. Let Γ be a cocompact discrete subgroup of L. Then*

  1) The L-action on X is proper if and only if the Γ-action is properly discontinuous.

  2) $L\backslash X$ is compact if and only if $\Gamma\backslash X$ is compact.

Proof: 1) Suppose Γ acts properly discontinuously. Take a compact subset C in L so that $L = C \cdot \Gamma$ and $C = C^{-1}$. Then, for any compact subset S in X, $L_S = \{g \in L : g \cdot S \cap S \neq \emptyset\} \subset C \cdot \Gamma_{CS}$. Because Γ_{CS} is a finite set from the assumption, we conclude that the action of L is proper. In view of $\Gamma_S = \Gamma \cap L_S$, the "only if" part follows immediately from the definition. 2) Suppose $L \backslash X$ is compact. We take an open covering $X = \bigcup_\alpha U_\alpha$ by relatively compact sets U_α. Then $X = \bigcup_\alpha L U_\alpha$ gives an open covering of $L \backslash X$. By the compactness of $L \backslash X$, we can choose finitely many U_j among $\{U_\alpha\}$ so that $X = \bigcup_j L \cdot U_j = \bigcup_j (\Gamma \cdot C) \cdot U_j = \bigcup_j \Gamma \cdot (C \cdot U_j)$, showing $\Gamma \backslash X$ is compact. The converse statement is clear. $\square$

2.10   Problems in a continuous analog

In view of Observation 2.12, we pose the following analogous problems in a continuous setting.

Problem 2.13   *Let G be a Lie group and H and L closed subgroups.*

1) Find the criterion on the triplet (L, G, H) such that the action of L on G/H is proper.

2) Find the criterion on the triplet (L, G, H) such that the double coset $L \backslash G / H$ is compact in the quotient topology.

We will give a complete answer to Problem 2.13 in terms of Lie algebras in the following cases:

  i) Problem 2.13(1) when G is reductive (see Section 2)
 ii) Problem 2.13(2) when the groups G, H, L are real reductive (see Section 5).

In preparation for the next section we introduce here some notations that are useful for a further study of Problem 2.13(1).

2.11   Relations $\sim$ and $\pitchfork$

Suppose that H and L are subsets of a locally compact topological group G.

Definition 2.14 ([Ko10, Definition 2.1.1])   We denote by $H \sim L$ in G if there exists a compact set S of G such that $L \subset SHS^{-1}$ and $H \subset SLS^{-1}$. Here, $SHS^{-1} := \{ahb^{-1} \in G : a, b \in S, \ h \in H\}$. Then the relation $H \sim L$ in G defines an equivalence relation.

We say the pair (H, L) is *proper* in G, denoted by $H \pitchfork L$ in G, iff $SHS^{-1} \cap L$ is relatively compact for any compact set S in G.

Definition 2.14 is motivated by the following:

Observation 2.15   *Let H and L be closed subgroups of G, and Γ a discrete subgroup of G.*

1) *The action of L on the homogeneous manifold G/H is proper if and only if $H \pitchfork L$ in G.*

2) *The action of Γ on the homogeneous manifold G/H is properly discontinuous if and only if $H \pitchfork \Gamma$ in G.*

Here are some elementary properties of the relations $\sim$, $\pitchfork$:

Lemma 2.16 ([Ko10])   *Suppose G is a locally compact topological group and that H, H' and L are subsets of G.*

 i) $H \pitchfork L$ if and only if $L \pitchfork H$.
 ii) If $H \sim H'$ and if $H \pitchfork L$ in G, then $H' \pitchfork L$ in G.

Now we are ready to give a reformulation of Problem 2.13(1) as:

Problem 2.17 (a reformulation of Problem 2.13(1))   *Let G be a Lie group, and H and L subsets of G. Find the criterion on the pair (L, H) (or on the pair of their equivalence classes with respect to $\sim$) such that $L \pitchfork H$ in G.*

2.12   Property (CI)

If a discrete group Γ acts on X properly discontinuously, then every isotropy subgroup is finite and every Γ orbit is closed (Lemma 2.3). The latter condition corresponds to the fact that each point is closed in the quotient topology of $\Gamma \backslash X$. In general, the converse implication does not hold (cf. Example 2.8).

We have a similar picture in a continuous setting. In fact, let H, L be closed subgroups of a locally compact topological group G. If L acts properly on G/H, then any L-orbit $LgH \simeq L/L \cap gHg^{-1} \subset G/H$ is a closed subset, and each isotropy subgroup $L \cap gHg^{-1}$ is compact. In general, the converse implication is not true. However, we focus our attention on the latter property, that is, each isotropy subgroup is compact.

Definition 2.18 ([Ko4, Ko6])   Suppose that H and L are subsets of a locally compact topological group G. We say that the pair (L, H) has *the property* (CI) in G if and only if $L \cap gHg^{-1}$ is compact for any $g \in G$.

Here (CI) stands for that the action of L has a <u>c</u>ompact <u>i</u>sotropy subgroup $L \cap gHg^{-1}$ at each point $gH \in G/H$, or stands for that L and gHg^{-1} has a <u>c</u>ompact <u>i</u>ntersection $(g \in G)$ (see also [Lip]).

If $H \pitchfork L$ in G, then the pair (L, H) has the property (CI) in G. The point is to understand how and to what extent the property (CI) implies the proper action.

Problem 2.19 ([Ko6, Problem 2]; see Open problem 6(ii))   *For which Lie groups, does the following equivalence (2.1) hold?*

$$H \pitchfork L \text{ in } G \Leftrightarrow \text{ the pair } (L, H) \text{ has the property (CI) in } G. \qquad (2.1)$$

In the next section, we shall see that the equivalence (2.1) holds if G, H, L are real reductive algebraic groups (see Theorem 3.18). R. Lipsman [Lip] has pointed out that it is likely that the equivalence (2.12.3) holds as well if G is a simply connected nilpotent group. There are some further cases, especially in the context of a continuous analog of the Auslander conjecture (see the remark following Example 1.9), where the equivalence (2.1) is known to hold. See Example 5 and Proposition A.2.1 in [Ko4, Ko6]; Theorem 3.1 and Theorem 5.4 in [Lip]. However, we should note that the equivalence (2.1) does not always hold. For instance, if $G = KAN$ is an Iwasawa decomposition of a real reductive group G and if we put $L := A$ and $H := N$, then (L, H) has the property (CI) in G, while $L \not\pitchfork H$ in G as we saw in Exercise 2.9 when $G = \mathrm{SL}(2, \mathbb{R})$.

3   Calabi–Markus phenomenon

3.1   Existence of infinite discontinuous groups

Let G be a Lie group and H a closed subgroup. In the section, we focus our attention on the first basic problem:

Problem 3.1 (see Problem 2.10(i))   *Which homogeneous manifold G/H admits an infinite discontinuous group?*

We first note that if M is a compact manifold, then there is no infinite discrete group $(\subset \mathrm{Diffeo}(M))$ acting on M properly discontinuously. So, we might expect that the answer to Problem 3.1 should be related to certain aspect of *compactness* of a homogeneous manifold G/H with respect to the transformation group $G \subset \text{Diffeo}(M)$. One observes that a criterion given in the reductive case (Theorem 3.6) has such an aspect.

Our main interest here is the (non-)existence of an infinite discontinuous group in the reductive case. However, we include a quick review for some other typical cases as well.

i)   relativistic spherical space form (Section 3.2),

ii) G/H with H compact (Section 3.3),

iii) G/H with G solvable (Section 3.4),

iv) G/H with G reductive (Section 3.5– Section 3.10).

3.2   Relativistic spherical space form

In 1962, Calabi and Markus discovered a surprising phenomenon on the fundamental group π_1 of a Lorentz manifold with constant curvature:

Theorem 3.2 ([CM])   *Every relativistic spherical space form is noncompact and has a finite fundamental group π_1 (see Example 1.10 for definition).*

As we saw in Example 1.10, the Calabi–Markus theorem is reformulated in a group language as:

Theorem 3.3   *If $n \geq 3$, then there does not exist a discrete subgroup Γ of $O(n,1)$ acting on $O(n,1)/O(n-1,1)$ properly discontinuously and freely such that the fundamental group $\pi_1(\Gamma \backslash G/H)$ is infinite.*

We note that if $n = 2$, then G/H is diffeomorphic to a cylinder and has the fundamental group $\simeq \mathbb{Z}$. Theorem 3.3 is also reformulated as

Theorem 3.4   *Any discontinuous group acting (see Definition 2.5) on a homogeneous manifold $SO(n,1)/SO(n-1,1)$ $(n \geq 2)$ is finite.*

This result will be generalized in Section 3.5 for homogeneous spaces of reductive type (Definition 3.12).

Setting the main subject aside for a moment, we mention a related problem in non-Riemannian geometry based on the following observation:

Observation 3.5   *i) Any complete Riemannian manifold M^n $(n \geq 2)$ with constant sectional curvature $+1$ is compact with finite fundamental group.*

ii) Any complete Lorentz manifold M^n $(n \geq 3)$ with constant sectional curvature $+1$ is noncompact with finite fundamental group.

We recall again in a group language that $M \simeq \Gamma \backslash O(n+1)/O(n)$ in the first case, whereas $M \simeq \Gamma \backslash O(n,1)/O(n-1,1)$ in the second case. A classical theorem due to Myers may be interpreted as a "perturbation" of Riemannian metric in the first statement:

Theorem 3.6 ([My])   *If the Ricci curvature of a complete Riemannian manifold M satisfies $K(U,U) \geq c > 0$ for all unit vectors, then M is compact with finite fundamental group.*

The author does not know a result concerning a "perturbation" of the Lorentz metric in Observation 3.5(ii). So we pay an attention on the following problem in pseudo-Riemannian geometry:

Problem 3.7   *Suppose M is a complete Lorentz manifold (or more generally a complete manifold equipped with an indefinite Riemannian metric). Find a sufficient condition in terms of a local property of M that assures that M is noncompact with a finite fundamental group. In particular, is there a sufficient condition given by some positiveness of the curvature of M?*

3.3 H is compact

Suppose that H is compact. Then it follows immediately from the definition of properly discontinuous actions that

The action of Γ on G/H is properly discontinuous
$\Leftrightarrow$ Γ is a discrete subgroup of G.

Moreover, if Γ is a torsion free discrete subgroup of G, then the action of Γ on G/H is properly discontinuous and free whenever H is compact. Therefore, we have the following:

Proposition 3.8   *If G is a noncompact connected linear Lie group and H is a compact subgroup, then there always exists an infinite discrete subgroup Γ of G acting on G/H properly discontinuously and freely on G/H.*

Proof: It suffices to show that any noncompact linear Lie group G contains an infinite torsion free discrete subgroup.

Assume first that G is a noncompact semisimple Lie group; we write $G = KAN$ for an Iwasawa decomposition. We note that G is noncompact if and only if $\mathbb{R}$-rank $G = \dim A > 0$. So we can take a lattice of A, which is isomorphic to $\mathbb{Z}^{\dim A}$ and is an infinite and torsion free discrete subgroup of G.

Assume second that G is a noncompact solvable Lie group. Let $\widetilde{G}$ be the universal covering group of G, and Z the kernel of the covering map $\widetilde{G} \to G$. It follows from Theorem 1 and Remark of [Ch] that we find $(0 \le)$ $r \le n = \dim G$ and a basis $X_1, \dots, X_n$ of the Lie algebra of G which has the following properties:

i) $\mathbb{R}^n \to \widetilde{G}$, $(t_1, \dots, t_n) \mapsto \exp(t_1 X_1) \dots \exp(t_n X_n)$ is a surjective homeomorphism.

ii) Z is isomorphic to the free abelian group generated by the elements $\exp X_1, \dots, \exp X_r$.

We note that G is noncompact if and only if $r < n$. Therefore, if G is noncompact, we put $\Gamma := \{\exp(n X_{r+1}) \in G : n \in \mathbb{Z}\}$, which is an infinite and torsion free discrete subgroup of G.

Now, we suppose G is a noncompact connected linear Lie group. Let $\mathfrak{g} = \mathfrak{l} + \mathfrak{s}$ is a Levi decomposition, where $\mathfrak{s}$ is the radical of $\mathfrak{g}$ and $\mathfrak{l}$ is a semisimple Lie algebra. We write L, S for the analytic subgroups of G with Lie algebra $\mathfrak{l}, \mathfrak{s}$, respectively. Then, L and S are closed subgroups of G. If G is noncompact, then L or S must be noncompact. In either case, we can find an infinite and torsion free discrete subgroup of G. $\square$

3.4   Solvable groups

In the case of simply connected solvable homogeneous spaces, it turns out that the Calabi–Markus phenomenon does not occur. The following result is proved based on a structural result of a simply connected solvable groups due to Chevalley [Ch].

Theorem 3.9, ([Ko8, Theorem 2.2]; see also [Lip])   *Suppose G is a solvable Lie group and H is a proper closed subgroup of G. Then there exists a discrete subgroup Γ of G acting on G/H properly discontinuously and freely such that the fundamental group $\pi_1(\Gamma \backslash G/H)$ is infinite.*

3.5   Reductive cases

The Calabi–Markus phenomenon in the reductive case has been studied by Calabi, Markus, Wolf, Kulkarni and Kobayashi, and it has been settled completely in terms of a rank condition:

Theorem 3.10, ([CM; Wo1; Wo2; Wo3; Ku; Ko2]) *Let G/H be a homogeneous space of reductive type. The following statements are equivalent:*

1) *Any discontinuous group acting on G/H is finite.*
2) $\mathbb{R}$-rank $G = \mathbb{R}$-rank H.

We will explain some terminology used here in Section 3.6 Section 3.6, and then give some examples and a sketch of the proof in Sections 3.7 and 3.8.

3.6   Homogeneous spaces of reductive type

We set up some notation of real reductive groups. Good references for what we need are [He, Chapter VI] and [War, Chapter I]. Let G be a connected real linear reductive Lie group.

Fix a maximal compact subgroup K in G. We write $\mathfrak{g} = \mathfrak{k} + \mathfrak{p}$ for the corresponding Cartan decomposition of the Lie algebra of G. Then, the homogeneous space G/K equipped with a G-invariant Riemannian metric is said to be a *Riemannian symmetric space*. We fix a maximal abelian subspace $\mathfrak{a} \subset \mathfrak{p}$. Then, $\mathfrak{a}$ is said to be *a maximally split abelian subspace for G*. If we want to emphasize the group G, we write $\mathfrak{a}_G$ for $\mathfrak{a}$. We write A for the corresponding connected Lie subgroup. We define

$\Sigma(\mathfrak{g}, \mathfrak{a})$: the restricted root system,

W_G : the Weyl group associated to $\Sigma(\mathfrak{g}, \mathfrak{a})$,

rank G := the dimension of any maximal semisimple abelian
            subspace of $\mathfrak{g}$,

c-rank G := rank K,

$\mathbb{R}$-rank G := dim $\mathfrak{a}$ = rank G/K,

$d(G)$:= dim $\mathfrak{p}$ = dim G/K.

We note that

$$c\text{-rank}\, G \ \leq \ \text{rank}\, G \geq \ \mathbb{R}\text{-rank}\, G \ \leq \ d(G).$$

Example 3.11 Let $G = \mathrm{SO}_0(p,q)$ $(p \geq q)$ be the identity component of the indefinite orthogonal group of signature (p,q). Then $K \simeq \mathrm{SO}(p) \times \mathrm{SO}(q)$, $A \simeq \mathbb{R}^q$, $c\text{-rank}\,G = [\frac{p}{2}] + [\frac{q}{2}]$, $\mathrm{rank}\,G = [\frac{p+q}{2}]$, $\mathbb{R}\text{-rank}\,G = q$, $d(G) = pq$, and $W_G \simeq \mathfrak{S}_q \ltimes (\mathbb{Z}/2\mathbb{Z})^q$. Here, $\mathfrak{S}_q$ denotes the qth symmetric group.

Definition 3.12 Suppose that H is a closed subgroup in G with at most finitely many connected components. If there exists a Cartan involution of G that stabilizes H, then H is said to be *reductive in G* and G/H is said to be *a homogeneous space of reductive type*.

If G/H is a homogeneous space of reductive type, then G and H have a realization in $\mathrm{GL}(n,\mathbb{R})$ such that $H \subset G \subset \mathrm{GL}(n,\mathbb{R})$ are closed subgroups and that $H = {}^tH$ and $G = {}^tG$. Here ${}^tG := \{{}^tg : g \in G\}$, the transposed set of G in $\mathrm{GL}(n,\mathbb{R})$.

Example 3.13 Suppose G is a real reductive linear group.

i) If H is compact, then G/H is a homogeneous space of reductive type.

ii) ([Yo]) If H is semisimple, then G/H is a homogeneous space of reductive type.

iii) Let σ be an involutive automorphism of G. If H is an open subgroup of $G^\sigma := \{g \in G : \sigma g = g\}$, then G/H is a homogeneous space of reductive type. The homogeneous space G/H is said to be a *reductive symmetric space*. If G is semisimple, G/H is said to be a *semisimple symmetric space*.

iv) If $X \in \mathfrak{g}$ is a semisimple element, namely, $\mathrm{ad}(X) \in \mathrm{End}(\mathfrak{g})$ is semisimple, then the semisimple orbit $G/Z_G(X) \simeq \mathrm{Ad}(G) \cdot X \,(\subset \mathfrak{g})$, where $Z_G(X) := \{g \in G : \mathrm{Ad}(g)X = X\}$, is a homogeneous space of reductive type.

v) ([Mos]) If $G_0 \supset G_1 \supset \cdots \supset G_n$ and if G_{i-1}/G_i $(1 \leq i \leq n)$ are all homogeneous spaces of reductive type, then so is G_0/G_n.

Suppose G/H is a homogeneous space of reductive type. There is a nondegenerate $\mathrm{Ad}(H)$-invariant bilinear form B on $\mathfrak{g}$ that is positive definite on $\mathfrak{p}$ and negative definite on $\mathfrak{k}$. (Actually, there exists B that is $\mathrm{Ad}(G)$-invariant.) Because $\mathfrak{h}$ is θ-stable, the restriction of B to $\mathfrak{h}$ is also nondegenerate. Therefore, B induces a nondegenerate $\mathrm{Ad}(H)$-invariant bilinear form $\mathfrak{g}/\mathfrak{h} \simeq T_o(G/H)$, the tangent space of G/H at $o = eH \in G/H$.

It induces a G-invariant (indefinite-)Riemannian metric on G/H by the left translation. The signature of this metric is $(\dim(\mathfrak{p}/\mathfrak{p} \cap \mathfrak{h}), \dim(\mathfrak{k}/\mathfrak{k} \cap \mathfrak{h}))$.

In analogy with the polar coordinate in the Euclidean space $\mathbb{R}^n$, there is a polar coordinate in a Riemannian symmetric space G/K. In group language, this means that a real reductive linear Lie group G has a Cartan decomposition

$$G = KAK.$$

In this decomposition, there is an element $a(g) \in A$, unique up to conjugation by W_G, such that $g \in Ka(g)K$ for each $g \in G$.

Definition 3.14 ([Ko2], [Ko10])   For each subset L of G, we define:

$$A(L) := A \cap KLK = \{w \cdot a(g) : w \in W_G, g \in L\} \subset A,$$
$$\mathfrak{a}(L) := \log A(L) \subset \mathfrak{a}.$$

Here, $\log\colon A \to \mathfrak{a}$ is the inverse of the diffeomorphism $\exp\colon \mathfrak{a} \to A$.

If L is a subgroup of G that is reductive in G, then we can take a maximal compact subgroup K of G such that $L \cap K$ is also a maximal compact subgroup of a reductive Lie group L. Let $\mathfrak{a}_L$ be a maximally split abelian subspace for L. Then there exists an element g of G such that $\operatorname{Ad}(g)\mathfrak{a}_L \subset \mathfrak{a}_G$. Then $\mathfrak{a}(L)$ is a finite union of subspaces in $\mathfrak{a}_G$:

$$\mathfrak{a}(L) = W_G \cdot \operatorname{Ad}(g)\mathfrak{a}_L \subset \mathfrak{a}_G.$$

In this case, the notation $\mathfrak{a}(L)$ coincides with that in [Ko2] as a subset of $\mathfrak{a}_G/W_G$.

Remark   Some remarks are in order.

i) If G/H is a homogeneous space of reductive type, then both G and H are real reductive Lie groups. Note that the converse statement is not always true.   ii) We avoid the terminology *reductive homogeneous space*, which is usually used in the following sense: the Lie algebra $\mathfrak{g}$ may be decomposed into a vector space direct sum of the Lie algebra $\mathfrak{h}$ and a H-stable subspace $\mathfrak{m}$ (see, for example, [KoN, Chapter X, Section 2]). This notion is wider than that of homogeneous spaces of reductive type in Definition 3.12. In particular, neither G nor H itself is required to be reductive in this usual definition of a reductive homogeneous space.

3.7   Examples of the Calabi–Markus phenomenon

We present here some examples of Theorem 3.10.

Example 3.15   Let $G/H = \mathrm{SO}(p+1,q)/\mathrm{SO}(p,q)$. Then $\mathbb{R}\text{-rank}\, G = \min(p+1,q)$ and $\mathbb{R}\text{-rank}\, H = \min(p,q)$. Therefore, we have

$$\text{there is no infinite discontinuous group acting on } G/H$$
$$\Leftrightarrow \quad \mathbb{R}\text{-rank}\, G = \mathbb{R}\text{-rank}\, H$$
$$\Leftrightarrow \quad q \leq p.$$

The case $q = 1$ is the original result of Calabi–Markus (see Theorem 3.4).

Example 3.16   i)Any para-Hermitian symmetric space does not admit an infinite discontinuous group (see [Lib] for the definition of para-complex structure; see also [KaKo] for the definition and a classification of irreducible para-Hermitian symmetric spaces).   ii) Any hyperbolic orbit does not admit an infinite discontinuous group. Here, we say a semisimple orbit $G/Z_G(X) \simeq \mathrm{Ad}(G)X$ (see Example 3.13(iv)) is a *hyperbolic orbit* if $X \in \mathfrak{g}$ is a hyperbolic element, that is, if all eigenvalues of $\mathrm{ad}(X) \in \mathrm{End}(\mathfrak{g})$ are real. We note that a para-Hermitian symmetric space is a reductive symmetric space that can be realized as a hyperbolic orbit.

Example 3.17   There does not exist an infinite discontinuous group acting on the following homogeneous manifolds:

i) reductive symmetric spaces

$$\mathrm{GL}(n,\mathbb{C})/\mathrm{GL}(n,\mathbb{R}), \quad \mathrm{SO}(2n+1,\mathbb{C})/\mathrm{SO}(2n,\mathbb{C}), \quad \mathrm{U}(m,n)/\mathrm{O}(m,n).$$

ii) para-Hermitian symmetric spaces

$$\mathrm{Sp}(n,\mathbb{R})/\mathrm{GL}(n,\mathbb{R}), \quad \mathrm{SO}^*(4n)/\mathrm{SU}^*(2n) \times \mathbb{R}.$$

iii) hyperbolic orbits

$$\mathrm{GL}(n_1 + \cdots + n_k, \mathbb{R})/\mathrm{GL}(n_1,\mathbb{R}) \times \cdots \times \mathrm{GL}(n_k,\mathbb{R}).$$

On the other hand, there exists a discontinuous group that is isomorphic to $\mathbb{Z}^n$ acting on the following homogeneous manifolds:

$$\mathrm{Sp}(2n,\mathbb{R})/\mathrm{U}(n,n), \quad \mathrm{GL}(2n,\mathbb{R})/\mathrm{GL}(n,\mathbb{C}), \quad \mathrm{O}(2m,2n)/\mathrm{U}(m,n)$$
$$(n \leq m).$$

3.8   Sketch of proof of the criterion for Calabi–Markus

First we explain a continuous analog, that is, Problem 2.13(1) for the criterion for a proper action in a general reductive setting (Theorem 3.18). The criterion for the Calabi–Markus phenomenon (Theorem 3.10) is obtained as a very special case of the continuous analog, combined with Observation 2.12.

3.9   Criterion of proper actions in a continuous setting

Theorem 3.18 ([Ko2])   *Let G/H, G/L be homogeneous spaces of reductive type. Then the following five conditions are equivalent:*

1) *L acts on G/H properly.*
1)′ *H acts on G/L properly.*
1)″ *$H \pitchfork L$ in G.*
2) *The triplet (L, G, H) has the property (CI). That is, for any $g \in G$, $L \cap gHg^{-1}$ is compact.*
2)′ *$\mathfrak{a}(L) \cap \mathfrak{a}(H) = \{0\}$.*

See Section 2.11 for the definition of $\pitchfork$, Section 2.12 for the property (CI), and Definition 3.14 for $\mathfrak{a}(L)$.

In the theorem, $(1) \Leftrightarrow (1)' \Leftrightarrow (1)''$, $(2) \Leftrightarrow (2)'$ and $(1) \Rightarrow (2)$ are easy. The nontrivial part is the implication $(2) \Rightarrow (1)$. The proof of $(2) \Rightarrow (1)$ is divided into two steps:

  i) Reduction to the case where H and L are abelian. This is an easy step that can be proved by using the Cartan decomposition.
 ii) Proof of the abelian case. This is done by looking at the infinite points in a Riemannian symmetric space based on some structural results on parabolic subgroups and nilpotent elements.

3.10   Criterion of proper actions in a discrete setting

Similar techniques lead us to a generalization to the case where H and L are not necessarily reductive:

Theorem 3.19 ([Ko10])   *Let H, L be subsets of a real reductive linear Lie group G.*

  i) $H \pitchfork L$ in $G \Leftrightarrow \mathfrak{a}(H) \pitchfork \mathfrak{a}(L)$ in $\mathfrak{a}$.
 ii) $H \sim L$ in $G \Leftrightarrow \mathfrak{a}(H) \sim \mathfrak{a}(L)$ in $\mathfrak{a}$.

If H and L are reductive in G, then it is easy to see that Theorem 3.19(i) implies Theorem 3.18. If $G = \mathrm{GL}(n, \mathbb{R})$ and $H = \mathrm{GL}(m, \mathbb{R})$, then Theorem 3.19(i) implies a result of Friedland ([Fr, Theorem 3.1]). For more details, we refer to [Ko2], [Ko10].

Example 3.20 Let $G = \mathrm{SO}(2m, 2n)$, $L = \mathrm{U}(m, n) \subset G$, and $H = \mathrm{SO}(p, q) \simeq 1_{2m-p} \times \mathrm{SO}(p, q) \times 1_{2n-q} \subset G$. Here, we assume $0 < q \leq p$, $0 < n \leq m$, $p \leq 2m$, and $q \leq 2n$ for simplicity. Then, with a suitable coordinate, we can identify $\mathfrak{a}_G$ with $\mathbb{R}^{2n}$ and $W_G \simeq \mathfrak{S}_{2n} \ltimes (\mathbb{Z}/2\mathbb{Z})^{2n}$ in $\mathrm{GL}(\mathfrak{a}_G)$. Up to the conjugacy by an element of W_G, we have

$$\mathfrak{a}(L) = \{(a_1, a_1, a_2, a_2, \ldots, a_n, a_n) : a_i \in \mathbb{R} \ (1 \leq i \leq n)\},$$
$$\mathfrak{a}(H) = \{(b_1, b_2, \ldots, b_q, 0, \ldots, 0) : b_i \in \mathbb{R} \ (1 \leq i \leq q)\}.$$

So the condition $(2)'$ in Theorem 3.18 amounts to $q = 1$. Therefore, we conclude that $\mathrm{U}(m, n)$ acts on $\mathrm{SO}(2m, 2n)/\mathrm{SO}(p, q)$ properly if and only if $q = 1$.

Exercise 3.21 Let $G/H = \mathrm{SO}(2p, 2q)/\mathrm{U}(i, j)$ $(i \leq p, \ j \leq q, \ i \leq j)$. Prove that if $i < \min(p, q)$ then there exists a discontinuous group $\Gamma \ (\subset G)$ acting on G/H such that Γ is isomorphic to $\pi_1(M_g)$. Hint: Show first that there is a subgroup of G isomorphic to $\mathrm{PSL}(2, \mathbb{R})$ that acts properly on G/H (use Theorem 3.18). Then use the fact that there exists a discrete subgroup Γ of $\mathrm{PSL}(2, \mathbb{R})$ such that $\Gamma \simeq \pi_1(M_g)$ (see Example 1.11).

3.11 Historical notes

Reductive case: The implication $(1) \Rightarrow (2)$ in Theorem 3.10 was first proved by E. Calabi and L. Markus in the case $G/H = \mathrm{SO}(n+1, 1)/\mathrm{SO}(n, 1)$ in [CM] (see Theorem 3.2). Then J. Wolf extended their result to the case $G/H = \mathrm{SO}(p + 1, q)/\mathrm{SO}(p, q)$ $(q \leq p)$ in [Wo1]. After finding some other similar results for symmetric spaces of rank one (e.g., [Wo2]), he finally obtained the sufficiency of the real rank condition in the case of semisimple symmetric spaces in the 1960s (see [Wo3]). His idea is also applicable to our more general setting. On the other hand, the proof of the necessity of the real rank condition given there was incomplete because of some confusion with the definition of properly discontinuous actions (see Example 2.8 for the difference between local Hausdorff topology and Hausdorff topology).

The converse implication (2) $\Rightarrow$ (1) in Theorem 3.10 is more difficult because we have to show the existence of an infinite discontinuous group acting on G/H if $\mathbb{R}$-rank $H < \mathbb{R}$-rank G. It took about 20 years before the converse implication was first proved in the rank one case $G/H = \mathrm{SO}(p+1,q)/\mathrm{SO}(p,q)$ $(q > p)$ by R. Kulkarni ([Ku, Theorem 5.7]).The method there is based on a study of quadratic forms of type $(p+1,q)$. with the aid of a lemma due to N. Wallach. The general case is due to T. Kobayashi [Ko2] as an application of Theorem 3.18 with dim $L = 1$.

Solvable case: It was proved by Kobayashi that there always exists a Clifford–Klein form of a homogeneous manifold G/H with infinite fundamental group (Theorem 3.9) if G is a solvable group and $H \neq G$ ([Ko8]). Lipsman made a further study of properly discontinuous actions and the property (CI) in [Lip].

General case: It is still an open problem to find a criterion on (G, H) for a general Lie group G such that a homogeneous space G/H admits an infinite discontinuous group.

4 Generalized Hirzebruch's proportionality principle and its application

4.1 Hirzebruch's proportionality principle

In 1956, Hirzebruch proved

Theorem 4.1([Hi]) *Let D be a bounded Hermitian symmetric domain, Γ a torsion free discrete cocompact subgroup of the automorphism group $\mathrm{Aut}(D)$ of D, and M the compact Hermitian symmetric space dual to D. Then there is a real number $A = A(\Gamma)$ such that $c^{\alpha}(\Gamma \backslash D)[\Gamma \backslash D] = Ac^{\alpha}(M)[M]$ for any c^{α}, where $\alpha = (\alpha_1, \ldots, \alpha_k)$ is a multi-index and $c^{\alpha} = c_1^{\alpha_1} \cup \cdots \cup c_k^{\alpha_k}$ is a monomial of Chern classes.*

In this section, we shall clarify this principle by eliminating unnecessary conditions. The setting is generalized as follows:

Hermitian symmetric spaces $\longrightarrow$ homogeneous spaces of reductive type

tangent bundles $\longrightarrow$ homogeneous vector bundles

characteristic numbers $\longrightarrow$ characteristic classes

As an application, we find an obstruction to the existence of compact Clifford–Klein forms of homogeneous spaces of reductive type by means of the Euler–Poincaré class (see Corollary 4.12).

4.2   Sketch of idea

It has been a classical and standard technique, in particular, in representation theory, to compare two objects through a holomorphic continuation: For instance,

i) Weyl's unitary trick — finite dimensional representations of reductive groups

ii) Flensted-Jensen duality — eigenspaces on semisimple symmetric spaces (e.g., [Fl])

The argument in this section lies in the same line:

iii) (generalized) Hirzebruch's proportionality principle — characteristic classes of homogeneous spaces of reductive type

Let us explain the idea briefly in the case of Theorem 4.1, taking $G = \mathrm{SL}(2, \mathbb{R})$ as an example. In this case

$$D = \mathrm{SL}(2, \mathbb{R})/\mathrm{SO}(2) \qquad \text{(the Poincaré upper half plane)},$$
$$M = \mathrm{SU}(2)/\mathrm{SO}(2)\,(\simeq \mathbb{C}P^1) \qquad \text{(the projective space)}.$$

There are two ways to compare the two manifolds D (or $\Gamma \backslash D$) and M:

i) $D \subset M$ (the Borel embedding),

ii) $D \subset \mathrm{SL}(2, \mathbb{C})/\mathrm{SO}(2, \mathbb{C}) \supset M$ (complexification).

The proportionality principle for Hermitian symmetric spaces (Theorem 4.1) can be proved based on the Borel embedding (i) as well as based on the complexification (ii). However, we will see that the argument using (ii) has a wider application, even when there is no natural map between D and M. Returning to our special example, we note that $M_{\mathbb{C}} := \mathrm{SL}(2, \mathbb{C})/\mathrm{SO}(2, \mathbb{C})$ is a common complexification of $M = \mathrm{SU}(2)/\mathrm{SO}(2)$ and $D = \mathrm{SL}(2, \mathbb{R})/\mathrm{SO}(2)$ in (ii), if we *forget* the original complex structures and look upon M and D simply as real manifolds. Now, we can compare differential forms that represent characteristic classes of two manifolds D and M through the holomorphic continuation on $M_{\mathbb{C}}$. This is the main idea of the proof of a generalized Hirzebruch's proportionality principle, which we discuss in Sections 4.3 and 4.4.

4.3   Complexification and associated Riemannian spaces of compact type

Let us introduce the setting to realize the idea in Section 3.2. Let G be a connected real reductive linear group and H a connected subgroup reductive in G. We assume that there is a connected complex reductive Lie group $G_{\mathbb{C}}$ and a connected closed subgroup $H_{\mathbb{C}}$ with Lie algebras $\mathfrak{g}_{\mathbb{C}}$ and $\mathfrak{h}_{\mathbb{C}}$ respectively, such that $G \subset G_{\mathbb{C}}$. Take a Cartan involution θ of G such that $\theta H = H$. We write the corresponding Cartan decomposition as $\mathfrak{g} = \mathfrak{k} + \mathfrak{p}$. Let G_U be the connected subgroup of $G_{\mathbb{C}}$ with Lie algebra $\mathfrak{g}_U := \mathfrak{k} + \sqrt{-1}\mathfrak{p}$, and H_U the connected subgroup of $G_{\mathbb{C}}$ with Lie algebra $\mathfrak{h}_U := \mathfrak{h} \cap \mathfrak{k} + \sqrt{-1}\mathfrak{h} \cap \mathfrak{p}$. Then, G_U is a maximal compact subgroup of $G_{\mathbb{C}}$, and so is H_U of $H_{\mathbb{C}}$. The homogeneous manifold G_U/H_U is said to be an *associated Riemannian space of compact type*. In summary, we have the following setting:

$$
\begin{array}{ccccc}
G & \subset & G_{\mathbb{C}} & \supset & G_U \\
\cup & & \cup & & \cup \\
H & \subset & H_{\mathbb{C}} & \supset & H_U
\end{array}
$$

Remark   The argument in this section is still valid for H with finitely many connected components, if we replace $H_{\mathbb{C}}$ by $H_{\mathbb{C}}(H \cap K)$ and H_U by $H_U(H \cap K)$.

Example 4.2   Here are some typical examples of $G_{\mathbb{C}}/H_{\mathbb{C}}$ and G_U/H_U.
  i) Let $G/H = \mathrm{SL}(n, \mathbb{R})/\mathrm{SO}(p, n-p)$.
     Then $G_{\mathbb{C}}/H_{\mathbb{C}} = \mathrm{SL}(n, \mathbb{C})/\mathrm{SO}(n, \mathbb{C})$ and $G_U/H_U = \mathrm{SU}(n)/\mathrm{SO}(n)$.
 ii) Let $G/H = \mathrm{GL}(n, \mathbb{R})/(\mathbb{R}^{\times})^n$.
     Then $G_{\mathbb{C}}/H_{\mathbb{C}} = \mathrm{GL}(n, \mathbb{C})/(\mathbb{C}^{\times})^n$ and $G_U/H_U = \mathrm{U}(n)/\mathrm{U}(1)^n$.
iii) Let $G/H = \mathrm{Sp}(2n, \mathbb{R})/\mathrm{Sp}(n, \mathbb{C})$.
     Then $G_{\mathbb{C}}/H_{\mathbb{C}} = \mathrm{Sp}(2n, \mathbb{C})/\mathrm{Sp}(n, \mathbb{C}) \times \mathrm{Sp}(n, \mathbb{C})$ and $G_U/H_U = \mathrm{Sp}(2n)/\mathrm{Sp}(n) \times \mathrm{Sp}(n)$.

Example 4.3   The foregoing assumption (i.e., closedness of $H_{\mathbb{C}}$) is satisfied for (G, H) in Example 3.13(i)–(iv). We recall that G/H is a reductive symmetric space in Example 3.13(iii); G/H is a semisimple orbit in Example 3.13(iv). The corresponding associated Riemannian space of compact type G_U/H_U is

$$G_U/H_U = \text{a compact symmetric space} \quad \text{for Example 3.13(iii)},$$
$$G_U/H_U = \text{a generalized flag variety} \quad \text{for Example 3.13(iv)}.$$

4.4   A homomorphism between cohomology rings

In the setting of Section 4.3, suppose a discrete subgroup Γ of G acts on G/H properly discontinuously and freely, so that $\Gamma \backslash G/H$ is a smooth manifold. (We do not require that $\Gamma \backslash G/H$ is compact. Γ is allowed to be the trivial subgroup $\{e\}$.)

We define

$$M_U := G_U/H_U \subset M_{\mathbb{C}} := G_{\mathbb{C}}/H_{\mathbb{C}} \supset M_{\mathbb{R}} := G/H.$$

Then, to compare characteristic classes of the two manifolds M_U and $M_{\mathbb{R}}$, we use the following restriction maps:

$$\mathcal{O}\left(\bigwedge^p TM_{\mathbb{C}}\right)$$

$$\swarrow \text{rest.} \qquad\qquad \searrow \text{rest.}$$

$$\mathcal{A}\left(\bigwedge^p{}_{\mathbb{R}} TM_U\right) \qquad\qquad\qquad \mathcal{A}\left(\bigwedge^p{}_{\mathbb{R}} TM_{\mathbb{R}}\right)$$

The groups G_U and G acting on M_U and $M_{\mathbb{R}}$, respectively, have the common complexification $G_{\mathbb{C}}$, which acts holomorphically on $M_{\mathbb{C}}$. Because G_U is compact, we can find a G_U-invariant differential form as a representative of an arbitrary element of the de Rham cohomology group $H^*(G_U/H_U; \mathbb{C})$. This element extends to a holomorphic differential form on $G_{\mathbb{C}}/H_{\mathbb{C}}$, and then we can restrict to G/H and to $\Gamma \backslash G/H$. Thus, the diagram induces a homomorphism on the cohomology level:

Theorem 4.4   *We have a natural $\mathbb{C}$-algebra homomorphism*

$$\eta \colon H^*(G_U/H_U; \mathbb{C}) \to H^*(\Gamma \backslash G/H; \mathbb{C}).$$

If $\Gamma \backslash G/H$ is compact and H is connected, then η is injective.

The last statement follows from the Poincaré duality.

4.5   Real homogeneous vector bundles

We review the definition of an associated vector bundle. Let

$$\rho\colon H \to \mathrm{GL}_{\mathbb{R}}(V)$$

be a representation of H on a real vector space V. Suppose a discrete subgroup Γ acts on a homogeneous space G/H properly discontinuously. Associated to the principal H-bundle

$$H \to \Gamma\backslash G \to \Gamma\backslash G/H$$

is a real homogeneous vector bundle over $\Gamma\backslash G/H$,

$$^{\Gamma}E := \Gamma\backslash G \times_{\rho} V,$$

defined by the set of equivalence classes with respect to the action of H, that is,

$$(\overline{g}, v), (\overline{g'}, v') \in \Gamma\backslash G \times V \text{ are equivalent}$$
$$\Leftrightarrow \overline{g} = \overline{g'}h \text{ and } v = \rho(h^{-1})v' \text{ for some } h \in H.$$

The projection on the first component $\Gamma\backslash G \times V \to \Gamma\backslash G$ gives rise to the projection $^{\Gamma}E \equiv \Gamma\backslash G \times_{\rho} V \to \Gamma\backslash G/H$ with typical fiber V. Similarly, an associated real vector bundle $E_U := G_U \times_{\rho_U} V_U$ over G_U/H_U is defined if a representation $\rho_U\colon H_U \to \mathrm{GL}_{\mathbb{R}}(V_U)$ is given.

4.6   Weyl's unitary trick

The idea of Weyl's unitary trick is a holomorphic continuation of finite dimensional representations. Here, we set up some notation of finite dimensional representations that fits into Hirzebruch's proportionality principle.

Let $H \subset H_{\mathbb{C}} \supset H_U$ be as in Section 4.3, namely, $H_{\mathbb{C}}$ is a complex reductive Lie group, H is a real form, and H_U is a compact real form. Let

$$\rho\colon H \to \mathrm{GL}_{\mathbb{R}}(V),$$
$$\rho_U\colon H_U \to \mathrm{GL}_{\mathbb{R}}(V_U)$$

be finite dimensional representations over $\mathbb{R}$. We say that *the complexifi-cations of ρ and ρ_U are isomorphic* if there are

  i) a complex vector space $V_{\mathbb{C}}$
 ii) a holomorphic representation $\rho_{\mathbb{C}} \colon H_{\mathbb{C}} \to \mathrm{GL}(V_{\mathbb{C}}, \mathbb{C})$
iii) isomorphisms $\psi \colon V \otimes \mathbb{C} \xrightarrow{\sim} V_{\mathbb{C}}$ and $\psi_U \colon V_U \otimes \mathbb{C} \xrightarrow{\sim} V_{\mathbb{C}}$

such that the following diagram commutes:

$$
\begin{array}{ccccc}
H & \xhookrightarrow{\iota} & H_{\mathbb{C}} & \hookleftarrow & H_U \\[2pt]
{\scriptstyle\rho}\downarrow & & \downarrow{\scriptstyle\rho_{\mathbb{C}}} & & \downarrow{\scriptstyle\rho_U} \\[2pt]
\mathrm{GL}_{\mathbb{R}}(V) & \underset{\psi_\sharp}{\hookrightarrow} & \mathrm{GL}_{\mathbb{C}}(V_{\mathbb{C}}) & \underset{\psi_{U\,\sharp}}{\hookleftarrow} & \mathrm{GL}_{\mathbb{R}}(V_U)
\end{array}
$$

4.7   Hirzebruch's proportionality principle

Theorem 4.5 ([KoO])   *Retain the setting of Section 4.3. Let Γ be any discrete subgroup of G acting on G/H freely and properly discontinuously from the left. Suppose that the complexifications of $\rho : H \to \mathrm{GL}_{\mathbb{R}}(V)$ and $\rho_U : H_U \to \mathrm{GL}_{\mathbb{R}}(V_U)$ are isomorphic. Then the ith Pontryagin class satis-fies*

$$ \eta(p_i(E_U)) = p_i({}^{\Gamma}E) \in H^{4i}(\Gamma\backslash G/H; \mathbb{R}). $$

In particular, if there is a relation $\sum a_\alpha p^\alpha(E_U) = 0$ in $H^*(G_U/H_U; \mathbb{R})$, then the equation $\sum a_\alpha p^\alpha({}^{\Gamma}E) = 0$ in $H^*(\Gamma\backslash G/H; \mathbb{R})$ holds. Here $\alpha = (\alpha_1, \ldots, \alpha_k)$ is a multi-index and $p^\alpha = p_1^{\alpha_1} \cup \cdots \cup p_k^{\alpha_k}$ is a monomial of Pontryagin classes.

Sketch of Proof: We take an invariant connection (the canonical con-nection of the second kind on G/H in the sense of [No]) on the principal bundles $G \to G/H$, $G_U \to G_U/H_U$, respectively. Then the curvature forms are represented in terms of the Lie algebras. By the Chern–Weil the-ory (see [D], [KoN], [We]), characteristic classes are represented by using curvature forms. Now, we have the theorem by the usual Weyl's unitary trick for finite dimensional representations and by using the holomorphic continuation through η. $\square$

We note that in the case of Riemannian symmetric spaces (in particu-lar, Hermitian symmetric spaces), we can prove the theorem without using a holomorphic continuation. That is, we pull back directly the curvature forms via the $\mathbb{R}$-linear bijection $\mathfrak{g}_U/\mathfrak{h}_U \xrightarrow{\sim} \mathfrak{g}/\mathfrak{h}$. This map is different from η in Theorem 4.4 only by a constant multiple *depending on* degrees.

4.8   Tangent bundle

The characteristic classes of a manifold are, by definition, those of the tangent bundle. The tangent bundle $T(\Gamma\backslash G/H)$ is associated to the adjoint representation

$$\mathrm{Ad}_{|H}\colon H \to \mathrm{GL}_{\mathbb{R}}(\mathfrak{g}/\mathfrak{h}),$$

that is,

$$T(\Gamma\backslash G/H) \simeq \Gamma\backslash G \underset{\mathrm{Ad}_{|H}}{\times} \mathfrak{g}/\mathfrak{h},$$

and similarly

$$T(G_U/H_U) \simeq G_U \underset{\mathrm{Ad}_{|H_U}}{\times} \mathfrak{g}_U/\mathfrak{h}_U.$$

Since $\mathrm{Ad}_{|H}\colon H \to \mathrm{GL}_{\mathbb{R}}(\mathfrak{g}/\mathfrak{h})$ and $\mathrm{Ad}_{|H_U}\colon H_U \to \mathrm{GL}_{\mathbb{R}}(\mathfrak{g}_U/\mathfrak{h}_U)$ have isomorphic complexifications, we have now relations of Pontryagin classes (of the tangent bundle) between G_U/H_U and $\Gamma\backslash G/H$ as follows.

Corollary 4.6   *In the same setting as Theorem 4.5, we have*

$$\eta\left(p_i(G_U/H_U)\right) = p_i(\Gamma\backslash G/H) \in H^{4i}(\Gamma\backslash G/H;\mathbb{R}).$$

Furthermore, if H is connected, then we have a relation of Euler–Poincaré classes:

$$\eta\left(\chi(G_U/H_U)\right) = \chi(\Gamma\backslash G/H) \in H^{n}(\Gamma\backslash G/H;\mathbb{R}).$$

We note that both $\Gamma\backslash G/H$ and G_U/H_U are orientable if H is connected.

4.9   Complex homogeneous vector bundles

Let $\rho\colon H_{\mathbb{C}} \to \mathrm{GL}_{\mathbb{C}}(V_{\mathbb{C}})$ be a holomorphic representation on a finite dimensional vector space $V_{\mathbb{C}}$ over $\mathbb{C}$. Associated to the principal H-bundle

$$H \to \Gamma\backslash G \to \Gamma\backslash G/H,$$

we define a homogeneous complex vector bundle over $\Gamma\backslash G/H$ by

$$^{\Gamma}E_{\mathbb{C}} := \Gamma\backslash G \underset{\rho_{|H}}{\times} V_{\mathbb{C}}.$$

Similarly, we define a homogeneous complex vector bundle over G_U/H_U by

$$E_{U\mathbb{C}} := G_U \underset{\rho_{|H_U}}{\times} V_{\mathbb{C}}.$$

4.10   Chern classes

Theorem 4.7 ([KoO])   *Retain the same setting as before. Let Γ be any discrete subgroup of G acting on G/H freely and properly discontinuously from the left. Then the ith Chern class satisfies*

$$\eta(c_i(E_{U\mathbb{C}})) = c_i(^\Gamma E_\mathbb{C}) \in H^{2i}(\Gamma\backslash G/H; \mathbb{R}).$$

In particular, if there is a relation $\sum a_\alpha c^\alpha(E_U) = 0$ in $H^*(G_U/H_U; \mathbb{R})$, then the equation $\sum a_\alpha c^\alpha(^\Gamma E) = 0$ in $H^*(\Gamma\backslash G/H; \mathbb{R})$ holds. If $\Gamma\backslash G/H$ is compact and H is connected, the converse statement also holds.

Exercise 4.8   Formulate an analogous result to Corollary 4.6 in the case of Chern classes of homogeneous manifolds. (Hint: The assumption will be that G/H is an elliptic orbit. See Section 5.12. See also [KoO, Corollary 4].)

4.11   Some examples

Example 4.9   We consider a semisimple symmetric space

$$G/H = \mathrm{SO}(p, q)/\mathrm{SO}(p - 1, q).$$

All Pontryagin classes of any Clifford–Klein form $\Gamma\backslash G/H$ of a homogeneous manifold G/H vanish in $H^*(\Gamma\backslash G/H; \mathbb{R})$ because we know the corresponding result holds for $G_U/H_U \simeq S^{p+q-1}$. In particular, all Pontryagin classes of a Riemannian manifold of constant sectional curvature vanish (see also [Su]). We mention that there exist compact Clifford–Klein forms of $\mathrm{SO}(p, q)/\mathrm{SO}(p - 1, q)$ if $(p, q) = (1, n), (2, 2n), (4, 4n), (8, 8)$ $(n \in \mathbb{N})$ (see Example 5.14).

Example 4.10   We endow $\mathbb{C}^{p+q+1}$ with an (indefinite) Hermitian metric of type $(p + 1, q)$, that is,

$$(z, z) := z_1\overline{z_1} + \cdots + z_{p+1}\overline{z_{p+1}} - z_{p+2}\overline{z_{p+2}} - \cdots - z_{p+q+1}\overline{z_{p+q+1}}.$$

Let $X(p, q)$ be the open subset of the projective space $\mathbb{C}P^{p+q}$, which consists of the complex lines on which the restriction of the indefinite Hermitian metric is positive definite. Then $U(p+1,q)$ acts transitively on $X(p,q)$, so that we have a diffeomorphism

$$X(p,q) \simeq U(p+1,q)/U(1) \times U(p,q) =: G/H$$
$$\subset \mathbb{C}P^n \quad \simeq \ U(n+1)/U(1) \times U(n) \quad =: G_U/H_U.$$

Here, we put $n = p + q$. We note that $X(n,0) = \mathbb{C}P^n \simeq G_U/H_U$ and that $X(0,n)$ is the dual Hermitian symmetric domain of the noncompact type (see [He] for the terminology). Let Γ be a discrete subgroup of $U(p+1,q)$ acting on $X(p,q)$ freely and properly discontinuously, so that $\Gamma\backslash X(p,q)$ is a Clifford–Klein form of $X(p,q)$. Then, we have a relation among Chern classes:

$$c_j(\Gamma\backslash X(p,q)) = \left(\prod_{l=0}^{j-1} \frac{n+1-l}{n+1} \right) c_1(\Gamma\backslash X(p,q))^j \qquad (1 \leq j \leq n).$$

This follows from the corresponding fact for

$$X(n,0) = \mathbb{C}P^n \simeq U(n+1)/U(1) \times U(n),$$

that is, the total Chern class $c(\mathbb{C}P^n) = 1 + c_1(\mathbb{C}P^n) + \cdots + c_n(\mathbb{C}P^n)$ of the complex projective space $\mathbb{C}P^n$ is given by

$$c(\mathbb{C}P^n) \equiv (1+x)^{n+1} \quad \mod x^{n+1},$$

where x is the first Chern class of the hyperplane section bundle (see [BoHi], 15.1], [MiSt, Theorem 14.10]). If $\Gamma\backslash X(p,q)$ is a compact Clifford–Klein form of $X(p,q)$, then $c_j(\Gamma\backslash X(p,q)) \neq 0$ for any j with $1 \leq j \leq n$. There exists a compact Clifford–Klein form of $X(0,n), X(n,0)$ (Riemannian case), and $X(1,2r)$ (see Corollary 4.7), whereas any discrete subgroup acting properly discontinuously on $X(p,q)$ with $p \geq q$ is finite (see Theorem 3.10).

Example 4.12 Let M be a compact Clifford–Klein form of a complex manifold

$$SO(p,q+2)/SO(p,q) \times SO(2).$$

Then the Chern class $c_j(M)$, for any j with $1 \leq j \leq p+q$, of a complex manifold M does not vanish, because we know that the corresponding result holds for the Hermitian symmetric space $SO(n+2)/SO(n) \times SO(2)$. There exists a compact Clifford–Klein form for $(p,q) = (n,0)$ ($n \in \mathbb{N}$) and $(4,1)$ (see Corollary 5.6).

4.12   Compact Clifford–Klein form

We mentioned in Section 2.2 a third proof that $\mathrm{PSL}(2,\mathbb{R})/\mathrm{SO}(1,1)$ does not admit a compact Clifford–Klein form by using a nonvanishing vector field. Now we are ready to state a generalization of this result to a higher dimensional setting:

Corollary 4.13   *Let (G, H) be as in Section 4.3. Denote by K a maximal compact subgroup of G such that $H \cap K$ is also a maximal compact subgroup of H. If $\operatorname{rank} G = \operatorname{rank} H$ and $\dim K/H \cap K$ is odd, then G/H admits no uniform lattice, that is, there exists no compact Clifford–Klein form of G/H.*

Sketch of proof: We may and do assume H is connected. Then, G/H admits a G-invariant orientation, so that a Clifford–Klein form $\Gamma\backslash G/H$ is an orientable manifold. The tangent bundle $T(\Gamma\backslash G/H)$ splits according to the $H \cap K$-module decomposition $\mathfrak{q} = \mathfrak{q} \cap \mathfrak{k} + \mathfrak{q} \cap \mathfrak{p}$. Then, $\chi(\Gamma\backslash G/H) = 0$ because $\dim_{\mathbb{R}} \mathfrak{q} \cap \mathfrak{k} = \dim K/H \cap K$ is odd. On the other hand, as H is of maximal rank in G, so is H_U in G_U. As H is connected, so is H_U. Therefore, G_U/H_U is a compact orientable manifold with nonvanishing Euler number $\chi(G_U/H_U)$ ([HoS]). Now it follows from Theorem 4.4 that $\chi(G_U/H_U) \neq 0$ contradicts the fact that $\chi(\Gamma\backslash G/H) = 0$ if $\Gamma\backslash G/H$ is compact. Hence, $\Gamma\backslash G/H$ cannot be compact.   □

Example 4.13 (see Section 2.2)   We know that the hyperboloid of one sheet $G/H = \mathrm{SL}(2,\mathbb{R})/\mathrm{SO}(1,1)$ does not admit a compact Clifford–Klein form. This fact was explained by using nonvanishing vector fields in the item (iii) of Section 2.2, which is a geometric idea of Corollary 4.12. In this case, we have $\operatorname{rank} G = \operatorname{rank} H = 1$ and $\dim K/H \cap K = 1$, so that the assumptions in Corollary 4.12 are satisfied.

Exercise 4.14 (see also Example 5.14)   Prove that a semisimple symmetric space $\mathrm{SO}(i + k, j + l)/\mathrm{SO}(i,j) \times \mathrm{SO}(k,l)$ does not have a compact Clifford–Klein form if exactly one element among i, j, k, l is even.

At the end of this section, we pose the following conjecture:

Conjecture 4.15 (see [Ko3, Conjecture 6.4])   *Let G/H be a homogeneous space of reductive type. It is likely that the inequality*

$$\operatorname{rank} G + \operatorname{rank}(H \cap K) \geq \operatorname{rank} H + \operatorname{rank} K$$

holds if G/H admits a compact Clifford–Klein form.

The case with rank G = rank H can be proved based on Corollary 4.6 and on an argument of the cohomological dimension of a discrete group, as a slight improvement of Corollary 4.12 (see [Ko2, Corollary 5]). If H is compact or if G/H is a group manifold, then the inequality is obviously satisfied. As far as the author knows, the inequality in Conjecture 4.15 holds for all homogeneous spaces of reductive type that are proved to admit compact Clifford–Klein forms (cf. Corollary 5.6).

Notes 4.16   i) Most material in this section is taken from [KoO].

ii) To establish the holomorphic continuation, we calculate characteristic forms in terms of Lie algebras. Similar calculations are also carried out in [Bo2], [CGW] for Riemannian symmetric spaces.

iii) There exist different generalizations of Hirzebruch's proportionality principle, due to D. Mumford [Mu] for the noncompact Clifford–Klein forms of Hermitian symmetric spaces, and due to J. L. Dupont and W. Kamber [DK] for the secondary characteristic numbers.

iv) F. Labourie informed me that Corollary 4.12 is also valid for a manifold modeled on a homogeneous space G/H (by e-mail, 1994).

v) A similar idea (at least implicitly) to Corollary 4.12 is used in [Ku] based on the Gauss–Bonnet theorem in the case of rank 1 symmetric spaces $G/H = \mathrm{SO}(p + 1, q)/\mathrm{SO}(p, q)$ (cf. Exercise 4.13).

5   Compact Clifford–Klein forms of non-Riemannian homogeneous spaces

In Section 5, we focus our attention on compact Clifford–Klein forms of homogeneous spaces of reductive type. This section is organized as follows:

Subsections 5.1 – 5.9	Homogeneous spaces with compact Clifford–Klein forms.
Subsections 5.10 – 5.12	Homogeneous spaces without compact Clifford–Klein forms.

5.1   Uniform lattice, compact Clifford–Klein form

Definition 5.1   Suppose G is a Lie group and H is a closed subgroup. Suppose a discrete subgroup $\Gamma \subset G$ satisfies the following two conditions:

Γ acts properly discontinuously and freely on G/H.   (5.1a)

$\Gamma \backslash G/H$ is compact.   (5.1b)

Then Γ is said to be a *uniform lattice for the homogeneous space* G/H. The double coset space $\Gamma \backslash G/H$ is a *compact Clifford–Klein form of the homogeneous space* G/H.

Suppose there exists a G-invariant measure on G/H. (This is the case for a homogeneous space of reductive type.) Then it induces a measure on a Clifford–Klein form $\Gamma \backslash G/H$. The discrete subgroup Γ is said to be a *lattice for the homogeneous space* G/H provided both (5.1a) and

$\Gamma \backslash G/H$ is of finite volume   (5.2)

are satisfied.

If H is compact (in particular $H = \{e\}$ or $H = K$), Definition 5.1 coincides with a usual one.

5.2   Riemannian cases

If the isotropy subgroup H is compact, then any discrete subgroup of G acts properly discontinuously on G/H, as we saw in Section 3.3. This case, reffered to as the *Riemannian case*, allows a compact Clifford–Klein form. We recall the following important theorem due to Borel, Harish-Chandra, and Mostow–Tamagawa.

Theorem 5.2 ([Bo1],[BoHa],[MoTa])   *Suppose G is a real linear reductive Lie group and H is a compact subgroup of G. Then G/H admits compact Clifford–Klein forms. Furthermore, G/H has noncompact Clifford–Klein forms with finite volume if G contains a noncompact semisimple factor.*

We shall not prove this theorem, and instead we refer the reader to [Ra], [Zi1] and the original papers. We shall mention, however, some typical examples.

i) A compact Riemann surface M_g ($g \geq 2$) is a compact Clifford–Klein form of the Poincaré plane $\mathrm{PSL}(2, \mathbb{R})/\mathrm{SO}(2)$ (see Example 1.11).

ii) If $\Gamma = \mathbb{Z}^n \subset G = \mathbb{R}^n \supset H = \{0\}$, then $\Gamma \backslash G/H \simeq S^1 \times \cdots \times S^1$ is a compact Clifford–Klein form of $G/H \simeq \mathbb{R}^n$.

iii) If $\Gamma = \mathrm{GL}(n, \mathbb{Z}) \subset G = \mathrm{GL}(n, \mathbb{R}) \supset H = \{e\}$ ($n \geq 2$), then $\Gamma \backslash G/H$ is a noncompact Clifford–Klein form of $G/H \simeq \mathrm{GL}(n, \mathbb{R})$ with finite volume.

5.3   Moore's ergodicity theorem

A simple remark here is that if Γ is a uniform lattice for a group manifold $G = G/\{e\}$, then the quotient topology of the double coset space $\Gamma\backslash G/H$ is always compact. However, if H is noncompact, the double coset space $\Gamma\backslash G/H$ does not have a good topology in general. In fact, the action of Γ on G/H is not properly discontinuous. We leave it to the reader as an easy exercise:

Exercise 5.3   Prove that the action of Γ on G/H is not properly discontinuous if Γ is a uniform lattice of G and if H is noncompact.

Furthermore, the action of Γ on G/H can be ergodic:

Theorem 5.4 ([Moo], see also [Zi1, Chapter 2])   *Let G be a noncompact simple linear Lie group. If $\Gamma \subset G$ is a lattice and $H \subset G$ is a closed subgroup, then*

$$\text{The action of } \Gamma \text{ is ergodic on } G/H \quad \Leftrightarrow \quad H \text{ is noncompact.}$$

Here, we recall that the action of Γ is said to be ergodic if every Γ-invariant measurable set is either null or conull.

Thus, a uniform lattice for G/H must be *smaller* than a uniform lattice for G in some sense. In this respect, the cohomological dimension of an abstract group is a nice measure for the "size" of a discrete group (see [Ko2, Corollary 5.5]). We use it in the proof of Theorems 5.7 and 5.8. Basic references for the material we use on the cohomological dimension of groups are [Ser] and [Bi].

5.4   Indefinite-Riemannian case

Theorem 5.5 ([Ko2, Section 4])   *Suppose that G is a real reductive linear group and that H and L are both reductive in G. If the triplet (G, L, H) satisfies both conditions*

$$\mathfrak{a}(L) \cap \mathfrak{a}(H) = \{0\}, \tag{5.3a}$$

$$d(L) + d(H) = d(G), \tag{5.3b}$$

then G/H admits compact Clifford–Klein forms. Furthermore, G/H admits noncompact Clifford–Klein forms that have finite volume if L contains a noncompact semisimple factor.

The same results hold for G/L because of the symmetric role of H and L.

5.5 Well known examples (I) — Riemannian cases revisited

Suppose G/H is of reductive type. We know that there exist compact Clifford–Klein forms of G/H if H is compact (Theorem 5.4). This fact is explained in the context of Theorem 5.5 as follows.

Suppose H is compact. Then we have $\mathfrak{a}(H) = \{0\}$ and $d(H) = 0$. If we take $L = G$, then the conditions (5.3a,b) are satisfied. Thus, Theorem 5.4 (Riemannian case) is a special case of Theorem 5.5. But this does not give a new proof of Theorem 5.4 because we shall use Theorem 5.4 as a starting point for the proof of Theorem 5.5.

5.6 Well known examples (II) — group manifold cases

Let G' be a noncompact real reductive Lie group. Suppose $(G, H) = (G' \times G', \operatorname{diag} G')$, so that $G/H \simeq G'$ is a group manifold. This case is trivial because it is obvious that G/H admits a compact Clifford–Klein form by Theorem 5.4. But it is instructive to see how Theorem 5.5 is applied in the group manifold case because $H \simeq G'$ is noncompact.

Let $\mathfrak{g}' = \operatorname{Lie} G' = \mathfrak{k}' + \mathfrak{p}'$ be a Cartan decomposition, and $\mathfrak{a}_{G'} \subset \mathfrak{p}'$ a maximally abelian subspace. We write $W_{G'}$ for the Weyl group of the restricted root system $\Sigma(\mathfrak{g}, \mathfrak{a})$. Then, we have

$$\mathfrak{a}_G = \mathfrak{a}_{G'} \oplus \mathfrak{a}_{G'},$$
$$W_G = W_{G'} \times W_{G'}.$$

Let us take $L := G' \times 1$ (or $1 \times G'$). Then, we have

$$\mathfrak{a}(H) = \{(X, wX) : X \in \mathfrak{a}_{G'},\ w \in W_{G'}\},$$
$$\mathfrak{a}(L) = \{(X, 0) : X \in \mathfrak{a}_{G'}\},$$
$$d(H) = d(L) = d(G'), \qquad d(G) = 2 \dim G'.$$

Hence, the conditions (5.3a,b) are satisfied.

5.7   Examples of indefinite-Riemannian homogeneous spaces that have compact Clifford–Klein forms

As a corollary of Theorem 5.5 we have

Corollary 5.6   *The following homogeneous spaces (six series and six isolated ones) admit compact Clifford–Klein forms. Also, they admit noncompact Clifford–Klein forms of finite volume.*

i)	*a)* $\mathrm{SU}(2,2n)/\mathrm{Sp}(1,n)$,	*b)* $\mathrm{SU}(2,2n)/\mathrm{U}(1,2n)$,
ii)	*a)* $\mathrm{SO}(2,2n)/\mathrm{U}(1,n)$,	*b)* $\mathrm{SO}(2,2n)/\mathrm{SO}(1,2n)$,
iii)	*a)* $\mathrm{SO}(4,4n)/\mathrm{Sp}(1,n)$,	*b)* $\mathrm{SO}(4,4n)/\mathrm{SO}(3,4n)$,
iv)	*a)* $\mathrm{SO}(8,8)/\mathrm{SO}(8,7)$,	*b)* $\mathrm{SO}(8,8)/\mathrm{Spin}(8,1)$,
v)	*a)* $\mathrm{SO}(4,4)/\mathrm{SO}(4,1) \times \mathrm{SO}(3)$,	*b)* $\mathrm{SO}(4,4)/\mathrm{Spin}(4,3)$,
vi)	*a)* $\mathrm{SO}(4,3)/\mathrm{SO}(4,1) \times \mathrm{SO}(2)$,	*b)* $\mathrm{SO}(4,3)/G_{2(2)}$.

Here, (a) and (b) in each line forms a pair $(G/H, G/L)$ that satisfies the assumptions (5.3a,b) of Theorem 5.5. For example, the first line means $(G, H, L) = (\mathrm{SU}(2,2n), \mathrm{Sp}(1,n), \mathrm{U}(1,2n))$ satisfies (5.3a,b).

The signature of the indefinite metric on G/H (and also on Clifford–Klein forms) induced from the Killing form is given by $(4n, 3n^2 - 2n)$, $(4n, 2)$, $(2n, n^2 - n)$, $(2n, 1)$, $(12n, 7n^2 - 4n + 3)$, $(4n, 3)$, $(8, 7)$, $(56, 28)$, $(12, 3)$, $(4, 3)$, $(8, 2)$, and $(4, 3)$, respectively.

5.8   Sketch of proof

The idea of Theorem 5.5 is illustrated by the abelian case. Let $G = \mathbb{R}^n \supset H = \mathbb{R}^k$ $(n \geq k)$. Take a complementary subspace $L \simeq \mathbb{R}^{n-k}$ of H in G, and choose a lattice $\Gamma \simeq \mathbb{Z}^{n-k}$ of L. Then,

$$\Gamma \backslash G/H \simeq \mathbb{Z}^{n-k} \backslash \mathbb{R}^n / \mathbb{R}^k \simeq S^1 \times \cdots \times S^1$$

is a compact Clifford–Klein form of G/H. In this case (5.3a) is satisfied because $L \simeq \mathbb{R}^{n-k}$ is complementary to $H \simeq \mathbb{R}^k$, and (5.3b) is satisfied because $d(L) + d(H) = (n - k) + k = n = d(G)$.

For the general case, the proof of Theorem 5.5 is divided into the following three steps:

i) The criterion for proper actions (continuous analog of discontinuous groups) (Theorem 3.18).

ii) The criterion for compact quotient (continuous analog of uniform lattice) (Theorem 5.7).

iii) Existence of uniform lattice in the Riemannian case (Theorem 5.4).

If both (5.3a,b) are satisfied, then L acts properly on G/H by Theorem 3.18 and the quotient topology of $L\backslash G/H$ is compact by Theorem 5.7. On the other hand, Theorem 5.4 assures the existence of a torsion free cocompact discrete subgroup Γ of L. Then, Γ turns out to be a uniform lattice for G/H thanks to Observation 2.12. This shows the first half of Theorem 5.5. Similarly, a torsion free covolume finite discrete subgroup Γ of L is also a lattice for G/H.

5.9   Continuous analog

A continuous analog of a compact Clifford–Klein form $\Gamma\backslash G/H$ of a homogeneous manifold is a compact double coset space $L\backslash G/H$ in the quotient topology where L, H are closed subgroups of G such that L acts properly on G/H (see Definition 2.11).

Theorem 5.7   *Let G be a real reductive linear Lie group, H and L closed subgroups that are reductive in G. Under the equivalent conditions in Theorem 3.18 (i.e. $L \pitchfork H$ in G in the sense of Definition 2.14), the following two conditions are equivalent:*

$$L\backslash G/H \text{ is compact in the quotient topology.} \qquad (5.4a)$$
$$d(L) + d(H) = d(G). \qquad (5.4b)$$

A flavor of the proof: We recall that $d(G) = \dim_{\mathbb{R}} \mathfrak{p}$ if we write a Cartan decomposition as $\mathfrak{g} = \mathfrak{k} + \mathfrak{p}$. In view of the Cartan decomposition of Lie group $G \simeq K \times \exp(\mathfrak{p})$, we may regard

$$d(G) = \text{ the dimension of "noncompact part" of } G. \qquad (5.5)$$

A homogeneous space of reductive type G/H has a vector bundle structure over a compact manifold $K/H \cap K$ with typical fiber $\mathfrak{p}/\mathfrak{p} \cap \mathfrak{h}$ (e.g., [Ko2, Appendix]). In view of $\dim_{\mathbb{R}} \mathfrak{p}/\mathfrak{p} \cap \mathfrak{h} = d(G) - d(H)$, we may regard

$$d(G) - d(H) = \text{ the dimension of "noncompact part" of } G/H. \qquad (5.6)$$

Now, one might expect that

$$d(G) - d(H) - d(L) = \text{ the dimension of "noncompact part" of } L\backslash G/H, \qquad (5.7)$$

and that (5.7) would lead to Theorem 5.7. However, (5.7) is not always "true" if we do not assume that $L \pitchfork H$ in G. (We remark that the dimension of "noncompact part" of $L\backslash G/H$ is not defined yet.) The simplest and illustrative observation is when $G = \mathbb{R}^n$ and H, L are vector subspaces of G with dimension k, l, respectively. Then

$$d(G) - d(H) - d(L) = n - k - l,$$
$$L\backslash G/H \simeq \mathbb{R}^{n-k-l \; + \; \dim(H \cap L)}.$$

Therefore, if we assume $\dim(H \cap L) = 0$ (or, equivalently, $L \pitchfork H$ in G), then we have

$$d(G) - d(H) - d(L) = 0 \iff L\backslash G/H \text{ is compact}.$$

This explains why the assumption $L \pitchfork H$ is necessary in Theorem 5.7, and how (5.4a) and (5.4b) are related under this assumption.

For the general case where G is a real reductive linear group, the underlying idea is similar but we need some more work. This is carried out in a framework of a discrete analog of Theorem 5.7, where L is replaced by an arithmetic subgroup Γ of L and $d(L)$ is replaced by the cohomological dimension of Γ. See [Ko2] for details. $\square$

5.10   Necessary conditions for the existence of compact Clifford–Klein forms

For the existence of compact Clifford–Klein forms of homogeneous spaces of reductive type, we have presented two necessary conditions so far:

i) Calabi–Markus phenomenon (Theorem 3.10)
ii) Hirzebruch's proportionality principle (Corollary 4.12)

Here, we give a third necessary condition for the existence of compact Clifford–Klein forms of homogeneous spaces:

Theorem 5.8   Let G/H be a homogeneous space of reductive type. G/H does not admit a compact Clifford–Klein form if there exists a closed subgroup L reductive in G satisfying the following two conditions:

$$\mathfrak{a}(L) \subset \mathfrak{a}(H), \tag{5.8a}$$
$$d(L) > d(H). \tag{5.8b}$$

The proof of Theorem 5.8 is similar to that of Theorem 5.5 and Theorem 5.7, by using the cohomological dimension of an abstract group together with Theorem 3.18. See [Ko7] for details.

For an application of Theorem 5.8, we need to find a suitable L satisfying (5.8a,b), provided G/H is given. This is done systematically for some typical homogeneous space of reductive type in the next two subsections.

5.11   Semisimple symmetric spaces

To apply Theorem 5.8 to a semisimple symmetric space, we need some results on the root system for semisimple symmetric spaces. First of all, we give a brief review of the notion of ε-families introduced by T. Oshima and J. Sekiguchi [OS].

Let $\mathfrak{g}$ be a semisimple Lie algebra over $\mathbb{R}$, σ an involution of $\mathfrak{g}$, and θ a Cartan involution of $\mathfrak{g}$ commuting with σ. Then, $\sigma\theta$ is also an involution of $\mathfrak{g}$, because $(\sigma\theta)^2 = \sigma^2\theta^2 = 1$. Let $\mathfrak{g} = \mathfrak{k} + \mathfrak{p} = \mathfrak{h} + \mathfrak{q} = \mathfrak{h}^a + \mathfrak{q}^a$ be direct sum decompositions of eigenspaces with eigenvalues ± 1 corresponding to θ, σ and $\sigma\theta$ respectively. The symmetric pair $(\mathfrak{g}, \mathfrak{h}^a)$ is said to be the *associated symmetric pair of* $(\mathfrak{g}, \mathfrak{h})$. Note that $\mathfrak{h}^a = \mathfrak{k} \cap \mathfrak{h} + \mathfrak{p} \cap \mathfrak{q}$ and that $(\mathfrak{h}^a)^a = \mathfrak{h}$. Fix a maximally abelian subspace $\mathfrak{a}_{\mathfrak{p},\mathfrak{q}}$ of $\mathfrak{p} \cap \mathfrak{q}$. Then $\Sigma(\mathfrak{a}_{\mathfrak{p},\mathfrak{q}}) \equiv \Sigma(\mathfrak{g}, \mathfrak{a}_{\mathfrak{p},\mathfrak{q}})$ satisfies the axiom of root system (see [Ro, Theorem 5], [OS, Theorem 2.11]) and is called the *restricted root system of* $(\mathfrak{g}, \mathfrak{h})$. As $\sigma\theta \equiv \mathrm{id}$ on $\mathfrak{a}_{\mathfrak{p},\mathfrak{q}}$, we have a direct sum decomposition of the root space $\mathfrak{g}(\mathfrak{a}_{\mathfrak{p},\mathfrak{q}}, \lambda) = (\mathfrak{g}(\mathfrak{a}_{\mathfrak{p},\mathfrak{q}}, \lambda) \cap \mathfrak{h}^a) + (\mathfrak{g}(\mathfrak{a}_{\mathfrak{p},\mathfrak{q}}, \lambda) \cap \mathfrak{q}^a)$. We define a map

$$(m^+, m^-)\colon \Sigma(\mathfrak{a}_{\mathfrak{p},\mathfrak{q}}) \to \mathbb{N} \times \mathbb{N}, \qquad (5.9)$$

by $m^+(\lambda) := \dim(\mathfrak{g}(\mathfrak{a}_{\mathfrak{p},\mathfrak{q}}, \lambda) \cap \mathfrak{h}^a)$, $m^-(\lambda) := \dim(\mathfrak{g}(\mathfrak{a}_{\mathfrak{p},\mathfrak{q}}, \lambda) \cap \mathfrak{q}^a)$. Note that if $(\mathfrak{g}, \mathfrak{h})$ is a Riemannian symmetric pair, then $\mathfrak{h}^a = \mathfrak{g}$ and $m^- \equiv 0$. A map $\varepsilon\colon \Sigma(\mathfrak{a}_{\mathfrak{p},\mathfrak{q}}) \cup \{0\} \to \{1, -1\}$ is said to be a *signature of* $\Sigma(\mathfrak{a}_{\mathfrak{p},\mathfrak{q}})$ if ε is a semigroup homomorphism with $\varepsilon(0) = 1$. It is easy to see that a signature is determined by its restriction to Ψ, a fundamental system for $\Sigma(\mathfrak{a}_{\mathfrak{p},\mathfrak{q}})$, and that any map $\Psi \to \{1, -1\}$ is uniquely extended to a signature.

To a signature ε of $\Sigma(\mathfrak{a}_{\mathfrak{p},\mathfrak{q}})$, we associate an involution σ_ε of $\mathfrak{g}$ defined by $\sigma_\varepsilon(X) := \varepsilon(\lambda)\sigma(X)$ if $X \in \mathfrak{g}(\mathfrak{a}_{\mathfrak{p},\mathfrak{q}}; \lambda)$, $\lambda \in \Sigma(\mathfrak{a}_{\mathfrak{p},\mathfrak{q}}) \cup \{0\}$. Then, σ_ε defines a symmetric pair $(\mathfrak{g}, \mathfrak{h}_\varepsilon)$. The set

$$F((\mathfrak{g}, \mathfrak{h})) := \{(\mathfrak{g}, \mathfrak{h}_\varepsilon) : \varepsilon \text{ is a signature of } \Sigma(\mathfrak{a}_{\mathfrak{p},\mathfrak{q}})\}$$

is said to be an ε-*family of symmetric pairs* ([OS, Section 6]). Among an ε-family, there is a distinguished symmetric pair, called *basic*, characterized by

$$m^+(\lambda) \geq m^-(\lambda) \qquad \text{for any } \lambda \in \Sigma(\mathfrak{a}_{\mathfrak{p},\mathfrak{q}}) \text{ such that } \frac{\lambda}{2} \notin \Sigma(\mathfrak{a}_{\mathfrak{p},\mathfrak{q}}).$$

It is known that there exists a basic symmetric pair of $F = F\left((\mathfrak{g}, \mathfrak{h})\right)$ unique up to isomorphisms ([OS, Proposition 6.5]).

Example 5.9

i) $\{(\mathfrak{sl}(n, \mathbb{R}), \mathfrak{so}(p, n - p)) : 1 \leq p \leq n\}$ is an ε-family for which the pair $(\mathfrak{sl}(n, \mathbb{R}), \mathfrak{so}(n))$ is basic. This family is an example of the so-called K_ε-family, which is a special case of an ε-family.

ii) For $\mathbb{F} = \mathbb{R}, \mathbb{C}$ or $\mathbb{H}$ (a quaternionic number field), we write $\mathrm{U}(p, q; \mathbb{F})$ for $\mathrm{O}(p, q)$, $\mathrm{U}(p, q)$, and $\mathrm{Sp}(p, q)$, respectively. We fix $r < q < p$. Then

$$\{(\mathfrak{u}(r, p + q - r; \mathbb{F}), \mathfrak{u}(k, p - k; \mathbb{F}) + \mathfrak{u}(r - k, q - r + k; \mathbb{R}) : 0 \leq k \leq r\}$$

is an ε-family with $(\mathfrak{u}(r, p + q - r; \mathbb{F}), \mathfrak{u}(r, p - r; \mathbb{F}) + \mathfrak{u}(q; \mathbb{R}))$ basic.

Now we are ready to state an application of Theorem 5.8 to symmetric spaces:

Corollary 5.10 ([Ko7, Theorem 1.4])   *If a semisimple symmetric space G/H admits a compact Clifford–Klein form, then the associated symmetric pair $(\mathfrak{g}, \mathfrak{h}^a)$ is basic in the ε-family $F\left((\mathfrak{g}, \mathfrak{h}^a)\right)$.*

Sketch of proof: We apply Theorem 5.8 with "H":$= H_\varepsilon^a$ and "L":$= H^a$. Then the assumptions (5.8a,b) follow from the following lemma (see [Ko7, Lemma 4.5.2]). $\square$

Lemma 5.11   *With notations as before, let $(\mathfrak{g}, \mathfrak{h})$ be basic in the ε-family $F = F\left((\mathfrak{g}, \mathfrak{h})\right)$ and $(\mathfrak{g}, \mathfrak{h}_\varepsilon)$ be not basic in F. Then we have*

i) $\mathfrak{a}(H^a) = \mathfrak{a}(H_\varepsilon{}^a)$,

ii) $d(H^a) > d(H_\varepsilon{}^a)$.

Here are some examples of symmetric pair $(\mathfrak{g}, \mathfrak{h})$ such that the associated pair $(\mathfrak{g}, \mathfrak{h}^a)$ is basic in its ε-family.

Example 5.12   As usual, we write $(\mathfrak{g}, \mathfrak{k})$ for a Riemannian symmetric space if $\mathfrak{g}$ is a Lie algebra of noncompact reductive Lie group G. In the following table, $\mathbb{F}$ denotes $\mathbb{R}$, $\mathbb{C}$, or $\mathbb{H}$.

$(\mathfrak{g}, \mathfrak{h})$	$(\mathfrak{g}, \mathfrak{h}^a)$ is basic
$(\mathfrak{g}, \mathfrak{g})$	$(\mathfrak{g}, \mathfrak{k})$
$(\mathfrak{g}, \mathfrak{k})$	$(\mathfrak{g}, \mathfrak{g})$
$(\mathfrak{g} + \mathfrak{g}, \operatorname{diag} \mathfrak{g})$	$(\mathfrak{g} + \mathfrak{g}, \operatorname{diag} \mathfrak{g})$
$(\mathfrak{g}_{\mathbb{C}}, \mathfrak{k}_{\mathbb{C}})$	$(\mathfrak{g}_{\mathbb{C}}, \mathfrak{g})$
$(\mathfrak{u}(p, q; \mathbb{F}), \mathfrak{u}(m, q; \mathbb{F}) + \mathfrak{u}(p - m; \mathbb{F}))$	$(\mathfrak{u}(p, q; \mathbb{F}), \mathfrak{u}(m; \mathbb{F}) + \mathfrak{u}(p - m, q; \mathbb{F}))$

An obvious remark is that the associated symmetric pair $(\mathfrak{g}, \mathfrak{h}^a)$ of a Riemannian symmetric pair $(\mathfrak{g}, \mathfrak{k})$ is $(\mathfrak{g}, \mathfrak{g})$, which is obviously basic, and that of a group manifold $(\mathfrak{g}' + \mathfrak{g}', \operatorname{diag} \mathfrak{g}')$ is again a group manifold, which is also basic. This is consistent with the fact that both Riemannian symmetric spaces and group manifolds admit compact Clifford–Klein forms (cf. Section 5.5 and Section 5.6). See [Ko7, Table 4.4] (and also [Ko5, Table 5.4.3]) for a list of semisimple symmetric spaces that are proved by this method not to admit compact Clifford–Klein forms.

5.12   Semisimple orbits

Another typical example of a homogeneous space of reductive type is a semisimple orbit $\operatorname{Ad}(G) \cdot X \simeq G/Z_G(X)$, where $X \in \mathfrak{g}$ is a semisimple element (Example 3.13(iv)). We apply Theorem 5.8 to semisimple orbits.

Corollary 5.13 ([Ko7], [BL]; see also [Ko4, Ko6])   *Let G be a real reductive linear Lie group, and X a semisimple element of $\mathfrak{g}$. If $G \cdot X \simeq G/Z_G(X)$ admits a compact Clifford–Klein form, then the orbit $G \cdot X$ carries a G-invariant complex structure.*

Before giving a sketch of proof, we remark that a simple group G cannot be a complex group if there is a nonzero semisimple element $X \in \mathfrak{g}$ such that $\mathrm{Ad}(G) \cdot X \simeq G/Z_G(X)$ admits a compact Clifford–Klein form. In fact, since $\mathrm{rank}\, G = \mathrm{rank}\, Z_G(X)$, we have $\mathbb{R}\text{-rank}\, G = \mathbb{R}\text{-rank}\, Z_G(X)$ if G is a complex reductive Lie group. Then there is no infinite discontinuous group acting on $G/Z_G(X)$ by Theorem 3.10 (the Calabi–Markus phenomenon). Therefore $G/Z_G(X)$ does not admit a compact Clifford–Klein form. This means that G itself is not a complex group. Nevertheless Corollary 5.13 asserts that the homogeneous space $G/Z_G(X)$ carries a G-invariant complex structure. This is because $G/Z_G(X)$ is realized as an elliptic orbit, which is a crucial point of the proof.

Sketch of proof: Let $X = X_e + X_h$ be a decomposition of a semisimple element $X \in \mathfrak{g}$, where X_e is elliptic and X_h is hyperbolic such that $[X_e, X_h] = 0$. Applying Theorem 5.8 with $L := Z_G(X_e)$ and $H := Z_G(X)$, we obtain Corollary 5.13. An alternative proof is given in [BL] based on symplectic structure. $\square$

Let X be an elliptic element of $\mathfrak{g}$. We give here a quick review of the rich geometric structure of an elliptic orbit $\mathrm{Ad}(G) \cdot X \simeq G/Z_G(X)$. It is well known that an elliptic orbit $G/Z_G(X)$ carries a G-invariant complex structure via the generalized Borel embedding into the generalized flag variety $G_U/Z_{G_U}(X)$:

$$G/Z_G(X) \subset G_U/Z_{G_U}(X).$$

See, for example, [KoO, Appendix] for a proof. (This embedding is a usual Borel embedding if $G/Z_G(X)$ is a Hermitian symmetric space, equivalently, if $G/Z_G(X)$ is a symmetric space and if $Z_G(X)$ is compact; see [He].) $G/Z_G(X)$ is said to be *dual manifolds of Kähler C-space* in the sense of Griffiths–Schmid if $Z_G(X)$ is compact; $G/Z_G(X)$ is said to be a $\frac{1}{2}$-*Kähler symmetric space* in the sense of Berger if $G/Z_G(X)$ is a symmetric space and if $Z_G(x)$ is noncompact.)

A G-invariant symplectic structure on the elliptic orbit is induced from the one on the coadjoint orbit through the identification $\mathfrak{g} \simeq \mathfrak{g}^*$, which is given by a nondegenerate G-invariant bilinear form B on $\mathfrak{g}$ (e.g., the Killing form if $\mathfrak{g}$ is semisimple). The orbit $G/Z_G(X)$ also carries a G-invariant (indefinite) Kähler structure induced by B. The indefinite Kähler structure is then compatible with the symplectic structure.

We note that Clifford–Klein forms of an elliptic orbit inherit these structures. Corollary 5.6 asserts that the homogeneous manifolds

$$\mathrm{U}(2,2n)/\mathrm{U}(1) \times \mathrm{U}(1,2n),$$

$$\mathrm{SO}(2,2n)/\mathrm{U}(1,n),$$

$$\mathrm{SO}(4,3)/\mathrm{SO}(2) \times \mathrm{SO}(4,1)$$

admit compact Clifford–Klein forms. Furthermore, these are realized as elliptic orbits of the adjoint action. So we obtain new examples of compact indefinite-Kähler manifolds.

Finally, in sharp contrast, recall that a hyperbolic orbit does not admit an infinite discontinuous group (Example 3.11).

5.13   Some examples

We give here a number of examples of homogeneous spaces that are proved not to have (or to have) Clifford–Klein forms by the method of this section. These examples may help to reveal the applications and limitations of various methods known so far (e.g., [Ko1], [Ko2], [Ko5], [KoO], [Ko4, Ko6], [Ko7], [BL], [Zi2], [LaMZ], [Ko11], [Co]) in studying the existence problem of compact Clifford–Klein forms, which has not yet found a final answer (see also Notes in Section 5.14).

Example 5.14   First, we consider a semisimple symmetric space

$$G/H = \mathrm{SO}(i+j, k+l)/\mathrm{SO}(i,k) \times \mathrm{SO}(j,l) \qquad (i \le j, k, l).$$

It follows from Theorem 5.8 (or Corollary 5.10) that if G/H admits a compact Clifford–Klein form, then G/H is compact, or H is compact, or $0 = i < l \le j - k$. Moreover, we have $jkl \equiv 0 \mod 2$ as an application of Hirzebruch's proportionality principle in Section 4 (see Corollary 4.12 and a remark after Conjecture 4.15).

Conversely, if $(i,j,k,l) = (0,2n,1,1), (0,4n,1,3), (0,4,2,1), (0,8,1,7)$, or if $i = l = 0$, then there exists a compact Clifford–Klein form of G/H (see Corollary 5.6).

Example 5.15 A semisimple symmetric space

$$G/H = \mathrm{SO}^*(2n)/\mathrm{U}(l, n - l)$$

does not admit a compact Clifford–Klein form if

$$3l \leq 2n \leq 6l$$

and if $n \geq 3$. We explain this in Example 5.18 together with similar examples.

It admits compact Clifford–Klein forms if

$$(n, l) = (4, 1),\ (4, 3),\ (3, 1),\ (3, 2),\ (2, 1),\ l = 0,\ \text{ or }\ l = n.$$

In fact, we have a local isomorphism $\mathrm{SO}^*(8)/\mathrm{U}(1,3) \approx \mathrm{SO}(2,6)/\mathrm{U}(1,3)$, which is proved to admit compact Clifford–Klein forms in Corollary 5.6. It is trivial in the cases $(3, 1)$, $(3, 2)$, and $(2, 1)$ because G/H is then compact. G/H is a Riemannian symmetric space in the cases $l = 0$ or $l = n$ (see Theorem 5.2).

Example 5.16 A semisimple symmetric space

$$G/H = \mathrm{SO}^*(4n)/\mathrm{SO}^*(4p + 2) \times \mathrm{SO}^*(4n - 4p - 2)$$

does not admit compact Clifford–Klein forms if $1 \leq p \leq n - 2$, by Corollary 5.13. We note that $\mathrm{SO}^*(8)/\mathrm{SO}^*(6) \times \mathrm{SO}^*(2)$ (namely, in the case $(n, p) = (2, 1)$) admits compact Clifford–Klein forms. Similarly, there exist compact Clifford–Klein forms in the case $(n, p) = (2, 0)$. In fact, we have a local isomorphism $\mathrm{SO}^*(8)/\mathrm{SO}^*(2) \times \mathrm{SO}^*(6) \approx \mathrm{SO}(2, 6)/\mathrm{U}(1, 3)$, which is in the list of Corollary 5.6.

Example 5.17 Suppose that $G_{\mathbb{C}}/H_{\mathbb{C}}$ is a complex irreducible semisimple symmetric space. Then, we have a conjecture that $G_{\mathbb{C}}/H_{\mathbb{C}}$ admits compact Clifford–Klein forms if and only if $G_{\mathbb{C}}/H_{\mathbb{C}}$ is locally isomorphic to a group manifold. One can prove that $G_{\mathbb{C}}/H_{\mathbb{C}}$ does not have compact Clifford–Klein forms unless $G_{\mathbb{C}}/H_{\mathbb{C}}$ is locally isomorphic to either a group manifold, $\mathrm{SO}(2n + 2, \mathbb{C})/\mathrm{SO}(2n + 1, \mathbb{C})$, $\mathrm{SL}(2n, \mathbb{C})/\mathrm{Sp}(n, \mathbb{C})$ $(n \geq 2)$, or $E_{6,\mathbb{C}}/F_{4,\mathbb{C}}$ (see [Ko7, Example 1.9]).

Example 5.18

G/H does not have a compact Clifford–Klein form		
G	H	Range of parameters
$SL(n, \mathbb{R})$	$Sp(l, \mathbb{R})$	$0 < 2l \leq n - 2$
$SL(n, \mathbb{C})$	$Sp(l, \mathbb{C})$	$0 < 2l \leq n - 1$
$SL(n, \mathbb{C})$	$SO(l, \mathbb{C})$	$0 < l \leq n$
$SO^*(2n)$	$U(l, n - l)$	$3l \leq 2n \leq 6l, \; n \geq 3$
$Sp(n, \mathbb{R})$	$Sp(l, \mathbb{C})$	$0 < 2l \leq n$
$SU^*(2n)$	$SO^*(2l)$	$1 < l \leq n$
$SL(2n, \mathbb{C})$	$SU(p, q)$	$p + q < n$ or $p = q \; (pq > 0)$
$SL(2n, \mathbb{R})$	$SO(p, q)$	$p + q < n$ or $p = q \; (pq > 0)$

These examples, except for $Sp(n, \mathbb{R})/Sp(l, \mathbb{C})$ with $n = 2l$, are proved by Theorem 5.8. In applying Theorem 5.8, the choice of "L" is not unique. Here are examples of the choice of "L" for G/H:

$$
\begin{array}{ll}
SO(l, n - l) & \text{for} \quad SL(n, \mathbb{R})/Sp(n, \mathbb{R}), \\
U(l, n - l) & \text{for} \quad SL(n, \mathbb{C})/Sp(l, \mathbb{C}),, \\
U([\tfrac{l}{2}], n - [\tfrac{l}{2}]) & \text{for} \quad SL(n, \mathbb{C})/SO(l, \mathbb{C}), \\
SO^*(4l + 2) & \text{for} \quad SO^*(2n)/U(l, n - l), \\
U(l, n - l) & \text{for} \quad Sp(n, \mathbb{R})/Sp(l, \mathbb{C}), \\
Sp([\tfrac{l}{2}], n - [\tfrac{l}{2}]) & \text{for} \quad SU^*(2n)/SO^*(2l), \\
U(p, 2n - p) \text{ or } Sp(p, \mathbb{C}) & \text{for} \quad SL(2n, \mathbb{C})/U(p, q), \\
SO(p, 2n - p) \text{ or } Sp(p, \mathbb{R}) & \text{for} \quad SL(2n, \mathbb{R})/SO(p, q).
\end{array}
$$

The proof for the non-existence of compact Clifford–Klein form of the symmetric space $G/H := Sp(2l, \mathbb{R})/Sp(l, \mathbb{C})$ is different. In this case we have rank $G = 2l = $ rank H and can apply an argument of Hirzebruch's proportionality principle for the Euler–Poincaré class (see Section 3). Then, in view of $K/H \cap K = U(2l)/Sp(l)$, the rank condition $\operatorname{rank} K = 2l > \operatorname{rank}(H \cap K) = l$ implies the non-existence of compact Clifford–Klein form of G/H (see [Ko2, Example (4.11)] for details).

For more examples of the foregoing type, we refer to Tables 4.4 and 5.3 in [Ko7].

Example 5.19

G/H does not have a compact Clifford–Klein form			
G	H	Range of parameters	
$(n > m > 1)$		m is even	m is odd
$SL(n, \mathbb{R})$	$SL(m, \mathbb{R})$	$n > \frac{3}{2}m$	$n > \frac{3}{2}m + \frac{3}{2}$
$SU^*(2n)$	$SU^*(2m)$	$n > \frac{3}{2}m - 1$	$n > \frac{3}{2}m + \frac{1}{2}$
$SO^*(2n)$	$SO^*(2m)$	$n > \frac{3}{2}m - 1$	$n > \frac{3}{2}m - \frac{1}{2}$
$Sp(n, \mathbb{R})$	$Sp(m, \mathbb{R})$	$n > \frac{3}{2}m + 1$	$n > \frac{3}{2}m + \frac{3}{2} + \delta_{m,3}$
$SL(n, \mathbb{C})$	$SL(m, \mathbb{C})$	$n > \frac{3}{2}m - 1$	$n > \frac{3}{2}m + \frac{1}{2}$
$SO(n, \mathbb{C})$	$SO(m, \mathbb{C})$	$n > \frac{3}{2}m - 1$	$n > \frac{3}{2}m - \frac{1}{2}$
$Sp(n, \mathbb{C})$	$Sp(m, \mathbb{C})$	$n > \frac{3}{2}m$	$n > \frac{3}{2}m + \frac{1}{2} + \delta_{m,3}$

Table A

We indicate the proof of Table A for $G = SL(n, \mathbb{R}) \supset H = SL(m, \mathbb{R})$ with m even. We want to show that G/H does not admit a compact Clifford–Klein form if $\frac{2}{3}n > m$. We take a subgroup

$$L := SO\left(\frac{m}{2}, n - \frac{m}{2}\right) \subset G.$$

We identify $\mathfrak{a}(G)$ with $\mathbb{R}^n$ by choosing a suitable coordinate of $\mathfrak{a}_G$, so that the action of the Weyl group $W_G \simeq \mathfrak{S}_n$ is given by permutation of coordinates. Then, we have

$$\mathfrak{a}(L) = \left\{ (y_1, \dots, y_{\frac{1}{2}m}, -y_1, \dots, -y_{\frac{1}{2}m}, 0, \dots, 0) : y_j \in \mathbb{R} \ (1 \le j \le \frac{1}{2}m) \right\},$$

$$\mathfrak{a}(H) = \left\{ (x_1, \dots, x_m, 0, \dots, 0) : x_j \in \mathbb{R}, \ (1 \le j \le m), \ \sum_{j=1}^{m} x_j = 0 \right\},$$

regarded as subsets in $\mathfrak{a}_G/W_G$. Therefore, we have $\mathfrak{a}(L) \subset \mathfrak{a}(H)$.

On the other hand,

$$d(L) - d(H) = \frac{m}{2}(n - \frac{m}{2}) - (\frac{1}{2}m(m+1) - 1)$$

$$= \frac{m}{2}(n - \frac{3}{2}m - 1) + 1 > 0.$$

Now Theorem 5.8 shows that G/H does not admit a compact Clifford–Klein form.

Other cases in Table A are proved similarly from Theorem 5.8. A choice of L for each (G, H) is listed in Table B.

A choice of L for Table A			
G	H	\multicolumn{2}{c}{L}	
$(n > m > 1)$		m is even	m is odd
$\mathrm{SL}(n, \mathbb{R})$	$\mathrm{SL}(m, \mathbb{R})$	$\mathrm{SO}(\frac{m}{2}, n - \frac{m}{2})$	$\mathrm{SO}(\frac{m-1}{2}, n - \frac{m-1}{2})$
$\mathrm{SU}^*(2n)$	$\mathrm{SU}^*(2m)$	$\mathrm{Sp}(\frac{m}{2}, n - \frac{m}{2})$	$\mathrm{Sp}(\frac{m-1}{2}, n - \frac{m-1}{2})$
$\mathrm{SO}^*(2n)$	$\mathrm{SO}^*(2m)$	$\mathrm{U}(\frac{m}{2}, n - \frac{m}{2})$	$\mathrm{U}(\frac{m-1}{2}, n - \frac{m-1}{2})$
$\mathrm{Sp}(n, \mathbb{R})$	$\mathrm{Sp}(m, \mathbb{R})$	$\mathrm{U}(\frac{m}{2}, n - \frac{m}{2})$	$\mathrm{U}(\frac{m-1}{2}, n - \frac{m-1}{2})$
$\mathrm{SL}(n, \mathbb{C})$	$\mathrm{SL}(m, \mathbb{C})$	$\mathrm{U}(\frac{m}{2}, n - \frac{m}{2})$	$\mathrm{U}(\frac{m-1}{2}, n - \frac{m-1}{2})$
$\mathrm{SO}(n, \mathbb{C})$	$\mathrm{SO}(m, \mathbb{C})$	$\mathrm{SO}(\frac{m}{2}, n - \frac{m}{2})$	$\mathrm{SO}(\frac{m-1}{2}, n - \frac{m-1}{2})$
$\mathrm{Sp}(n, \mathbb{C})$	$\mathrm{Sp}(m, \mathbb{C})$	$\mathrm{Sp}(\frac{m}{2}, n - \frac{m}{2})$	$\mathrm{Sp}(\frac{m-1}{2}, n - \frac{m-1}{2})$

Table B

Example 5.20

G/H does not have a compact Clifford–Klein form		
G	H	range of parameters
$(j \geq i > 0,\ p \geq i,\ q \geq j)$		
$\mathrm{O}(p,q)$	$\mathrm{O}(i,j)$	$j \neq q$ or $p > q$ or $(p+1)iq \equiv 1 \mod 2$
$\mathrm{U}(p,q)$	$\mathrm{U}(i,j)$	$j \neq q$ or $p > q$
$\mathrm{Sp}(p,q)$	$\mathrm{Sp}(i,j)$	$j \neq q$ or $p > q$

For $\mathbb{F} = \mathbb{R}, \mathbb{C}$ or $\mathbb{H}$ (a quaternionic number field), we write $\mathrm{U}(p,q;\mathbb{F})$ for $\mathrm{O}(p,q), \mathrm{U}(p,q)$ and $\mathrm{Sp}(p,q)$, respectively.

Sketch of Proof: To prove the results in the table, we first choose $L = \mathrm{U}(i,q;\mathbb{F})$ ($\mathbb{F} = \mathbb{R}$, $\mathbb{C}$, $\mathbb{H}$). Then $\mathfrak{a}(H) = \mathfrak{a}(L)$ because $i \leq j \leq q$, and $d(H) \leq d(L)$. Here, the equality holds if and only if $j = q$. Therefore, it follows from Theorem 5.8 that $\mathrm{U}(p,q;\mathbb{F})/\mathrm{U}(i,j;\mathbb{F})$ admits a compact Clifford–Klein form only if $j = q$. Now, $\mathrm{U}(p,q;\mathbb{F})/\mathrm{U}(i,q;\mathbb{F})$ admits a compact Clifford–Klein form if and only if so does a reductive symmetric space $\mathrm{U}(p,q;\mathbb{F})/\mathrm{U}(p-i;\mathbb{F}) \times \mathrm{U}(i,q;\mathbb{F})$. Then, by Corollary 5.10, we have $p > q$. See Example 5.14 for the condition $(p+1)iq \equiv 1 \mod 2$ in the case where $\mathbb{F} = \mathbb{R}$. $\square$

Example 5.21 i) Suppose $G = \mathrm{SL}(n,\mathbb{R}) \supset H = \mathrm{SL}(2,\mathbb{R}) \times \cdots \times \mathrm{SL}(2,\mathbb{R})$ (k times) ($n \geq 2k > 0$). If $n > 3$, then there does not exist a compact Clifford–Klein form of G/H.
  ii) Suppose $G = \mathrm{SL}(n,\mathbb{C}) \supset H = \mathrm{SL}(2,\mathbb{C}) \times \cdots \times \mathrm{SL}(2,\mathbb{C})$ (k times) ($n \geq 2k > 0$). If $n \geq 3$, then there does not exist a compact Clifford–Klein form of G/H. In particular, $\mathrm{SL}(3,\mathbb{C})/\mathrm{SL}(2,\mathbb{C})$ does not admit a compact Clifford–Klein form ([Ko6, Example7]).

Proof: i) Take $L = \mathrm{SO}(k, n-k)$. Then $\mathfrak{a}(H) = \mathfrak{a}(L)$ and $d(L) - d(H) = k(n-k) - 2k = k(n-k-2) > 0$ if $n > 4$ and if $n \geq 2k > 0$. Take $L = \mathrm{Sp}(2,\mathbb{R})$ for $n = 4$.

ii) Take $L = \mathrm{SU}(k, n - k)$. Then $\mathfrak{a}(H) = \mathfrak{a}(L)$ and $d(L) - d(H) = 2k(n - k) - 3k = k(2n - 2k - 3) > 0$ if $n \geq 2k$. $\square$

For more examples of homogeneous spaces without compact Clifford–Klein forms, we refer to [Ko7, Table 5.3].

5.14 Notes

The existence problem of a compact Clifford–Klein form of a homogeneous space G/H has its origin in the uniformization theorem of Riemann surfaces due to Klein, Poincaré, and Koebe (Example 1.11). In this case, the homogeneous space is the Poincaré plane $G/H = \mathrm{PSL}(2, \mathbb{R})/\mathrm{SO}(2)$, which is the simplest example a Riemannian symmetric space of the non-compact type.

At the beginning of the 1960s, the existence problem of a compact Clifford–Klein form of any Riemannian symmetric space was settled affirmatively by Borel, Harish-Chandra, Mostow, and Tamagawa ([Bo1], [BoHa], [MoTa]; see Theorem 5.2). Contrary to this, around the same time, it was found by Calabi, Markus, and Wolf that certain pseudo-Riemannian symmetric spaces (of rank 1) do not admit compact Clifford–Klein forms (Theorem 3.2; [CM], [Wo1], [Wo2]; see also Section 3.10). Clifford–Klein forms of the real hyperbolic space $\mathrm{SO}(p, q)/\mathrm{SO}(p - 1, q)$ were studied by R. Kulkarni in the begining of 1980s, and in particular, the Calabi–Markus phenomenon of this special case was settled [Ku].

It was in the late 1980s that a systematic study of the existence problem of compact Clifford–Klein forms was basically begun by T. Kobayashi [Ko1–8] for a general homogeneous space of reductive type, which is a wide class of homogeneous spaces containing Riemannian symmetric spaces, reductive group manifolds, pseudo-Riemannian symmetric spaces and semisimple orbits of the adjoint action. An overview is given in [Ko4, Ko6]. A sufficient condition for the existence of compact Clifford–Klein forms (Theorem 5.5) was proved in [Ko2]. The proof rests on an argument in the continuous setting, namely the criterion for proper actions (Theorem 3.18) and the criterion for the compactness of a double coset space (Theorem 5.7). Among six series and six isolated homogeneous spaces that admit compact Clifford–Klein forms in Corollary 5.6, (ii-b) and (iii-b) (i.e., in the case $\mathrm{SO}(p + 1, q)/\mathrm{SO}(p, q)$) were first proved by Kulkarni ([Ku]) and other cases were by Kobayashi (see [Ko2] for (i-a), (i-b), (ii-a), (iii-a); [Ko4, Ko6] for (vi-a) and (vi-b); and [Ko11] for (iv-a), (iv-b), (v-a), (v-b)).

Conversely, necessary conditions for the existence of compact Clifford–Klein forms have been also studied since the late 1980s by various approaches. That is,

  i) The Calabi–Markus phenomenon and the criterion for proper actions ([Ko2], [Fr], [Ko10])
 ii) Hirzebruch's proportionality principle ([KoO])
iii) Comparison theorem ([Ko7])
 iv) Construction of symplectic forms ([BL])
  v) Ergodic theory ([Zi2], [LaMZ])

We have explained (i), (ii), and (iii) in Section 3 (e.g., Theorem 3.10, 3.18, and 3.19), Section 4 (e.g., Corollary 4.12), and Section 5 (e.g., Theorem 5.8, Corollary 5.10, Corollary 5.13), respectively. We have not dealt with (iv) and (v) here, which are quite different from other methods. However, examples obtained so far by other methods (e.g. [BL], [Zi2], [LaMZ], [Co]) are not necessarily new in the reductive case; most of them are also proved (sometimes in a stronger form) by (i), (ii), and (iii).

To clarify the applications and limitations of various methods, we will examine some typical classes of homogeneous spaces. We note that the setting of the foregoing results due to Benoist, Labourier, Zimmer, Mozes, Corlete ([BL], [Zi2], [LaMZ], [Co]) requires either

$$Z_G(H) \text{ contains } \mathbb{R} \qquad (5.10)$$

or

$$Z_G(H) \text{ contains a semisimple Lie group with } \mathbb{R}\text{-rank} \geq 2. \qquad (5.11)$$

First, suppose that G/H is a semisimple symmetric space; we assume G/H is irreducible for simplicity. We have seen that the methods (i), (ii), and (iii) give rise to necessary conditions for the existence of compact Clifford–Klein forms, namely, Theorem 3.10, Corollary 4.12 and Corollary 5.10. On the other hand, no new results are obtained for semisimple symmetric spaces if we assume (5.10) or (5.11). In fact, the assumption (5.10) is satisfied if and only if G/H is a para-Hermitian symmetric space. And a para-Hermitian symmetric space does not admit compact Clifford–Klein forms (Example 3.15(i)) by the Calabi–Markus phenomenon. The assumption (5.11) is never satisfied in the symmetric case.

Second, suppose that G/H is a semisimple orbit. Then, the assumption (5.10) is satisfied. In this case, Benoist and Labourie, using the method (iv), gave an alternative proof of Corollary 5.13 (the method (iii)) in a strengthened form ([BL, Theorem 1]).

Typical examples of homogeneous spaces satisfying the assumption (5.11) are those listed in some parts of Examples 5.18–5.20 (these are not symmetric spaces as we already remarked.) Zimmer proved in particular that $SL(n, \mathbb{R})/SL(m, \mathbb{R})$ (which is referred as the " basic test case" in [Zi2]) does not admit a compact Clifford–Klein form if $\frac{1}{2}n > m$. His approach is based on Ratner's theorem and ergodic theory ([Zi2, Corollary 1.3]). Labourier, Mozes, and Zimmer simplified and extended Zimmer's approach to manifolds locally modeled on homogeneous spaces ([LaMZ]). Their result allows the case $\frac{1}{2}n = m$ in this example. On the other hand, our method (iii) (see Section 5; [Ko4, Ko6], [Ko7]) applied to this special case shows that $SL(n, \mathbb{R})/SL(m, \mathbb{R})$ does not admit a compact Clifford–Klein form if $\frac{2}{3}n > m$ for even m (or if $\frac{2}{3}n > m+1$ when m is odd), which gives a sharper result in most cases (when n and m are large enough; see Example 5.19). Similarly, the indefinite Stiefel manifolds $U(p, q; \mathbb{F})/U(i, j; \mathbb{F})$ ($\mathbb{F} = \mathbb{R}, \mathbb{C}, \mathbb{H}$) ($j \geq i > 0$, $p \geq i$, $q \geq j$) do not admit compact Clifford–Klein forms if $j \neq q$ or if $p > q$, as we explained in Example 5.20 as another example of our method (iii).

Again, the very special case with $(p, q, i, j) = (n, 2, m, 1)$, $n > 2m$, and $\mathbb{F} = \mathbb{H}$ implies a result that $Sp(n, 2)/Sp(m, 1)$ ($n > 2m$) does not admit a compact Clifford–Klein form, which was announced by Corlette in ICM-94 ([Co, Theorem 12]). Thus, Corlette's results were not new, but his approach apparently differs from methods (i)–(v).

Since the beginning of the 1990s, the existence problem of compact Clifford–Klein forms of homogeneous spaces with noncompact isotropy subgroups has interacted with other branches of mathematics. For example, in the classification of certain Anosov flows, Benoist, Foulon, and Labourier [BFL] encountered the nonexistence problem of compact Clifford–Klein forms of G/H where G is real reductive and H is a semisimple part of a Levi subgroup of a maximal parabolic subgroup. Also, symplectic geometry [BL], ergodic theory and Ratner's theory [Zi2], [LaMZ], harmonic maps [Co], restriction of unitary representations [Ko9] have come to be interacted with the existence problem of compact Clifford–Klein forms.

Many basic questions about Clifford–Klein forms of non-Riemannian homogeneous spaces have not yet found a final answer. As observed in the recent developments just mentioned, it is fascinating that different areas of mathematics seem to be closely related to the existence problem of compact Clifford–Klein forms.

6   Open problems

We collect some open problems.

i) (Problem 2.10(ii))   Find a criterion for the Calabi–Markus phenomenon for a general Lie group.

ii) (Problem 2.17)   Find a criterion for $L \pitchfork H$ in G if L, H are closed subgroups of G.

iii) (Problem 2.19)   In which class of Lie groups, does the following equivalence hold?

$$H \pitchfork L \text{ in } G \Leftrightarrow \text{ the pair } (L, H) \text{ has the property (CI) in } G.$$

iv) (Problem 3.7)   Is there a sufficient condition on some positiveness of the curvature of a pseudo-Riemannian manifold M (in particular, Lorentz manifold) assuring that M is noncompact with a finite fundamental group?

v) (Problem 2.10 (ii))   Find a criterion on G/H that admits a compact Clifford–Klein form. Also, find a criterion on G/H that admits a noncompact Clifford–Klein form of finite volume.

vi) Suppose that G/H is a homogeneous space of reductive type that admits a compact Clifford–Klein form. Is there a subgroup L reductive in G which satisfies both of the conditions (5.3a) and (5.3b)?

vii) (Conjecture 4.15)   Does the inequality

$$\operatorname{rank} G + \operatorname{rank}(H \cap K) \geq \operatorname{rank} H + \operatorname{rank} K$$

hold if G/H admits a compact Clifford–Klein form?

viii) Is there a Teichmüller theory for a compact Clifford–Klein form?

ix) If there exists a compact Clifford–Klein form of G/H, then does there also exist a noncompact Clifford–Klein form of G/H of finite volume, and vice versa?

x) ([Wal]) Is there an analog of Eisenstein series that describes the decomposition of $L^2(\Gamma\backslash G/H)$ if $\Gamma\backslash G/H$ is a compact Clifford–Klein form or if $\Gamma\backslash G/H$ is a noncompact Clifford–Klein form of finite volume?

We have already explained most of these problems. Here are some short comments for the convenience of the reader.

Problem (i) is solved for the reductive case and the simply connected solvable case as we explained in Section 3.

Problem (ii) is solved if G is a reductive group.

Problem (iii) is a subproblem of (ii), and in particular, the property (CI) might be the final solution for a simply connected nilpotent group (the conjecture of Lipsman).

Problem (iv) is a question in pseudo-Riemannian geometry. It is regarded as a "perturbation" of the Calabi–Markus phenomenon, and should be in a good contrast to a classical theorem of Myers [My].

Problem (v) has been one of the main subjects of this chapter. For example, $\mathrm{SO}(i+j, k+l)/\mathrm{SO}(i,j) \times \mathrm{SO}(k,l)$ remains open (see Example 4.13.1 for partial results obtained so far).

Problems (vi) and (vii) are subproblems of (v). Problem (vi) asks if a converse of Theorem 5.5 holds. This is a delicate problem because the Zariski closure of a uniform lattice for G/H of reductive type does not always satisfy the conditions (5.3a,b).

We have found that an analog of the Weil rigidity does not always hold for semisimple symmetric spaces of higher ranks, and that an analog of the Mostow rigidity does not always hold for semisimple symmetric spaces of higher dimensions [Ko8]. There seems to be much to study in the deformation of uniform lattices, in which area we pose Problem (viii).

Regarding Problem (x), we recall that an existence result for compact Clifford–Klein forms (or noncompact ones of finite volume) of Riemannian symmetric spaces (Theorem 4.2) has opened a theory of Eisenstein series in harmonic analysis on square integrable functions over the double coset space $\Gamma\backslash G/H$. It is natural to expect that an existence result for pseudo-Riemannian symmetric spaces could open a theory of harmonic analysis on such nice double coset spaces.

Acknowledgment

The author would like to express his sincere gratitude to Professor Bent Ørsted and Professor Henrik Schlichtkrull for inviting me to give lectures and for their heartful hospitality in European School of Group Theory, August 1994.

Postscript December 21 1995

There has been some recent progress on the open problems in Section 6.

Y. Benoist gave a different proof of Theorem 3.19(i) and proved a nonexistence theorem of compact Clifford–Klein forms of some homogeneous spaces such as $SO(4n, \mathbb{C})/SO(4n - 1, \mathbb{C})$ (see Example 5.16) in [1] (see Open Problem (v)).

É. Ghys studied a fine structure of the deformation of a lattice for a group manifold $G' \times G'/\operatorname{diag} G'$ with $G' = SL(2, \mathbb{C})$ in [2], and T. Kobayashi studied for which homogeneous manifolds of reductive type local rigidity of a uniform lattice fails in [3] (see Open Problem (viii)).

[1] Y. Benoist, *Actions propres sur les espaces homogenes reductifs*, Preprint.

[2] É. Ghys, *Déformations des structures complexes sur les espaces homogènes de* $SL(2, \mathbb{C})$, J. reine angew. Math. **468** (1995), 113–138.

[3] T. Kobayashi, *Deformation of compact Clifford–Klein forms of indefinite-Riemannian homogeneous manifolds*, Preprint, Mittag–Leffler Institute 95-96, No. 32.

References

[AhSa] L. V. Ahlfors and L. Sario, *Riemann Surfaces*, Princeton Mathematics Series **26**, Princeton Univesity Press, Princeton, New Jersey, 1960.

[Au] L. Auslander, *The structure of compact locally affine manifolds*, Topology **3** (1964), 131–139.

[BL] Y. Benoist and F. Labourie, *Sur les espaces homogenes modeles de varietes compactes*, I.H.E.S. Publ. Math. **76** (1992), 99–109.

[BFL]   Y. Benoist, P. Foulon, F. Labourie, *Flots d'Anosov à distributions stanble et instable différentiables*, J. Amer. Math. Soc. **5** (1992), 33–74.

[Bi]   R. Bieri, *Homological dimension of discrete groups*, Mathematics Notes, Queens Mary College, 1976.

[Bo1]   A. Borel, *Compact Clifford–Klein forms of symmetric spaces*, Topology **2** (1963), 111–122.

[Bo2]   ———, *Sur une généralisation de la formule de Gauss-Bonnet*, An. Acad. Bras. Cienc. **39** (1967), 31–37.

[BoHa] A. Borel and Harish-Chandra, *Arithmetic subgroups of algebraic groups*, Ann. of Math. **75** (1962), 485–535.

[BoHi] A. Borel and F. Hirzebruch, *Characteristic classes and homogeneous spaces I*, Amer. Math. J. **80** (1958), 458–538.

[Bou]   N. Bourbaki, *Éléments de mathématique, Topologie générale*, Hermann, Paris, 1960.

[CM]   E. Calabi and L. Markus, *Relativistic space forms*, Ann. of Math. **75** (1962), 63–76.

[Car]   H. Cartan, *Quotients of analytic spaces*, Contributions to functional theory, Bombay, 1960.

[CGW]   R. S. Cahn, P. B. Gilkey, and J. A. Wolf, *Heat equation, proportionality principle, and volume of fundamental domains*, pp. 43–45, in Differential Geometry and Relativity (M. Cahen and M. Flato, eds.), Reidel, Dordrecht, 1976.

[Ch]   C. Chevalley, *On the topological structure of solvable groups*, Ann. of Math. **42–43** (1941), 668–675.

[Co]   K. Corlette, *Harmonic maps, rigidity and Hodge theory* (1994), ICM-94, invited talk.

[D]   J. L. Dupont, *Curvatures and Characteristic Classes*, Lecture Notes in Math. **640**, Springer-Verlag, Berlin/Heidelberg/New York, 1978.

[DK]   J. L. Dupont and F. W. Kamber, *Cheeger–Chern–Simons classes of transversally symmetric foliations: dependence relations and eta-invariants*, Math. Ann. **295** (1993), 449–468.

[Fl]   M. Flensted-Jensen, *Analysis on Non-Riemannian Symmetric Spaces*, C. B. M. S., Regional Conference Series **61**, Amer. Math. Soc., 1986.

[Fr]   Friedland, *Properly discontinuous groups on certain matrix homogeneous spaces*, Preprint, 1994.

[GK]   W. Goldman and Y. Kamishima, *The fundamental group of a compact flat space form is virtually polycyclic*, J. Differential Geometry **19** (1984), 233–240.

[He]   S. Helgason, *Differential Geometry, Lie Groups and Symmetric Spaces*, Pure and Applied Mathematics **80**, Academic Press, New York/London, 1978.

[Hi]   F. Hirzebruch, *Automorphe Formen und der Satz von Riemann–Roch*, Symposium Internacional de Topologia algebraica (1956), 129–144.

[HoS]   H. Hopf and H. Samelson, *Ein Satz über die Wirkungsräume geschlossener Liescher Gruppen*, Comment. Math. Helv. **13** (1940–41), 240–251.

[KaKo]   S. Kaneyuki and M. Kozai, *Paracomplex structures and affine symmetric spaces*, Tokyo J. Math. **8** (1985), 81–98.

[Ko-S]   S. Kobayashi, *Transformation Groups in Differential Geometry*, Ergebnisse der Mathematik und ihrer Grenzgebiete **70**, Springer, Berlin/Heidelberg/New York, 1972.

[KoN]   S. Kobayashi and K. Nomizu, *Foundations of Differential Geometry, Vol. 2*, Interscience, John Wiley & Sons, Inc., New York/London, 1969.

[Ko1]   T. Kobayashi, *Properly discontinuous actions on homogeneous spaces of reductive type*, Seminar Reports of Unitary Representation Theory **8** (1988), 17–22. (Japanese)

[Ko2]   ———, *Proper action on a homogeneous space of reductive type*, Math. Ann. **285** (1989), 249–263.

[Ko3]   ———, *Discontinuous groups and homogeneous spaces with indefinite metric*, Proceedings of the 36th Symposium of Geometry (1989), 104–116. (Japanese)

[Ko4]   ———, *Discontinuous groups acting on homogeneous spaces of reductive type*, Seminar Reports of Unitary Representation Theory

10 (1990), 41–45.

[Ko5]   ———, *Discontinuous group in a non-Riemannian homogeneous space*, R. I. M. S. Kokyuroku, Kyoto University **737** (1990), 1–24.

[Ko6]   ———, *Discontinuous groups acting on homogeneous spaces of reductive type*, Proceedings of the Conference on Representation Theory of Lie Groups and Lie Algebras held in 1990 August–September at Fuji-Kawaguchiko (ICM-90 Satellite Conference) (1992), World Scientific, Singapore/New Jersey/London, 59–75.

[Ko7]   ———, *A necessary condition for the existence of compact Clifford–Klein forms of homogeneous spaces of reductive type*, Duke Math. J. **67** (1992), 653–664.

[Ko8]   ———, *On discontinuous groups acting on homogeneous spaces with noncompact isotropy subgroups*, J. Geometry and Physics **12** (1993), 133–144.

[Ko9]   ———, *Discrete decomposability of the restriction of $A_q(\lambda)$ with respect to reductive subgroups and its applications*, Invent. Math. **117** (1994), 181–205.

[Ko10]   ———, *Criterion of proper actions on homogeneous space of reductive groups*, J. Lie Theory (to appear).

[Ko11]   ———, *Compact Clifford–Klein forms and Clifford modules*, in preparation.

[KoO]   T. Kobayashi and K. Ono, *Note on Hirzebruch's proportionality principle*, J. Fac. Soc. University of Tokyo **37** (1) (1990), 71–87.

[Ku]   R. S. Kulkarni, *Proper actions and pseudo-Riemannian space forms*, Advances in Math. **40** (1981), 10–51.

[LaMZ]   F. Labourie, S. Mozes and R. J. Zimmer, *On manifolds locally modelled on non-Riemannian homogeneous spaces*, preliminary version (1994).

[Lib]   R. Libermann, *Sur le structures presque paracomplexes*, C. R. Acad. Sci. Paris **234** (1952), 2517–2519.

[Lip]   R. Lipsman, *Proper actions and a compactness condition*, J. Lie Theory **5** (1995), 25–39.

[Ma]   G. A. Margulis, *Free completely discontinuous groups of affine transformations*, Soviet Math. Dokl. **28** (2) (1983), 435–439.

[Mi]   J. Milnor, *On fundamental groups of complete affinely flat mani-fold*, Advances in Math. **25** (1977), 178–187.

[MiSt]   J. W. Milnor and J. D. Stasheff, *Characteristic Classes*, Annals of Mathematics Studies **76**, Princeton University Press, Princeton, New Jersey, 1974.

[Moo]   C. C. Moore, *Ergodicity of flows on homogeneous spaces*, Amer. J. Math. **88** (1966), 154–178.

[Mos]   G. D. Mostow, *Self-adjoint groups*, Ann. of Math. **62** (1955), 44–55.

[MoTa]   G. D. Mostow and T. Tamagawa, *On the compactness of arith-metically defined homogeneous spaces*, Ann. of Math. **76** (1962), 446–463.

[Mu]   D. Mumford, *Hirzebruch's proportionality theorem in the noncom-pact case*, Invent. Math. **42** (1977), 239–272.

[My]   S. B. Myers, *Riemannian manifolds with positive mean curvature*, Duke Math. J. **8** (1941), 401–404.

[No]   K. Nomizu, *Invariant affine connections on homogeneous spaces*, Amer. J. Math. **76** (1954), 33–65.

[OS]   T. Oshima and J. Sekiguchi, *The restricted root system of a semisimple symmetric pair*, Advanced Studies in Pure Math. **4** (1984), 433–497.

[Pa]   R. S. Palais, *On the existence of slices for actions of noncompact Lie groups*, Ann. of Math. **73** (1961), 295–323.

[Ra]   M. Raghunathan, *Discrete Subgroups of Lie Groups*, Ergebnisse der Mathematik und ihrer Grenzgebiete **68**, Springer, Berlin Hei-delberg New York, 1972.

[Ro]   W. Rossmann, *The structure of semisimple symmetric spaces*, Canad. J. Math. **31** (1979), 156–180.

[Sa]   I. Satake, *On a generalization of the notion of manifold*, Proc. Nat. Acad. Sci. USA **42** (1956), 359–363.

[Sel]   A. Selberg, *On discontinuous groups in higher-dimensional sym-metric spaces*, Contributions to functional theory, Bombay, 1960, pp. 147–164.

[Ser]   J. P. Serre, *Cohomologie des groupes discrètes*, Annals of Math. Studies **70**, Princeton University Press, Princeton, New Jersey., 1971, pp. 77–169.

[Sp]   G. Springer, *Introduction to Riemann Surfaces*, Addison Wesley, 1957.

[Su]   D. Sullivan, *A Generalization of Milnor's inequality concerning affine foliations and affine manifolds*, Comment. Math. Helv. **51** (1976), 183–189.

[To]   G. Tomanov, *The virtual solvability of the fundamental group of a generalized Lorentz space form*, J. Differential Geometry **32** (1990), 539–547.

[Wal]   N. R. Wallach, *Two problems in the theory of automorphic forms*, Open Problems in Representation Theory, 1988, pp. 39–40, Proceedings at Katata, 1986.

[War]   G. Warner, *Harmonic Analysis on Semisimple Lie Groups* I, II, Springer Verlag, Berlin, 1972.

[We]   R. O. Wells, Jr., *Differential Analysis on Complex Manifolds*, Graduate Texts in Mathematics **65**, Springer-Verlag, 1980.

[Wo1]   J. A. Wolf, *The Clifford–Klein space forms of indefinite metric*, Ann. of Math. **75** (1962), 77–80.

[Wo2]   _____ , *Isotropic manifolds of indefinite metric*, Comment. Math. Helv. **39** (1964), 21–64.

[Wo3]   _____ , *Spaces of Constant Curvature*, 5th ed., Publish or Perish, Inc., Wilmington, Delaware, 1984.

[Yo]   K. Yosida, *A theorem concerning the semisimple Lie groups*, Tohoku Math. J. **44** (1938), 81–84.

[Zi1]   R. J. Zimmer, *Ergodic Theory and Semisimple Groups*, Birkhäuser, Boston, 1984.

[Zi2]   _____ , *Discrete groups and non-Riemannian homogeneous spaces*, J. Amer. Math. Soc. **7** (1994), 159–168.

Chapter 4

The Method of Stationary Phase and Applications to Geometry and Analysis on Lie Groups

V. S. VARADARAJAN

University of California, Los Angeles

CONTENTS

Introduction

It is well known that many problems arising in optics and electromagnetic theory, number theory, probability theory, and quantum physics lead to questions involving integrals of rapidly oscillating functions, especially the asymptotic behavior of these integrals as the frequency parameter becomes large. The method used in these studies is invariably what is known as the *method of stationary phase* that goes back to *Stokes* and *Kelvin*. It is equally the case, although less widely known, that oscillatory integrals and their asymptotics play an important role in many more modern parts of mathematics such as topology of manifolds, geometry and analysis on homogeneous and locally homogeneous spaces, differential equations on complex manifolds, and so on. In these notes, we shall attempt to give a brief introduction to oscillatory integrals and the method of stationary phase, and explain how it relates to the various areas of mathematics we have mentioned.

1   Oscillatory integrals. Introduction and examples

1.1 The principle of stationary phase

Let us begin with a classical situation in optics, namely the problem of determining the radiation at a point x in space, coming from a surface S that is emitting monochromatic radiation of very short wavelength isotropically from each of its points (see [GS]). If the frequency of radiation is τ, it is easy to conclude by looking at spherically symmetric outgoing wave solutions of the wave equation that the total radiation at x (at a fixed time t, which we shall promptly forget) is a superposition of the radiations associated to the various points of the surface and so can be written as an integral

$$I(x : \tau) = \int_S \frac{e^{i\tau|x-y|}}{|x-y|} a(y) dy.$$

The problem is that of examining the behavior of this integral as $\tau \longrightarrow \infty$. In optics, this is the problem of determining the intensity $|I(x : \tau)|^2$ of the radiation at x in the regime when wave optics is replaced by geometrical optics. The integral $I(x : \tau)$ is an example of an *oscillatory integral* because the function

$$e^{i\tau|x-y|}$$

oscillates very rapidly when τ is very large. The real-valued function

$$\varphi_x(y) = |x-y| \qquad (y \in S, \ x \notin S)$$

is called the *phase*, and the possibly complex-valued function

$$a_x(y) = \frac{a(y)}{|x-y|} \qquad (y \in S, \ x \notin S)$$

is called the *amplitude*. Note that both the phase and amplitude, especially the phase, depends on x. Analytically, x plays the role of a *parameter* that governs, in a way that we shall explain now, the behavior of this integral.

The basic method of dealing with this integral is to appeal to the principle of stationary phase of Stokes and Kelvin, which asserts the following: The principal contributions to $I(x : \tau)$ come from those points of S that are *critical points* of the phase φ_x, namely, those points y at which the first order derivatives of φ_x (along S) all vanish. If, for instance, there are no critical points, then for amplitudes that are smooth and compactly supported, the integral is a rapidly decreasing function of τ; this depends of course on the location of x, and such x are said to be in the *shadow* or the *dark zone*:

$$I(x : \tau) = O(\tau^{-r}) \quad \forall \, r \geq 1 \qquad (x \text{ in shadow}).$$

Suppose now that there are critical points in S for φ_x but that these are all *nondegenerate* in the sense that the *Hessian matrix*, namely the matrix of second derivatives (along S again) at each critical point, is nonsingular. It is then possible to show that the radiation is of the order of magnitude of τ^{-1} at these points x. Since the intensity of light at these points x is much larger than at the points in the shadow, these points are said to form the *light zone*. More precisely (this is the technical content of the principle of stationary phase),

$$I(x : \tau) \sim \gamma \tau^{-1} \qquad (x \text{ in light zone}).$$

where γ is a constant depending on the amplitude a_x and is not identically 0. However, as x varies in space, it often happens that there will be points for which the associated phase φ_x has *degenerate* critical points. Typically, this has the consequence that the radiation at x is of the order τ^{-d} where $d < 1$. Such points form a surface C in space and have neighboring points that are in the light zone; since $d < 1$, the intensity at their location is very much higher than at the neighboring points of the light zone. Light *burns* at these points, and so they have been known traditionally as *caustics*. The surface C is called the *caustic surface* of the phase φ_x.

The foregoing classification of points of space into light, dark, and caustic zones (see [AGV]) can be refined a little more if we assume that the surface is *opaque* to the radiation. The total radiation at x is then an integral, not on the entire surface S but only on the part visible from x. This is a region R_x, which now has possibly a *boundary*, and there will be locations x at which $I(x : \tau)$ is of the order $\tau^{-3/2}$. These are called points in the *twilight zone*:

$$I(x : \tau) \sim \gamma \tau^{-3/2} \qquad (x \text{ in twilight zone}).$$

In this example, it is easy to check that the time dependent radiation is

$$e^{i\tau t} I(x : \tau),$$

and that it satisfies the wave equation *exactly*; in fact, it was derived from this point of view. One can turn this around and ask, for very general hyperbolic equations, whether asymptotic solutions can be expressed as such integrals, where we now allow amplitudes that depend on τ and have asymptotic expansions in τ. Here, asymptotic solution means that the integral satisfies the equation in question to greater and greater accuracy as τ becomes large. Thus, the study of oscillatory integrals is important from the point of view of classical hyperbolic partial differential equations.

Summarizing the preceding discussion, we may say the following. In problems of high frequency optics and classical partial differential equations, one comes across integrals of the form

$$I(x : \tau) = I(x : a : \tau) = \int e^{i\tau \varphi(x,y)} a(x, y, \tau) dy.$$

Here, τ is called the *frequency*; a is the *amplitude*, which is a smooth complex valued function, defined on $X \times Y \times \mathbb{R}_+$ where $X \subset \mathbb{R}^k, Y \subset \mathbb{R}^n$ are open sets and $\mathbb{R}_+$ is the set of positive real numbers; φ is the *phase*, which is a smooth real function defined on $X \times Y$; and there is a compact set $K \subset Y$ such that $a(x, y, \tau) = 0$ if $y \notin K$. The problem is to understand the limiting behavior of the integrals $I(x : \tau)$ when τ is very large. This is mathematically formulated as the problem of studying the behavior of $I(x : \tau)$ as $\tau \longrightarrow \infty$. Since for large τ the function $e^{i\tau \varphi(x,y)}$ oscillates rapidly, the integrals such as I are called *oscillatory integrals*. Typically, the amplitudes do not depend on τ. But if they do, it is natural to assume that their dependence on τ is expressed by asymptotic expansions

$$a \sim \sum_{r \geq 0} a_r(x, y) \tau^{\mu - r},$$

$$a_{pq} \sim \sum_{r \geq 0} a_{r,p,q}(x, y) \tau^{\mu - r}.$$

Here, $\sim$ refers to asymptotics as $\tau \longrightarrow \infty$ and means that for any integer $N \geq 0$, the difference of the left side and the sum up to $r = N$ of the right side is $O(\tau^{\mu_N})$, where $\mu_N \longrightarrow \infty$ with N, and the O is locally uniform in x and α; the subscripts p, q denote differentiations with respect to x (p times) and y (q times). The simplest special case of such integrals are those of the form

$$I(\tau) = \int e^{i\tau\varphi(y)} a(y) dy \qquad (a \in C_c^\infty(K))$$

where there are no parameters.

Using partitions of unity in the amplitudes, it is clear that we need to study this problem only *locally*, i.e., at each point of Y for amplitudes supported within arbitrarily small neighborhoods of this point. The *Principle of stationary phase* in this context asserts that *the essential contribution to the integrals comes from the stationary (critical) points of the phase function φ; the other points contribute only terms that are rapidly decreasing in τ.* We can state this in more precise language (see [D2]):

Principle of stationary phase: *Let φ be a phase function, and let*

$$S_\varphi = \{z \in K \mid d\varphi(z) = 0\}.$$

Then, modulo rapidly decreasing functions of τ, the behavior, as $\tau \longrightarrow \infty$, of the integral

$$I(\tau) = \int e^{i\tau\varphi(y)} a(y) dy$$

depends only on the behavior of the amplitude a in arbitrarily small neighborhoods of S_φ. Moreover, if a is localized at a critical point of φ that is nondegenerate in the sense that its Hessian matrix (the matrix of second derivatives) is nonsingular, and if n is the number of integration variables, the integral is asymptotic to

$$\left(\frac{2\pi}{\tau}\right)^{n/2} |\det H|^{-1/2} e^{i\pi/4 \, \mathrm{sgn}(H)} e^{i\tau\varphi(0)} a(0),$$

where $\mathrm{sgn}(H)$ is the signature of H.

Note that there is a *change of phase* in the asymptotic form of the oscillatory integral as the signature of the Hessian changes. In particular, from the minimum point ($\mathrm{sgn} = n$) to the maximum point ($\mathrm{sgn} = -n$) the phase changes by $n\pi/2$; for $n = 2$, this is a phase shift of $180°$, the classic case in geometric optics.

1.2 Morse theory

In spite of the rather specialized nature of the origins of the principle of stationary phase, it has become one of the main tools in modern analysis and its applications have become extraordinarily diverse. In this subsection we shall try to elucidate this a little more.

First of all, the study of the behavior of oscillatory integrals leads directly to questions involving the *local structure* of smooth functions and maps. Indeed, since integrals transform very simply under diffeomorphisms, we see that there is a direct relationship to the problem of describing the simplest possible forms of functions and maps when we are allowed to make arbitrary C^∞ change of coordinates. The first such theorems were proved by *Marston Morse* in his epochmaking paper [Mo] when he showed that in the neighborhood of a *nondegenerate critical point* a smooth function can be transformed to a quadratic form, namely to its *Hessian form*, by a C^∞ change of variables. As is well known, Morse used this result as a stepping stone to his famous theorems relating the structure of critical points of a function to the topology of the space on which the function is defined, and he used these in turn to obtain his fundamental theorems on the calculus of variations in the large. Since then, *Morse theory* has become a fundamental tool in analysis and geometry, and it has been shown to have many interesting applications to questions of geometry of homogeneous spaces.

While Morse theory deals with the relationship between the topology of a manifold (both finite and infinite dimensional) and the structure of the critical points of a generic function defined on the manifold, the optical example discussed at the beginning focuses on an entirely different theme! Here we have a manifold M on which is defined somehow a *family* of real smooth functions $\varphi(x, m)$ $(x \in X, m \in M)$, X being the space of *parameters*. For generic values of x, the function $\varphi_x := \varphi(x, \cdot)$ has only nondegenerate critical points on M (is a *Morse function*, as it is nowadays called); but for special values of x, it may happen that the critical points become *degenerate*. The behavior of the oscillatory integral

$$I(x : \tau) = \int_M e^{i\tau\varphi(x,m)} a(m) dm$$

changes dramatically as x passes through these points, which are called, in analogy with the optical example, *caustics*.

As long as we are dealing with a single function, it is not an essential loss of generality to assume that it is a Morse function. This is due to the result that the Morse functions on a compact smooth manifold M form a dense open set; thus, a small perturbation (which can be attributed to imperfect measurement) will change any given function into a Morse function. This is true locally also. Thus, the function x^3 on $\mathbb{R}$ is not Morse, as its critical point at 0 is degenerate; but the neighboring function $x^3 - 3\varepsilon^2 x$ for small real but nonzero ε is Morse, because its two critical points are at $\pm\varepsilon$ and are both nondegenerate. We thus see that a small perturbation has made the degenerate critical point 0 of x^3 split into a pair of nondegenerate critical points. We speak of this as a *Morsification* of the function x^3. This might lead one to suppose that there is no physical need to study the phenomena of degenerate critical points. But to do so would be an error. The point is that, as in the optical example previously discussed, phase functions of physical problems often arise in a *family* depending on a parameter of physical significance (such as the location of a point in space), and one has to perturb *this family* to see if the degenerate critical points will disappear. It can be shown that this cannot always be done. Thus, in the example given just now, the family $x^3 - 3\varepsilon^2 x$ has a degenerate critical point at 0 when $\varepsilon = 0$; but, in every neighboring family, there will be a parameter value for which the phase has a degenerate critical point. This is not surprising, since, in the optical example, the degenerate critical points correspond to the caustics and so have real physical significance.

In this way, one is led to the problem of describing and classifying, up to local and global diffeomorphisms, smooth functions and families of smooth functions, and to study how the degenerate members of the classification change into generic members under perturbations. This is nothing but the modern theory of singularities of smooth functions and maps, to which many mathematicians, most notably, *Whitney, Thom, Malgrange, Mather, Arnold, Gel'fand, Bernstein,* and many others, have made remarkable and profound contributions, both conceptually and technically.

1.3 Some examples

We shall illustrate the preceding remarks with some examples.

i) *Shadow*   The oscillatory integral is

$$I(\tau) = \int_{\mathbb{R}^n} e^{i\tau x_1} a(x) dx \qquad (a \in C_c^\infty(\mathbb{R}^n)).$$

This is rapidly decreasing in τ by the fundamental result of Fourier transform theory. The phase has no critical points.

ii) *Light zone* The classic example is that of the *Fresnel integral*, namely

$$I(\tau) = \int_{\mathbb{R}} e^{i\tau x^2/2} a(x) dx \qquad (a \in C_c^\infty(\mathbb{R})).$$

This is evaluated using Fourier transforms. For any Schwartz function f on $\mathbb{R}$, we define its Fourier transform $\mathcal{F}f$ by

$$\mathcal{F}f(\xi) = \int_{\mathbb{R}} f(x) e^{ix\xi} dx \qquad (\xi \in \mathbb{R}).$$

The Fourier inversion formula says that for two Schwartz functions f, g we have

$$\int_{\mathbb{R}} f(x) g(x) dx = \frac{1}{2\pi} \int_{\mathbb{R}} \mathcal{F}f(-\xi) \mathcal{F}g(\xi) d\xi.$$

We now take $f(x) = \exp(-\sigma x^2/2)$ where $\sigma > 0$ and get, for our amplitude a,

$$\int_{\mathbb{R}} e^{-\sigma x^2/2} a(x) dx = \frac{\sigma^{-1/2}}{\sqrt{2\pi}} \int_{\mathbb{R}} e^{-\xi^2/2\sigma} \mathcal{F}a(\xi) d\xi.$$

Now, both sides are analytic functions of σ in the domain $\Re(\sigma) > 0$, and so this equality persists in this domain; the function $\sigma^{-1/2}$ is defined as $r^{-1/2} e^{i\theta/2}$ in polar coordinates with $\sigma = re^{i\theta}$, $|\theta| < \pi/2$. If we now let σ converge to $-i\tau$ ($\tau > 0$), we get (using the dominated convergence theorem on both sides)

$$\int_{\mathbb{R}} e^{i\tau x^2/2} a(x) dx = \frac{\tau^{-1/2} e^{i\pi/4}}{\sqrt{2\pi}} \int_{\mathbb{R}} e^{-i\xi^2/2\tau} \mathcal{F}a(\xi) d\xi. \tag{1.1}$$

The right side has τ in the denominator in the exponential, and so, on expanding the exponential we get an *asymptotic expansion* in powers of τ^{-1}. From the Taylor expansion of e^{iu} ($u \in \mathbb{R}$), we get

$$\left| e^{iu} - \left(1 + iu + \frac{(iu)^2}{2!} + \cdots + \frac{(iu)^m}{m!} \right) \right| \leq \frac{|u|^{m+1}}{(m+1)!},$$

whereas the formula $\mathcal{F}(a^{(2m)})(\xi) = (-1)^m \xi^{2m} \mathcal{F}a(\xi)$ gives

$$a^{(2m)}(0) = \frac{(-1)^m}{2\pi} \int_{\mathbb{R}} \xi^{2m} \mathcal{F}a(\xi) d\xi.$$

Putting these together and integrating the expansion

$$e^{-i\xi^2/2\tau} = \sum_{k\geq 0} \frac{(-i)^k}{2^k k!}\tau^{-k}\xi^{2k},$$

we get

$$I(\tau) = \left(\frac{2\pi}{\tau}\right)^{1/2} e^{i\pi/4} \sum_{0\leq k\leq m-1} \frac{i^k}{2^k k!}a^{(2k)}(0)\tau^{-k} + O(\tau^{-(m+(1/2))}).$$

In particular, taking $m = 1$ we get

$$I(\tau) = \left(\frac{2\pi}{\tau}\right)^{1/2} e^{i\pi/4}a(0) + O(\tau^{-3/2}) \qquad (\tau \longrightarrow \infty).$$

If $E_m(\tau : a)$ is the error term in the first expansion, it is easy to show from our remarks that we have the estimate

$$|E_m| \leq \frac{1}{\sqrt{2\pi}2^m}||\mathcal{F}(a^{(2m)})||_1\, \tau^{-(m+1)/2}.$$

iii) *Twilight zone*   We now consider a Fresnel type integral taken over a *domain with boundary*, namely, the half-space $\{x \geq 0\}$ in the plane $\mathbb{R}^2$ with coordinates x, y:

$$I(\tau : a) = \int\int_{x\geq 0} e^{i\tau(x+(1/2)y^2)}a(x,y)dxdy \qquad (a \in C_c^\infty(\mathbb{R}^2)).$$

We now integrate by parts with respect to x to obtain

$$I(\tau : a) = \frac{i}{\tau}\int_{\mathbb{R}} e^{i\tau y^2/2}a(0,y)dy + \frac{i}{\tau}\int\int_{x\geq 0} e^{i\tau(x+(1/2)y^2)}a_x(x,y)dxdy,$$

which leads to

$$I(\tau : a) = i\sqrt{2\pi}\tau^{-3/2}e^{i\pi/4}a(0,0) + O(\tau^{-5/2})$$

and

$$I(\tau : a) = i\sqrt{2\pi}\tau^{-3/2}e^{i\pi/4}\left[a(0,0) + \frac{i}{\tau}(a_x(0,0) + \frac{1}{2}a_{yy}(0,0))\right] + O(\tau^{-7/2}),$$

and so on. We leave it to the reader to verify that the general expansion is

$$I(\tau : a) = i\sqrt{2\pi}\tau^{-3/2}e^{i\pi/4} \sum_{0\leq r\leq m-1} i^r A_r(a)\tau^{-r} + O(\tau^{-(m+3/2)})$$

with

$$A_r(a) = \sum_{0\leq k\leq r} 2^{-k} a^{(r-k,2k)}(0,0).$$

iv) *Oscillatory integral associated to x^k in $\mathbb{R}$: caustics* This is the most basic case in which we encounter degenerate critical points. The family of phase functions is

$$\varphi(\alpha : x) = x^3 + \alpha x \qquad (\alpha, x \in \mathbb{R}).$$

For $\alpha > 0$, the phase is strictly increasing and has no critical points; for $\alpha < 0$, there are two critical points, at $\pm\sqrt{|\alpha|/3}$, both of which are nondegenerate, where the Hessians have opposite signs. For very small α, these are very close to each other. When $\alpha = 0$, the critical point is 0 and is degenerate. From our classification, we see that the region $\alpha > 0$ is the shadow or the dark zone, the region $\alpha < 0$ is the light zone, and the origin $\alpha = 0$ is the caustic. The oscillatory integral is

$$\int_{\mathbb{R}} e^{i\tau(x^3+\alpha x)}a(x)dx. \qquad (1.2)$$

It was first considered by *Airy* in 1836 in a famous paper [Ai], perhaps the first to investigate the behavior of light in the neighborhood of a caustic. Airy mentions that the study of this integral is at the foundation of the theory of the rainbow, and that he has been unable to integrate it in terms of the functions known at that time. We now know not too much more, except that we call it the *Airy function* and denote it by Ai(x):

$$\text{Ai}(t) = \int_{\mathbb{R}} e^{i(x^3+tx)}dx. \qquad (1.3)$$

The singularity of x^3 at 0 is called a *fold*, and it splits into two nondegenerate singularities $\pm\sqrt{|\alpha|/3}$ when we go over to the nearby function $x^3 + \alpha x$. The family $x^3 + \alpha x$ is thus called an *unfolding* (or a *Morsification* in this case) of the function x^3. There are asymptotic expansions of the integral (1.2) uniform in the parameter α, but they will not be in powers of τ^{-1}; rather they are in terms of the Airy function (1.3) (see [D3]). In terms of powers of τ, one can get only uniform *estimates*

$$\left| \int_{\mathbb{R}} e^{i\tau(x^3 + \alpha x)} a(x) dx \right| \leq \mu(a) \tau^{-1/3} \qquad (a \in C_c^\infty(\mathbb{R})),$$

where μ is a continuous seminorm on the Schwartz space of $\mathbb{R}$. Of course, for (1.3) there will be an asymptotic expansion in powers of τ^{-1}. We shall derive this now; it goes back, as far as I know, to Erdelyi [Er].

We consider

$$I(\tau : a) = \int_{\mathbb{R}} e^{i\tau x^k} a(x) dx \qquad (a \in C_c^\infty(\mathbb{R})).$$

To treat the neighborhood of 0 separately, we introduce a function $b \in C_c^\infty(\mathbb{R})$ that is 1 in a neighborhood of the origin, and we write

$$I = I_1 + I_2,$$

where

$$I_1 = \int e^{i\tau x^k} a(x) b(x) dx, \qquad I_2 = \int e^{i\tau x^k} a(x)(1 - b(x)) dx.$$

The amplitude $a(1 - b)$ in I_2 vanishes in a neighborhood of the origin, and so we can integrate by parts using the differential operator D:

$$D = \frac{1}{kx^{k-1}} \frac{d}{dx}, \qquad D\left(e^{i\tau x^k} \right) = (i\tau)^{-1} e^{i\tau x^k}.$$

Let us write $^t D$ for the transpose of the operator D. Then, $^t D^N(a(1 - b))$ is still in $C_c^\infty(\mathbb{R})$, so that

$$I_2(\tau : a) = (i\tau)^{-N} \int_{\mathbb{R}} e^{i\tau x^k} (^t D^N(a(1 - b))) dx,$$

which shows that I_2 is rapidly decreasing in τ:

$$|I_2(\tau : a)| \leq \mu_N(a) \tau^{-N} \qquad \forall\, N \geq 1,$$

where μ_N are continuous seminorms on the Schwartz space of $\mathbb{R}$.

The treatment of I_1 requires more care. Again we integrate by parts, integrating the function $e^{i\tau x^k}$ repeatedly. We split the integral into two, $I_1^{\pm}$, defined by the integrations over the positive and the negative semiaxes, and rewrite the integral I_1^- as an integral over the positive semiaxis. Thus,

$$I_1(\tau : a) = \int_0^{\infty} e^{i\tau x^k} a(x)b(x)dx + \int_0^{\infty} e^{(-1)^k i\tau x^k} a(-x)b(-x)dx.$$

Note that for k even the integrals are of the same nature, whereas for k odd they have τ entering with opposite signs. The basic idea in [Er] is to define the repeated integrals of the exponential as an integral in the complex domain; the line of integration depends on the sign in front of τ. For any integer $m \geq 0$, we define

$$f_{m+1}^{\pm}(u) = (-1)^{m+1} \int_u^{u+e^{\pm i\pi/2k}\infty} \frac{(z-u)^m}{m!} e^{\pm i\tau z^k} dz \qquad (u > 0, \ \tau > 0),$$

where the notation means that the integration is over the ray

$$\{v \longmapsto u + ve^{\pm i\pi/2k} \mid v \geq 0\}.$$

It is clear that

$$f_{m+1}^{(m+1)}(u) = e^{i\tau u^k} \qquad (u > 0).$$

Moreover,

$$f_{m+1}(0) = (-1)^{m+1} e^{\pm i(m+1)\pi/2k} \frac{1}{km!} \Gamma\left(\frac{m+1}{k}\right) \tau^{-(m+1)/k}.$$

To estimate the function f_{m+1} itself, we note that

$$\pm i\tau \left(u + ve^{\pm i\pi/2k}\right)^k = \pm i\tau \left(\pm iv^k + \cdots\right),$$

where the ellipses indicate terms the imaginary part of which are of the form $\pm iw \sin(s\pi/2k)$ with $w > 0$ and $0 \leq s \leq k$. Hence, on the line of integration, we have

$$\left| e^{\pm i\tau z^k} \right| \leq e^{-\tau|z-u|^k}.$$

Hence,

$$|f_{m+1}(u)| \leq \frac{1}{km!}\tau^{-(m+1)/k}\Gamma\left(\frac{m+1}{k}\right) \qquad (u > 0).$$

After these preliminaries, we can now integrate by parts the integral I_1 and get the following: If

$$I_1^{\pm}(\tau : a) = \int_0^{\infty} e^{\pm i\tau x^k} a(x)b(x)dx,$$

then

$$I_1^{\pm}(\tau : a) = \sum_{0 \leq s < m} \frac{1}{ks!} e^{\pm i(s+1)\pi/2k}\Gamma\left(\frac{s+1}{k}\right) a^{(s)}(0)\tau^{-(s+1)/k} + E_m(\tau : a),$$

where

$$|E_m(\tau : a)| \leq \mu_m(a)\tau^{-(m+1)/k},$$

μ_m being a continuous seminorm on the Schwartz space of $\mathbb{R}$.

Write now

$$I^k(\tau : a) = \int_{\mathbb{R}} e^{i\tau x^k} a(x)dx \qquad (a \in C_c^{\infty}(\mathbb{R})).$$

We consider the case of even and odd k separately. The μ_m denote continuous seminorms on the Schwartz space of $\mathbb{R}$.

k **even**   We have

$$I^k(\tau : a)$$
$$= \frac{2}{k} \sum_{0 \leq s < m} \frac{1}{(2s)!} e^{i(2s+1)/2k}\Gamma\left(\frac{2s+1}{k}\right) a^{(2s)}(0)\tau^{-(2s+1)/k} + E_m(\tau : a),$$

where

$$|E_m(\tau : a)| \leq \mu_m(a)\tau^{-(2m+1)/k}.$$

It is not difficult to verify that this agrees with the asymptotic expansion derived earlier by direct use of the Fourier transform when $k = 2$. We leave it to the reader to do this.

k odd   Define

$$\varepsilon_{ks} = e^{i(s+1)/2k} + (-1)^s e^{-i(s+1)/2k}.$$

Then,

$$I^k(\tau : a) = \frac{1}{k} \sum_{0 \le s < m} \frac{1}{s!} \Gamma\left(\frac{s+1}{k}\right) \varepsilon_{ks} a^{(s)}(0) \tau^{-(s+1)/k} + E_m(\tau : a),$$

where

$$|E_m(\tau : a)| \le \mu_m(a)\tau^{-(m+1)/k}.$$

In particular,

$$I^3(\tau : a) = \Gamma\left(\frac{4}{3}\right) a(0)\tau^{-1/3} + E, \quad |E| \le \mu(a)\tau^{-2/3}.$$

The leading term of the asymptotic expansion is now $\tau^{-1/3}$ instead of the $\tau^{-1/2}$ that is characteristic of the points for which the phase is a Morse function. The reader should also note that the derivation of the asymptotic expansion is considerably harder than in the generic case of Morse functions. This is fairly typical.

1.4 Phase functions on homogeneous spaces

When working in the geometric context of homogeneous spaces, there is a great deal of symmetry, and the phase functions that are significant display this symmetry in their definition. The typical situation is the following. We have a connected Lie group G and a linear representation of G in a finite dimensional vector space V over the real or the complex field. We suppose that there is a bilinear form $\langle \cdot, \cdot \rangle$ defined on $V \times V$ that is G-invariant. If G is compact, such a form always exists, and we may even suppose that it is positive definite. Let $v \in V$ be a point such that the orbit M_v of v under G is closed in V; if G_v is the stabilizer of v in G, we have an analytic diffeomorphism

$$G/G_v \approx M_v,$$

and so

$$\varphi(w : gG_v) := \langle g \cdot v, w \rangle \qquad (w \in V)$$

is a well-defined phase function on $G/G_v \approx M_v$. Let us write

$$L_w(u) = \langle w - u, w - u \rangle \qquad (u \in M_v).$$

Then, it is immediate that

$$L_w(g \cdot v) = \langle w, w \rangle + \langle v, v \rangle - 2\varphi(w : gG_v),$$

showing that, except for the possibility that $\langle u, u \rangle$ may be an *indefinite metric*, these are just the generalization of the phase functions coming from the optics example; and, in the case when G is compact and the metric is positive definite, they are exactly the phase functions defined by taking the distance from an arbitrary point.

When G is compact or semisimple, this method leads to some very interesting phase functions that have a lot of significant information that is useful in the topology and analysis on the homogeneous spaces G/G_v. Later, we shall study these cases in greater detail. We shall obtain many well-known results from this point of view: the conjugacy of maximal abelian subspaces of $\mathfrak{s}$, where $\mathfrak{g} = \mathfrak{k} \oplus \mathfrak{s}$ is the Cartan decomposition of a real semisimple Lie algebra $\mathfrak{g}$; the beautiful theorem of Kostant that the *Iwasawa projection* of $(\exp H) \cdot K$ in $\mathfrak{a}$ is the convex hull of the Weyl group transforms of H; computations of the cohomology of complex flag manifolds; estimates for the matrix elements of the principal series representations of a connected real semisimple Lie group with finite center; applications of these estimates to the problems of error estimation of spectra in locally symmetric spaces of negative curvature; and so on.

It may not be out of place to discuss briefly an interesting nonclassical aspect of the phase functions defined on the flag manifolds via the Iwasawa decomposition. Let G be a connected real semisimple Lie group with finite center, and let $G = KAN$ be an Iwasawa decomposition of G with $\mathfrak{a}$ the Lie algebra of A. We then have an analytic map

$$H : G \longrightarrow \mathfrak{a}, \qquad x \longmapsto H(x),$$

where $H(x)$ is defined by

$$x = u \exp H(x) n \qquad (u \in K, \ H(x) \in \mathfrak{a}, \ n \in N).$$

The matrix elements of the principal series representations of G are of the form

$$\int_K e^{i\tau(H(xk))} g(k)dk,$$

where $\tau \in \mathfrak{a}^*$. Although the integral is written over K, it is really over K/M, where M is the centralizer of A in K, so that we have *oscillatory integrals* on the flag manifold K/M. But there is a new element; the frequency variable is now *vector-valued* since it varies in $\mathfrak{a}^*$, and so the asymptotics are much more subtle than in the classical case where there is only one frequency variable. This is not very surprising; for example, there is a deep analogy between problems of scattering theory and problems of harmonic analysis on G/K; but when the rank of the symmetric space G/K is > 1, the past and future of conventional scattering are replaced by the chambers defined by the Weyl group, and the single scattering operator of the conventional theory is replaced by a family of scattering operators, going from an *initial* chamber to a *final* chamber. The situation in the case of oscillatory integrals with several frequency parameters is approximately similar. This is a situation (together with the scattering analogy just mentioned) that seems to merit some further examination from a general point of view.

2   Oscillatory integrals with nondegenerate phase functions

2.1 Local structure of smooth maps

We begin with a brief discussion of the local structure of smooth functions and maps between smooth manifolds. These are always assumed to be second countable and without boundary unless otherwise stated. Let M, N be smooth manifolds, and let f be a smooth map of M into N:

$$f: M \longrightarrow N;$$

and let

$$x \in M, \quad f(x) = y \in N.$$

A basic problem is to understand the structure of f in a neighborhood of x, especially the problem of describing the simplest possible forms of f that arise when we subject f to an arbitrary smooth change of coordinates in M and N. We write

$$f: (M, x) \longrightarrow (N, y)$$

and keep in mind that we are only interested in the *germ* of the map f. There is an obvious notion of isomorphism for such maps. Let G_x be the group of germs of diffeomorphisms $g: (M, x) \longrightarrow (M, x)$. If we write

$$h \circ f \circ g^{-1} \qquad (g \in G_x, \; h \in G_y)$$

for the (germ of the) map $t \longmapsto h(f(g^{-1}(t)))$, we see that

$$((h, g), f) \longmapsto h \circ f \circ g^{-1}$$

gives an action of $G_y \times G_x$ on the set of germs of the maps f. Maps in the same orbit are the ones that are isomorphic to each other. The problem is thus one of *classifying the orbits for this action*, or at least to describe *generic* orbits, where generic is to be interpreted appropriately and reasonably. Note that this problem makes sense in the *complex analytic category* also. Let $T_x(M)$ be the tangent space to M at x. We denote as usual by df_x the tangent map of f at x:

$$df_x: T_x(M) \longrightarrow T_y(N).$$

Write $m = \dim(M), n = \dim(N)$.

i) If df_x is *surjective* (then $m \geq n$, and f is called a *submersion at x*), then f is isomorphic to the projection

$$(x_1, \ldots, x_m) \longmapsto (x_1, \ldots, x_n).$$

ii) If df_m is *injective* (then $m \leq n$, and f is called an *immersion at x*), then f is isomorphic to the injection

$$(x_1, \ldots, x_m) \longmapsto (x_1, \ldots, x_m, 0, \ldots, 0).$$

Both (i) and (ii) are valid in the complex analytic category. If $m = n$, in either of these cases, f is a local diffeomorphism (or a local analytic isomorphism in the complex analytic case).

iii) By a *critical point* of a map $f(M \longrightarrow N)$ we mean a point x where f is neither immersive nor submersive, i.e., df_x is neither injective nor surjective; $f(x)$ is then called a *critical value*. If $N = \mathbb{R}$ so that f is a real function, these definitions agree with the usual definitions of critical points and critical values of a function. The fundamental result on critical values is *Sard's theorem*:

Theorem 2.1 (Sard)   *The set of values of a smooth map at points at which it is not submersive has measure zero.*

We shall not give a proof here but indicate it in the exercises. In local coordinates, measure zero is understood to mean measure zero in the Euclidean space defined by the coordinates; since this property is preserved under diffeomorphisms, it makes sense to speak of a set of measure zero in a coordinate neighborhood without specifying the chart; and as the manifold can be covered by a countable number of coordinate neighborhoods, it is clear that measure zero makes sense globally: A set is of measure zero if and only if its intersection with each coordinate open set has measure zero.

For arbitrary dimensions of M and N, even the local classification of maps under local diffeomorphisms is a very difficult problem. Nevertheless, a great deal of understanding has been gained in recent years on what happens generically. For *functions* (real smooth or complex analytic) the situation is much better. For a function f or a family of functions, it is a question of finding canonical forms in the neighborhood of a critical point. The most fundamental result of this type goes back, as we mentioned in the introduction, to Morse. It is called *Morse's lemma*. To formulate it we first define the notion of the *Hessian form* of a function at a critical point.

Definition–Proposition 2.2   Let f be a smooth function on M, and let x be a critical point of f. Then the *Hessian form* of f at x is the symmetric bilinear form H_x defined on $T_x(M) \times T_x(M)$ as follows: If $u, v \in T_x(M)$, and U, V are vector fields defined around x such that $U_x = u$, $V_x = v$, then

$$H_x(u, v) = (UVf)(x).$$

We must show that this makes sense. First, note that there is symmetry in u, v because the difference $(UVf)(x) - (VUf)(x)$ is $([U, V]f)(x) = 0$, since x is a critical point. To see that H_x is well defined, note that if $u = 0$, the expression for $H_x(u, v)$ is 0 because $U_x = 0$; if $v = 0$, the same conclusion is valid, because we can replace $(UVf)(x)$ by $(VUf)(x)$. If $t_1, \ldots, t_m$ are local coordinates around x taking x to 0, then the matrix of H_x in the basis $\partial/\partial t_j$ $(1 \le j \le m)$ is

$$\left(\frac{\partial^2 f}{\partial t_i \partial t_j} \right)_0,$$

where the subscript 0 means that the derivatives are calculated at 0.

Definition 2.3   Let $f(M \longrightarrow \mathbb{R})$ be a smooth function and $x \in M$ a critical point of f. Then x is said to be *nondegenerate* if the Hessian form H_x at x is nondegenerate. If all the critical points of f are nondegenerate, f is said to be a *Morse function*.

Proposition 2.4   *A nondegenerate critical point is always isolated. In particular, for a Morse function, the critical set is discrete, and so is finite if M is compact.*

Proof: The first assertion is local, and so we may assume that f is a map of a neighborhood of 0 in $\mathbb{R}^m$ and has 0 as a critical point. We then have the gradient map

$$\nabla f : (t_1, \ldots, t_m) \longmapsto \left(\frac{\partial f}{\partial t_1}, \ldots, \frac{\partial f}{\partial t_m} \right),$$

and the matrix of its tangent map at the origin is just the Hessian matrix

$$\left(\frac{\partial^2 f}{\partial t_i \partial t_j} \right)_0.$$

So if this matrix is nonsingular, which is the content of the assumption that 0 is a nondegenerate critical point, ∇f is a local diffeomorphism in a neighborhood of 0, and so, in that neighborhood, 0 is the only point that maps to 0. Since the points where $\nabla f = 0$ are precisely the critical points of f, it follows that there are no critical points in a sufficiently small neighborhood of 0. The remaining assertions are trivial. $\square$

2.2 Morse's lemma

We shall now state and prove Morse's fundamental lemma, which describes completely the local structure of a map in a neighborhood of a nondegenerate critical point: It says that in a neighborhood of a nondegenerate critical point, the map can be written as a quadratic form in a suitable chart. We shall actually treat the case with parameters for greater flexibility in applications.

Theorem 2.1 (Morse's lemma with parameters)   *Let $f \in C^\infty(X \times Y)$, where $X \times Y$ is an open neighborhood of $(0,0)$ in $\mathbb{R}^k \times \mathbb{R}^n$, and write $f_x(y) = f(x,y)$. Suppose that*

$$d_y f_0(0) = 0, \qquad d_y^2 f_0(0) \text{ is invertible}$$

(that is, 0 is a nondegenerate critical of f_0). Then:

i) *$\exists$ an open neighborhood $X_1 \times Y_1$ of $(0,0)$ and a smooth map $x \longmapsto y(x)$ of X_1 into Y_1 such that $y(0) = 0$ and $y(x)$ is the unique critical point of f_x in Y_1; moreover, the Hessian matrix $Q(x)$ (in the coordinates y_j) of f_x at $y(x)$ is invertible, so that this critical point is nondegenerate.*

ii) *$\exists$ a diffeomorphism*

$$(x, y) \longmapsto (x, u(x, y))$$

of $X_1 \times Y_1$ with $u(0,0) = 0$ such that, for all $(x, y) \in X_1 \times Y_1$,

$$f(x, y) = f(x, y(x)) + \frac{1}{2}\langle Q(x)u, u \rangle.$$

Proof: i) Consider the map

$$(x, y) \longmapsto (x, \nabla_y f_x(y)),$$

which takes $(0,0)$ into $(0,0)$. Its tangent map at $(0,0)$ has the matrix

$$\begin{pmatrix} I & 0 \\ 0 & Q(0) \end{pmatrix},$$

where $Q(0)$ is the Hessian matrix of f_0 at 0. Since $Q(0)$ is invertible, this map is a local diffeomorphism around $(0,0)$, mapping an open neighborhood $X_1 \times Y_1$ of $(0,0)$ onto an open neighborhood $W(X_1)$ of $(0,0)$ that projects onto X_1. By shrinking X_1 and Y_1, we may assume that the matrix

$$\left(\frac{\partial^2 f}{\partial y_i \partial y_j} \right)$$

is invertible throughout $X_1 \times Y_1$. Obviously, there is a unique point of the form $(x, y(x))$ that maps to $(x, 0)$, the dependence of $y(x)$ on x being C^∞. Clearly, $y(x)$ is the unique critical point of f_x in Y_1, and it is nondegenerate.

ii) The map $(x, y) \longmapsto (x, y - y(x))$ takes $(0, 0)$ to $(0, 0)$ and is a diffeomorphism around $(0, 0)$. If f goes to g under this map, it is clear that we need to prove the theorem with g in place of f (with smaller X_1 and Y_1). For g, 0 is the unique critical point of g_x in Y_1, and $Q(x)$ is the (invertible) Hessian matrix of g_x at 0.

For any function F of a real variable u defined on an interval containing 0, its Taylor expansion at 0 is

$$F(u) = F(0) + uF'(0) + u^2 \int_0^1 (1 - v)F''(vu)dv.$$

Applying this at $u = 1$ to the function $g(x, uy)$, we get, as 0 is a critical point for g_x for all $x \in X_1$,

$$g(x, y) = g(x, 0) + \frac{1}{2} \sum_{1 \le i,j \le n} y_i y_j h_{ij}(x, y),$$

$$h_{ij}(x, y) = 2 \int_0^1 (1 - v)g_{ij}(x, vy)dv,$$

with g_{ij} being written for the partial $\partial^2 g / \partial y_i \partial y_j$. Let $H(x, y)$ be the matrix with $h_{ij}(x, y)$ as its entries. We now search for a map of the form

$$(x, y) \longmapsto (x, R(x, y)y), \qquad R(0, 0) = I,$$

where R is a smooth $n \times n$ matrix and y is viewed as a column vector. This is a diffeomorphism around $(0, 0)$ for any R and sends $(x, 0)$ to $(x, 0)$. The idea is to see if we can choose R so that the requirements of Morse's lemma are met, i.e.,

$$g(x, y) = g(x, 0) + \frac{1}{2}\langle Q(x)Ry, Ry \rangle$$

for all (x, y) near $(0, 0)$. For this it is enough to have

$$^tR(x, y)Q(x)R(x, y) = H(x, y), \tag{2.1}$$

where ^tR is the transposed of the matrix R.

To this end, we consider the linear space M of real $n \times n$ matrices and its linear subspace S of symmetric matrices. Let s_0 be an invertible element of S, and let $x \longmapsto s(x)$ be a smooth map of X into S with $s(0) = s_0$. Let us now consider the map π of $X \times M$ into $X \times S$ given by $\pi(x, m) = (x, {}^t m s(x) m)$. We assert that this map is a submersion at $(0, I)$. For this, we must check that the tangent map at $(0, I)$ is surjective. If L is the tangent map in question, L takes $(0, Z)$ to $(0, {}^t Z s_0 + s_0 Z)$, and by choosing Z to be $\frac{1}{2} s_0^{-1} W$, where W is any $n \times n$ real symmetric matrix, we see that the range of L contains all elements of the form $(0, W)$. On the other hand, if e_j are the standard vectors in $\mathbb{R}^k$, L maps $(e_j, 0)$ into $(e_j, \partial s / \partial x_j(0))$ for any j. From this, it follows at once that L is surjective. Since submersions are isomorphic to projections, we can find a smooth map ρ of an open neighborhood $X_0 \times S_0$ of $(0, s_0)$ in $X \times S$ into an open neighborhood of $(0, I)$ in $X \times M$ such that $\pi \circ \rho$ is the identity map. Obviously, $\rho(x, W) = (x, r(x, W))$, and so we have the equation

$$ {}^t r(x, W) s(x) r(x, W) = W $$

for all $(x, W) \in X_0 \times S_0$.

This result is now immediately applicable to (2.1). We take $s(x) = Q(x)$, and then define

$$ R(x, y) = r(x, H(x, y)). $$

Then (2.1) is satisfied. This completes the proof of the theorem. $\square$

Remark This proof is the same as in [D1]. Morse's proof of it when there are no parameters [Mo] was different and was a variation of the usual proof of the reduction of a quadratic form to a linear combination with ± 1 coefficients of squares. In the exercises, we indicate a sketch of this proof.

We shall now treat the case of nondegenerate critical points in the setting of manifolds with a boundary. Locally, this comes down to working at a boundary point of a half-space. Let $x, y_1, \ldots, y_n$ be the coordinates in $\mathbb{R}^{n+1}$, and let H be the half-space $\{x \geq 0\}$. Let Y be the boundary of H, i.e., the subspace $x = 0$. We are given a phase function φ defined in an open neighborhood of 0 in H, and we shall investigate its structure near 0 under assumptions about its critical behavior at 0. To say that φ

is smooth means that there is a smooth function in a full neighborhood of
0 in $\mathbb{R}^{n+1}$ that coincides with φ when restricted to H; a similar remark
applies to maps into a manifold; in particular, by a diffeomorphism of H at
0 we mean a diffeomorphism on a neighborhood N of 0 in $\mathbb{R}^{n+1}$ such that
it fixes 0, takes $N \cap H$ into H, and is a diffeomorphism on $N \cap Y$ into Y.

Proposition 2.6   *i) Suppose φ is a smooth real function defined on a
neighborhood of 0 in H. If 0 is not a critical point of $\varphi|_Y$, then there is a
local diffeomorphism of the form*

$$(x, y) \longmapsto (x, u(x, y)), \qquad u = (u_1, \ldots, u_n), \ u(0,0) = 0,$$

such that
$$\varphi(x, y) = \varphi(0, 0) + u_1.$$

*ii) If 0 is a nondegenerate critical point of $\varphi|_Y$ but is not a critical
point of φ, there is a local diffeomorphism of the form*

$$(x, y) \longmapsto (t(x), u(x, y)),$$

with $u = (u_1, \ldots, u_n)$, $u(0,0) = 0$, $t(0) = 0$, $t'(0) > 0$, such that

$$\varphi(x, y) = \varphi(0, 0) + \varepsilon t + \frac{1}{2} \langle Qu, u \rangle,$$

*where $\varepsilon = \pm 1$ is a constant, and Q is a constant $n \times n$ real invertible
symmetric matrix.*

Proof: We note that in case (ii) the partial derivative $\partial_\xi \varphi(0, 0)$ in any
direction ξ directed *interior* to H is nonzero, and its sign does not depend
on the choice of ξ. It is clear that ε is this sign.

   i) We assume that φ is defined and smooth in a full neighborhood of 0.
By permuting the coordinates y_j, we may assume that $\partial\varphi/\partial y_1(0, 0) \neq 0$.
Then,
$$(x, y) \longmapsto (x, \varphi(x, y) - \varphi(0, 0), y_2, \ldots, y_n)$$

is a local diffeomorphism of H at 0 with the required property.

   ii) We regard φ as a function of y with x as a parameter and apply
Morse's lemma with parameter. Write $\varphi_x(y) = \varphi(x, y)$. Then, we can find
a local diffeomorphism at 0 of the form

$$(x, y) \longmapsto (x, v(x, y))$$

such that

$$\varphi(x, y) = \varphi(x, z(x)) + \frac{1}{2}\langle Q(x)v, v\rangle,$$

where $z(x)$ is a smooth map of a neighborhood of 0 in $\mathbb{R}$ into $\mathbb{R}^n$ taking 0 to 0 with the property that $z(x)$ is the unique critical point of φ_x near 0. The matrix $Q(x)$ is real, symmetric, and invertible, and so it has the same signature as $Q(0)$ for all small x. By elementary Lie theoretic arguments, it is easy to see that there is a smooth map o of a neighborhood of 0 in $\mathbb{R}$ into the orthogonal group of n variables such that $o(0) = I$ and ${}^{t}o(x)Q(0)o(x) = Q(x)$ for all x near 0. The diffeomorphism $(x, v) \longmapsto (x, w)$, $w = o(x)v$, has the property

$$\varphi(x, y) = \varphi(x, z(x)) + \frac{1}{2}\langle Q(0)w, w\rangle.$$

We now observe that as 0 is not a critical point of φ but of φ_0, the derivative of $\varphi(x, z(x))$ with respect to x at $x = 0$ is nonzero, so that we can write $\varphi(x, z(x)) - \varphi(0, 0) = \varepsilon t(x)$ where $\varepsilon = \pm 1$ and $t'(0) > 0$. The diffeomorphism

$$(x, w) \longmapsto (t(x), v)$$

is now seen to have the required property. $\square$

2.3 Construction and genericity of Morse functions

How can we produce Morse functions on arbitrary, or at least compact, manifolds? We shall presently indicate a very simple method of constructing Morse functions which shows that they are dense in the space of smooth functions. For another method of constructing Morse functions and showing that they are generic, see [GP, pp. 43–48]. Later on we shall see that for homogeneous spaces, the structure of their groups of motions allow us to define certain phase functions that are in the nicest possible position and so yield a great deal of information on the geometry and topology of these homogeneous spaces.

This method of constructing Morse functions goes back to the example of optics considered in the introduction. There we encountered the case of a surface M imbedded in $\mathbb{R}^3$ from which isotropic monochromatic radiation is issuing; the radiation at a point not on M is given by an oscillatory integral the phase function of which is the distance of the point from a variable point on the surface. We adapt this method to a general manifold by first imbedding it in a Euclidean space of sufficiently high dimension and working with the square of the distance rather than the distance itself for obvious reasons (to get a smooth function everywhere).

If M is a smooth manifold of dimension m, compact or not, we know by the famous theorem of H. Whitney that M can be imbedded as a regular closed submanifold of $\mathbb{R}^{2m+1}$. Hence, there is no loss of generality in supposing that M is contained as a closed submanifold of some $\mathbb{R}^n$. We write $||\cdot||$ for the usual norm on $\mathbb{R}^n$ and consider, for any $p \in \mathbb{R}^n$, the function L_p that is the square of the distance of p from a variable point of M:

$$L_p(q) = ||p - q||^2 \qquad (q \in M).$$

In the example from optics discussed in Section 1, it was a question of $L_p^{1/2}$ rather than L_p. However, for $p \notin M$, $L_p^{1/2}$ and L_p have the same critical set and proportional Hessians at the critical points. So we may work with the L_p, which are smoother.

This definition gives us a family of smooth functions on M parametrized by points of $\mathbb{R}^n$. It can be shown (see the exercises) that given p, a point q is critical for L_p if and only if the line $\overrightarrow{qp}$ is "normal to M at q," i.e., if and only if $q - p$ is orthogonal to the tangent space $T_q(M)$ regarded as a subspace of $\mathbb{R}^n$. Let N be the total space of the *normal bundle* of M, i.e., the set of all (q, v) such that $q \in M$ and v is orthogonal to $T_q(M)$. It can be shown that N is a submanifold of dimension n in $\mathbb{R}^{2n}$. We have the map E that takes (q, v) to the "endpoint" of the vector based at q in the direction v, i.e.,

$$E(q, v) = q + v \qquad (q \in M, \ v \perp T_q(M)).$$

This is obviously smooth, and it can be shown that for $(q, v) \in N$ and $p = q + v$, q is a degenerate critical point of L_p if and only if (q, v) is a critical point of E. In other words, if p is not a *critical value* of E, then L_p is a Morse function on M. By Sard's theorem, we now see that for almost all $p \in \mathbb{R}^n$ the function L_p is Morse on M. It is even possible to use the L_p to show that the Morse functions are dense in the space $C^\infty(M)$. A critical value of E is, heuristically, a point where *infinitesimally near normals intersect*. So they are called *focal points*. It can be shown that for a given $(q, v) \in N$, the focal points on the normal to M at q along v are the points $q + K_i^{-1} v$ where K_i are the principal radii of curvature (with multiplicities) of M at q in the direction of v. The set of focal points is called the *caustic variety* of M, and the foregoing is the method of constructing it.

It is clear from the preceding description of the caustic variety that for a curve M imbedded in $\mathbb{R}^2$ the caustic variety is the locus of the centers of curvatures at its points or the envelope of its normals, and so it is the so-called *evolute* of the curve M, the study of which goes back to *Huygens* (in connection with his studies on the theory of pendulum clocks). For example, for the ellipse with equation

$$\frac{x^2}{a^2} + \frac{y^2}{b^2} = 1,$$

the evolute is a curve called the *astroid*, which has 4 cusps. In parametric form, it is given by

$$x = A\cos^3\theta, \qquad y = B\sin^3\theta,$$

and it is easy to check that the cusps arise for $\theta = 0, \frac{\pi}{2}, \pi, \frac{3\pi}{2}$. At each of these points, the cusp is diffeomorphic locally to the cusp at the origin of the *semicubical parabola* given by the classical equation $y^2 = x^3$. It is remarked by Arnold et al. (see [AGV]) that this is a stable situation; if the ellipse is perturbed a little, its evolute is still a curve with four cusps of the type in a semicubical parabola that resembles the astroid. For the ellipsoid

$$\frac{x^2}{a^2} + \frac{y^2}{b^2} + \frac{z^2}{c^2} = 1,$$

the caustic surface was investigated by *Cayley* as far back as 1873 (see [AGV]). For the symmetric case of the sphere S when $a = b = c = 1$, the distance function becomes

$$L_p(q) = 1 + u^2 + v^2 + w^2 - 2p{\cdot}q \qquad (p = (u,v,w),\ q \in S).$$

In view of the rotational symmetry it is enough to consider the case

$$p_w = (0,0,w), \qquad w \neq 0, \pm 1.$$

The corresponding phase function is

$$\varphi_w(q) = 1 + w^2 - 2wz \qquad (q = (x,y,z)).$$

Thus, it is essentially the *height function* (see [Mi, p. 1]), which has two critical points $q^{\pm} = (0, 0, \pm 1)$, q^{+} being the maximum and q^{-} the minimum. So the oscillatory integral

$$I_w(\tau) = \int_S e^{i\tau |p_w - q|} a(q) dq \sim L\tau^{-1} \qquad (w \neq 0, \pm 1).$$

The normals at all points of S pass through the origin, so that $\{0\}$ is the caustic set. For $w = 0$ the phase is *constant* on S, and so

$$I_0(\tau) = \gamma(a)$$

is independent of τ. This is a good illustration of the increase in intensity at a caustic in a simple everyday example.

2.4 Oscillatory integrals with nondegenerate phase functions

We begin with some preliminary remarks on asymptotics. Let $F(\tau)$ be a complex-valued function on $(0, \infty)$. A statement such as

$$F(\tau) \sim \sum_{r \geq 0} c_r \tau^{-e_r} \qquad (\tau \longrightarrow \infty) \tag{2.2}$$

means the following:

i) The exponents e_r are real numbers and

$$e_1 < e_2 < \cdots < e_r < \cdots, \qquad e_r \longrightarrow \infty \text{ as } r \longrightarrow \infty.$$

ii) $c_r \in \mathbb{C}$ for all $r \geq 0$.

iii) There are positive numbers f_N such that $f_N \longrightarrow \infty$ as $N \longrightarrow \infty$ and

$$F(\tau) - \sum_{0 \leq r \leq N} c_r \tau^{-e_r} = O(\tau^{-f_N}) \qquad (\tau \longrightarrow \infty).$$

We refer to (2.2) as an *asymptotic expansion* of F. Given the exponents, it is clear that the coefficients c_r are determined uniquely and inductively by

$$c_r = \lim_{\tau \longrightarrow \infty} \tau^{e_r} \left(F(\tau) - \sum_{0 \leq j < r} c_j \tau^{-e_j} \right).$$

It is also obvious that the exponents corresponding to those c_r that are not zero are also uniquely determined. Note that if $G(\tau)$ is rapidly decreasing in τ, F and $F + G$ have the same asymptotic expansion.

Suppose now that $F(\tau)$ is a distribution on a smooth manifold Y for each τ, and write $F(\tau)(a) = F(\tau : a)(a \in C_c^\infty(Y))$. Then by the asymptotic expansion

$$F \sim \sum_{r \geq 0} c_r \tau^{-e_r} \qquad (\tau \longrightarrow \infty)$$

is meant the following:

i) For each $a \in C_c^\infty(Y)$, we have

$$F(\tau : a) \sim \sum_{r \geq 0} c_r(a) \tau^{-e_r} \qquad (\tau \longrightarrow \infty),$$

where the exponents e_r are independent of a; in this case, the c_r are distributions on Y.

ii) Given any compact set $K \subset Y$, there are numbers f_N $(N \geq 1)$ and continuous seminorms μ_N on $C_c^\infty(K)$ such that $f_N \longrightarrow \infty$ as $N \longrightarrow \infty$, and for all $a \in C_c^\infty(K)$ and $\tau \geq \tau_0$ we have

$$\left| F(\tau : a) - \sum_{0 \leq r \leq N} c_r(a) \tau^{-e_r} \right| \leq \mu_N(a) \tau^{-f_N}. \qquad (2.3)$$

Suppose, finally, that F depends in addition on $x \in X$, $X \subset \mathbb{R}^k$ an open set such that, for each $x \in X$,

$$F(x : \tau : a) \sim \sum_{r \geq 0} c_r(x : a) \tau^{-e_r} \qquad (\tau \longrightarrow \infty),$$

where the exponents are independent of a and x. The asymptotic expansion is said to be *locally uniform at* x_0 if there is a compact neighborhood N of x_0 such that (2.3) is valid for suitable f_N, μ_n, τ_0 independent of $x \in N$. If this is so for each $x_0 \in X$, we say that the asymptotic expansion is locally uniform in X.

In the special case when all the c_r are zero, F is rapidly decreasing and we write it as

$$F \sim 0,$$

with the obvious interpretations when F is a distribution for each τ, depending possibly on additional parameters.

We now study the asymptotics of integrals of the form

$$I(x:\tau:a) = \int_{\mathbb{R}^n} e^{i\tau\varphi(x,y)} a(y) dy \qquad (x \in X, \ \tau > 0),$$

where X is open in $\mathbb{R}^k$.

Theorem 2.7   *Suppose* $(x_0, y_0) \in K \times \mathbb{R}^n$ *is such that* $d_y\varphi(x_0, y_0) \neq 0$. *Then we have the asymptotic expansion*

$$I(x:\tau:\cdot) \sim 0 \qquad (\tau \longrightarrow \infty),$$

locally uniformly on X.

Proof: We may clearly assume, by permuting the coordinates if necessary, that $u := \partial\varphi/\partial y_1$ is bounded away from 0 on the closure of $X_0 \times K_0$ for a suitable neighborhood $X_0 \times K_0$ of (x_0, y_0). Hence if $L := u^{-1}\partial/\partial y_1$, then L is a smooth first order differential operator on $X_0 \times K_0$ with bounded coefficients, and $L\varphi = 1$ on $X_0 \times K_0$. So

$$e^{i\tau\varphi} = (i\tau)^{-1} L(e^{i\tau\varphi}) = \cdots = (i\tau)^{-r} L^r(e^{i\tau\varphi})$$

for all integers $r \geq 1$. Integration by parts now gives us, with $L^\dagger$ denoting the formal transpose of L,

$$I(x:\tau:a) = (i\tau)^{-r} \int e^{i\tau\varphi} (L^{\dagger^r} a) dy.$$

Since the coefficients of $L^{\dagger^r}$ are bounded on $X_0 \times K_0$, we have, for all $a \in C_c^\infty(K_0)$, $x \in X_0$,

$$|I(x:\tau:a)| \leq C\mu(a)\tau^{-r},$$

where $\mu(a) = \sum_{0 \leq j \leq r} \sup |\partial^j a/\partial y_1|$. $\square$

We shall now treat the neighborhoods of points where the phase has a nondegenerate critical point. In view of the application of Morse's lemma, the key step is to calculate the Fourier transform of

$$e^{i\tau\langle Qx, x\rangle/2},$$

where Q is a nonsingular symmetric $n \times n$ matrix. Now this is a bounded continuous function and so defines a tempered distribution. It is the limit (in the sense of bounded convergence, hence in the weak topology of tempered distributions)

$$\lim_{z \longrightarrow -i\tau} e^{-z\langle Qy,y\rangle/2}.$$

The function of one variable

$$e^{-zy^2/2}$$

is in the Schwartz space for $\Re(z) > 0$; its Fourier transform for $z > 0$ is

$$(2\pi)^{1/2} z^{-1/2} e^{-\xi^2/2z}.$$

By analytic continuation, we therefore have

$$\mathcal{F}\left(e^{-zy^2/2}\right)(\xi) = (2\pi)^{1/2} z^{-1/2} e^{-\xi^2/2z},$$

where the square root is the continuous one defined for $\Re(z) \geq 0$, $z \neq 0$, in polar coordinates by

$$z^{-1/2} = r^{-1/2} e^{-i\theta/2} \qquad (z = re^{i\theta},\ r > 0,\ |\theta| \leq \frac{\pi}{2}).$$

Taking the distribution limit as $z \longrightarrow -i\tau q$ ($q \in \mathbb{R}$), we obtain (as $\tau > 0$)

$$\mathcal{F}\left(e^{i\tau q y^2/2}\right)(\xi) = (2\pi)^{1/2} |\tau q|^{-1/2} e^{(i\pi/4)\,\mathrm{sgn}(q)} e^{-i\xi^2/2\tau q}.$$

From this, we get at once the following n-variable formula:

$$\mathcal{F}\left(e^{i\tau \sum_j q_j y_j^2/2}\right)(\xi)$$
$$= \left(\frac{2\pi}{\tau}\right)^{n/2} |q_1 \ldots q_n|^{-1/2} e^{(i\pi/4) \sum_j \mathrm{sgn}(q_j)} e^{-i \sum_j \xi_j^2/2\tau}.$$

In $\mathbb{R}^n$ we have an orthogonal matrix R such that

$$Q = RDR^{-1}, \qquad D = \mathrm{diag}\,(q_1, \ldots, q_n),$$

and so finally we have

$$\mathcal{F}\left(e^{i\tau\langle Qy,y\rangle/2}\right)(\xi) = \left(\frac{2\pi}{\tau}\right)^{n/2} |\det(Q)|^{-1/2} e^{(i\pi/4)\,\mathrm{sgn}(Q)} e^{-i\langle Q^{-1}\xi,\xi\rangle/2\tau},$$

where

$$\mathrm{sgn}(Q) = n_+ - n_-$$

with n_+, n_- as the number of positive, respectively negative, eigenvalues of Q. Thus, for any function h in the Schwartz space of $\mathbb{R}^n$, we see that

$$\int e^{i\tau \langle Qy, y\rangle /2} h(y) dy$$

equals

$$\left(\frac{2\pi}{\tau} \right)^{n/2} |\det(Q)|^{-1/2} e^{(i\pi/4)\,\mathrm{sgn}(Q)} \int e^{-i\langle Q^{-1}\xi, \xi\rangle /2\tau} \mathcal{F}(h)(\xi) d\xi.$$

The exponential can now be expanded as a power series in τ^{-1} and leads us to an asymptotic expansion. By the Fourier inversion formula, we have, for any polynomial P,

$$\int P(\xi)\mathcal{F}(h)(\xi)d\xi = (P(i\nabla)h)(0).$$

Hence, we finally get the asymptotic expansion

$$\int e^{i\tau \langle Qy, y\rangle /2} h(y) dy$$

$$\sim \left(\frac{2\pi}{\tau} \right)^{n/2} |\det(Q)|^{-1/2} e^{(i\pi/4)\,\mathrm{sgn}(Q)} \sum_{k \geq 0} \frac{1}{k!} (R^k h)(0)\tau^{-k}$$

as $\tau \longrightarrow \infty$, R being the operator

$$R = \frac{i}{2}\langle Q^{-1}\nabla, \nabla \rangle.$$

The estimation of the error terms is not particularly difficult in this asymptotic expansion (see the analysis given in Section 1) and so is left as an exercise. We thus finally obtain the following key result.

Lemma 2.8   *Suppose $Q(x)$ is a family of nonsingular $n \times n$ symmetric real matrices depending continuously on x. Then, for all smooth $a \in C_c^\infty(K)$, we have, with*

$$I(x : \tau : a) = \int e^{i\tau \langle Q(x)y, y \rangle / 2} a(y) dy,$$

the asymptotic expansion

$$I \sim \left(\frac{2\pi}{\tau} \right)^{n/2} |\det Q(x)|^{-1/2} e^{(i\pi/4)\,\mathrm{sgn}(Q(x))} \sum_{k \geq 0} \frac{1}{k!} (R^k a)(0) \tau^{-k}$$

as $\tau \longrightarrow \infty$, locally uniformly in $x \in X$; here R is the second order constant coefficient (but x dependent) partial differential operator

$$R_y = \frac{i}{2} \langle Q(x)^{-1} \nabla_y, \nabla_y \rangle.$$

We are now in a position to formulate the asymptotics of the integrals

$$I = I(x : \tau : a) = \int e^{i\tau \varphi(x, y)} a(x, y) dy.$$

Theorem 2.9   *Suppose that*

$$d_y \varphi(x_0, y_0) = 0, \qquad d_y^2 \varphi(x_0, y_0) \text{ is invertible.}$$

Then there is a compact neighborhood $X_0 \times K_0$ of (x_0, y_0) such that for all amplitudes a with support in y contained in K_0, I has the asymptotic expansion

$$\left(\frac{2\pi}{\tau} \right)^{n/2} |\det H(x)|^{-1/2} e^{(i\pi/4)\,\mathrm{sgn}(H(x_0))}$$

$$\times e^{i\tau \varphi(x, y(x))} \sum_{k \geq 0} \frac{1}{k!} (R_y^k a)(x, y(x)) \tau^{-k}$$

as $\tau \longrightarrow \infty$, uniformly in $x \in X_0$; here, R_y is a second order partial differential operator, $y(x)$ is the unique critical point in K_0 of $\varphi(x, \cdot)$, and $H(x)$ is the Hessian matrix of $\varphi(x, \cdot)$ at the critical point $y(x)$.

Proof: This follows at once from Morse's lemma with parameters and Lemma 2.8, by making the x dependent change of the integration variables that changes the phase function to the Hessian quadratic form at its critical point $y(x)$. $\square$

There is a variation of the nondegenerate situation that is encountered not infrequently and is called *clean* by Bott [Bo1]. If φ is a smooth function on a smooth manifold M and C is its set of critical points, (φ, M) is *clean* if

i) C is a smooth, not necessarily connected, imbedded submanifold of M.
ii) At each point c of C, the Hessian of φ is *transversally nondegenerate*, i.e., nondegenerate on $T_c(M)/T_c(C)$.

One should note that φ is constant on C, and so the Hessian form vanishes if one of the vectors is from $T_c(C)$. Hence, the Hessian form descends to the quotient space $T_c(M)/T_c(C)$, and the condition (ii) is that this is nondegenerate.

Proposition 2.10   *If (φ, M) is clean, and $c \in C$, there is a chart (y, v) at c, $(y, v) = (y_1, \ldots, y_n, v_1, \ldots, v_p)$ at c, such that φ takes the form*

$$\varphi(x, v) = \varphi(c) + \frac{1}{2}\langle Q(v)y, y\rangle$$

in that chart, $Q(v)$ being a $n \times n$ real invertible symmetric matrix depending smoothly on v.

Proof: There is a chart $x = (u, v) = (u_1, \ldots, u_n, v_1, \ldots, v_p)$ at c such that C is locally described by the equations $u = 0$. Then $\varphi(0, v)$ is constant, thus equal to $\varphi(0, 0)$. We now treat $\varphi(u, v)$ as a family of phase functions in u depending on the parameter v and apply Morse's lemma with parameters. For any v we have 0 as a unique critical point for $\varphi(\cdot, v)$, and so there is a diffeomorphism of the form

$$(u, v) \longmapsto (y(u, v), v)$$

such that

$$\varphi(u, v) = \varphi(0, 0) + \frac{1}{2}\langle Q(v)y, y\rangle. \quad \square$$

Theorem 2.9 now leads to the following corollary.

Corollary 2.11   *We have, for a sufficiently small open neighborhood U of c, an asymptotic expansion for*

$$I(\tau : a) = \int e^{i\tau\varphi(x)}a(x)dx \qquad (a \in C_c^\infty(U))$$

of the form

$$I \sim \left(\frac{2\pi}{\tau}\right)^{n/2} \sum_{r \geq 0} c_r$$

as $\tau \longrightarrow \infty$, uniformly in $x \in X_0$, where the c_r are distributions of order $\leq 2r$ supported by C, c_0 being therefore a measure on C.

The local results lead to global results on compact manifolds in the usual way by localizing and using partitions of unity. We state the following result, which is an immediate consequence of the theorems that we have proved. We treat the case of Morse functions on a compact manifold M of dimension m. This is the generic case because it can be shown that they form an open set in $C^\infty(M)$ (see the exercises).

Before formulating the global result, we note that the oscillatory integrals on M will be obtained by integrating with respect to an m-form on M. In local coordinates, this will give rise to an integration in Euclidean space with respect to some density function, and it is important to make sure that the factors in front of the coefficients of the asymptotic expansion make sense globally. Suppose V is an m-dimensional real vector space and B is a nondegenerate symmetric bilinear form on $V \times V$. By $\det B$ we mean the determinant of the matrix $(B(v_i, v_j))$, where (v_i) is a basis of V. This, of course, depends on the basis. If, however, there is an exterior m-form L on V that is nonzero, and we write $|L|$ for $|L(v_1, \ldots, v_m)|$, it is obvious that $|\det B|^{-1/2}|L|$ is independent of the choice of the basis used in computing it, provided of course the same basis is used for both. We write it as

$$|\det B|^{-1/2}|L|.$$

Theorem 2.12   *Let M be a smooth oriented compact manifold of dimension m, ω a smooth m-form > 0 on M, X an open set in $\mathbb{R}^k$, and φ a real smooth function on $X \times M$ such that $\varphi_x(\cdot) := \varphi(x, \cdot)$ is a Morse function on M for each $x \in X$. Let*

$$I(\tau : a) = \int_M e^{i\tau\varphi(y)} a(y)\omega \qquad (a \in C^\infty(M)).$$

Then we have the following asymptotic expansion:

$$I(\tau : a) \sim \left(\frac{2\pi}{\tau}\right)^{m/2} \sum_{r \geq 0} c_r(a)\tau^{-r},$$

where the $c_r(\cdot)$ are distributions supported by the critical set C of φ of order $\leq 2r$, and c_0 is the measure given by

$$c_0 = \sum_{t \in C} |\det H(t)|^{-1/2} |\omega_t| e^{(i\pi/4)\,\mathrm{sgn}(H(t))} e^{i\tau\varphi(t)} \delta_t.$$

Here, $H(t)$ is the Hessian form of φ at t, and δ_t is the Dirac delta measure at t. If φ depends smoothly on a parameter, this expansion is locally uniform in the parameter. Finally, if φ is clean, the asymptotic expansion is valid again, but the c_r are now distributions of order $\leq 2r$ supported by the critical manifold, c_0 being therefore a measure on the critical manifold.

2.5 Manifolds with (smooth) boundary

We shall now consider the case when M has a *smooth boundary*. We write M^0 for the *interior* of M and ∂M for the *boundary* of M; we recall that ∂M is a smooth manifold without boundary. As before, m is the dimension of M. We assume that M is oriented and use the usual definition of a smooth m-form on M that is > 0. We refer the reader to [GP] for the basic definitions and results concerning manifolds with boundary.

A real-valued smooth function φ on M is said to be a *Morse function* if it has the following properties:

i) The critical points of φ in M^0 are all nondegenerate.

ii) φ has no critical points on ∂M.

iii) The restriction of φ to ∂M is a Morse function on ∂M.

It follows from our local study of phase functions on half-spaces in Section 2.2 that φ has only finitely many critical points in M^0. The same method of partitions of unity used in the boundaryless case now reduces the study of the integrals to the local case, which has one of the two following forms:

$$\int_{x \geq 0} e^{i\tau y_1} a(x,y)dy \quad \text{or} \quad \int_{x \geq 0} e^{i\tau[x+\langle Qy,y\rangle]} a(x,y)dy,$$

where Q is a real invertible $n \times n$ symmetric matrix. We thus easily obtain the following theorem.

Theorem 2.13   *Let φ be a Morse function on M, and let ω be a smooth m-form > 0 on M. Write*

$$I(\tau : a) = \int_M e^{i\tau\varphi(u)} a(u)\omega \qquad (a \in C^\infty(M)).$$

Then we have the asymptotics

$$I(\tau : a) \sim \left(\frac{2\pi}{\tau}\right)^{(m+1)/2} \sum_{r \geq 0} \tau^{-r}\alpha_r(a) + \left(\frac{2\pi}{\tau}\right)^{m/2} \sum_{r \geq 0} \tau^{-r}\beta_r(a),$$

where α_r and β_r are distributions, with α_r supported by the critical set of the restriction of φ to ∂M of order $\leq 2r$, while β_r are distributions supported by the critical set of φ in M^0, of order $\leq 2r$ there. The distributions α_0 and β_0 are of the form described in Theorem 2.12.

2.6 Exercises

SARD'S THEOREM ([GP, pp. 202–207])

1. Let $f(M \longrightarrow N)$ be a smooth map, and $C \subset M$ the set of x in M such that df_x is not surjective. The following steps are designed to lead to a proof that $f(C)$ has measure 0 in N. It is clear that we may suppose that M (resp., N) is an open neighborhood of the origin in $\mathbb{R}^m$ (resp., $\mathbb{R}^n$) and $f = (f_1, \ldots, f_n)$ with $f_i(0) = 0$. Let $C_0 = C$, and C_1 the set of all $x \in C_0$ such that all first order partial derivatives of the f_i vanish; more generally, let C_p for $p \geq 1$ be the set of all points of C where all partial derivatives of the f_i of order $\leq p$ vanish. There are two steps to the proof: (i) to prove that $f(C_p \setminus C_{p+1})$ has measure zero for all $p \geq 0$, and (ii) to prove that $f(C_p)$ has measure zero if p is large enough, for example, if $p + 1 > n/m$.

a) Suppose $x \in C_0 \setminus C_1$ and $\partial f_1 / \partial x_1(x) \neq 0$. Show that the map $(x_1, \ldots, x_m) \longmapsto (f_1(x), x_2, \ldots, x_m)$ is a diffeomorphism around x.

b) If f has the form $(x_1, \ldots, x_m) \longmapsto (x_1, f_2(x), \ldots, f_n(x))$, prove that $(x_1', x_2', \ldots x_m') \in C$ if and only if $(x_2', \ldots, x_m')$ is a critical point for he map $(x_2, \ldots, x_m) \longmapsto (f_2(x_1', x_2, \ldots, x_m), \ldots, f_m(x_1', x_2, \ldots, x_m))$.

c) Deduce from (a), (b), induction on m, and the Fubini theorem that $f(C_0 \setminus C_1)$ has measure zero.

d) Suppose $x \in C_p \setminus C_{p+1}$ and $\partial/\partial x_1(\partial^\beta f)(x) \neq 0$, where β is a multi-index with $|\beta| = p$. Show that $(x_1, \ldots, x_m) \longmapsto (\partial^\beta f, x_2, \ldots, x_m)$ is a diffeomorphism around x, and hence show that in the corresponding new set of coordinates at x the points of $C_p \setminus C_{p+1}$ are mapped into a subset of $y_1 = 0$, and hence deduce that $f(C_p \setminus C_{p+1})$ has measure zero.

e) Prove by Taylor series expansion that if $x \in C_p$ and $\varepsilon > 0$, $\|f(y) - f(x)\| \leq L\varepsilon^{p+1}$ for $\|y - x\| \leq \varepsilon$, where L is bounded as x varies in compact subsets of C_p.

f) Fix $\delta > 0$, and let R be a cube in M of side δ. For any integer $k \geq 1$, let R_j $(1 \leq j \leq k^m)$ be the cubes into which R is subdivided by dividing the sides into k equal segments. Prove that if D_p is a compact subset of C_p, the image under f of $R \cap D_p$ is contained in a union of k^m sets of diameter $\leq Lm^{p+1}(\delta/k)^{p+1}$ and so has measure $\leq L'k^m(\delta/k)^{p+1}$, where L' is another constant depending only on L and n.

g) Use (f) to show that $f(D_p)$ has measure $O(k^{-(n(p+1)-m)})$ for $k \geq 1$, and hence that if $n(p + 1) > m$ the set $f(D_p)$, hence also $f(C_p)$, has measure zero.

MORSE'S LEMMA ([Mi, pp. 5–9])

2. Use the Taylor expansion with remainders in integral form to prove that if φ is a smooth function defined around the origin in $\mathbb{R}^n$, then the following are true.

a) If $\varphi(0) = 0$, there exists an open neighborhood U of 0 and $\varphi_i \in C^\infty(U)$ such that $\varphi = \sum_{1 \leq i \leq n} x_i \varphi_i$ on U.

b) If $\varphi(0) = 0$, $d\varphi(0) = 0$, then $\exists U$ and $h_{ij} = h_{ji} \in C^\infty(U)$ as in (a) such that $\varphi = \sum_{i,j} x_i x_j h_{ij}$.

3. Suppose that $\varphi(0) = 0$, $d\varphi(0) = 0$, and that $d^2\varphi(0)$ is invertible.

a) If $n = 1$, write $\varphi = x^2 h$ where h is smooth and $h(0) \neq 0$. If ε is the sign of $h(0)$, show that $h = \varepsilon k^2$ where k is smooth and $k(0) \neq 0$. Check that $y = xk(x)$ defines a diffeomorphism around 0, and $\varphi = \varepsilon y^2$ around 0.

b) Assume $n \geq 2$ and that in terms of a chart (u_i) at 0 we can write

$$\varphi = \sum_{i<r} \varepsilon_i u_i^2 + \sum_{i,j \geq r} u_i u_j k_{ij},$$

where $k_{ij} = k_{ji}$ are smooth and $\varepsilon_i = \pm 1$. Show that the matrix $(k_{ij}(0))_{i,j \geq r}$ is invertible, and that after a linear transformation of the u_i it may be assumed that $k_{rr}(0) \neq 0$. Write $k_{rr} = \varepsilon_r g^2$, where g is smooth and $g(0) \neq 0$.

c) Define the v_i by

$$v_i = \begin{cases} u_i, & \text{if } i \neq r, \\ g\left[u_r + \sum_{i>r} u_i k_{ir} k_{rr}^{-1}\right], & \text{if } i = r. \end{cases}$$

Show that the (v_i) define a chart at the origin and that

$$\varphi = \sum_{i \leq r} \varepsilon_i v_i^2 + \sum_{i,j \geq r+1} v_i v_j t_{ij},$$

where $t_{ij} = t_{ji}$ are smooth.

d) Show by induction on r that (c) gives Morse's lemma without parameters.

PHASE FUNCTIONS ON IMBEDDED SUBMANIFOLDS
([Mi, pp. 32–38])

4. Let M be a smooth manifold of dimension m imbedded as a closed regular submanifold of $\mathbb{R}^n$. Let $N \subset \mathbb{R}^n$ be the *normal bundle* of M, namely, the set of points $(q, v) \in \mathbb{R}^{2n} = \mathbb{R}^n \times \mathbb{R}^n$ such that $q \in M$ and $v \perp T_q(M)$.

a) Let $u = (u_1, \ldots, u_m) \longmapsto x(u) = (x_1(u), \ldots x_m(u))$ be a diffeomorphism of some ball B in $\mathbb{R}^m$ onto an open set $M(B)$ in M. Show that if B is sufficiently small, there are smooth maps $w_\alpha (B \longrightarrow \mathbb{R}^n)$ ($1 \leq \alpha \leq n - m$) such that $w_\alpha \cdot w_\beta = \delta_{\alpha\beta}$ and $w_\alpha \perp T_{x(u)}(M)$ for all $u \in B$.

b) Show that if $N(B) = \{(q, v) \mid q \in M(B)\}$ the map

$$((u_i), (t_\alpha)) = (u, t) \longmapsto (x(u), \sum_{1 \le \alpha \le n} t_\alpha w_\alpha(u))$$

defines a chart on $N(B) \subset N$, and that these charts give the normal bundle N the structure of a closed submanifold of dimension n of $\mathbb{R}^{2n}$.

c) Show that in the coordinates (u, t) the endpoint map $E((q, v) \longmapsto q + v)$ becomes

$$E(u, t) = x(u) + \sum_\alpha t_\alpha w_\alpha.$$

Taking the scalar products of $\partial E / \partial u_i$ and $\partial E / \partial t_\alpha$ with $\partial x / \partial u_j$ and the w_β, show that the tangent map dE is singular at (u, t) if and only if

$$\det (g_{ij} - v \cdot h_{ij})_{1 \le i, j \le m} = 0,$$

where

$$g_{ij} = \frac{\partial x}{\partial u_i} \cdot \frac{\partial x}{\partial u_j}, \qquad h_{ij} = \frac{\partial^2 x}{\partial u_i \partial u_j}, \qquad v = \sum_\alpha t_\alpha w_\alpha.$$

Hint: Use the identity $h_{ij} \cdot w_\alpha = -\partial x / \partial u_j \cdot \partial w_\alpha / \partial u_i$ obtained by differentiating $\partial x / \partial u_j \cdot w_\alpha = 0$.

5. Let L_p for $p \in \mathbb{R}^n$ be the phase functions defined on $M \subset \mathbb{R}^n$ by the formula

$$L_p(q) = ||q - p||^2,$$

where $||\cdot||$ is the usual norm on $\mathbb{R}^n$.

a) Show that

$$\frac{\partial L_p}{\partial u_i} = 2 \left((x(u) - p) \cdot \frac{\partial x}{\partial u_i} \right),$$

and hence deduce that $x(u)$ is a critical point of L_p if and only if $\overrightarrow{qp} := p - x(u) \in T^{\perp}_{x(u)}$, that is, if and only if the line from $q = x(u)$ to p is normal to M at q.

b) Suppose that $x(u)$ is a critical point of L_p. Show that

$$\frac{\partial^2 L_p}{\partial u_i \partial u_j} = 2 \left(g_{ij} - \overrightarrow{qp} \cdot h_{ij} \right).$$

c) Hence deduce that $q = x(u)$ is a degenerate critical point for L_p if and
   only if p is a focal point associated to q.
d) Use Exercise 2 and the preceding in conjunction with Sard's theorem
   to show that for almost all p the functions L_p and $L_p^{1/2}$ are Morse
   functions on M.

FOCAL POINTS AND RADII OF CURVATURE

6. For any $v \in \mathbb{R}^n$, $(v{\cdot}h_{ij})$ is the so-called second fundamental form in
the direction v. Suppose $q = x(0)$ and $g_{ij}(0) = \delta_{ij}$. Show that the focal
points on the normal to M at q along v are the points $q + K_i^{-1}v$ (with
multiplicity) where K_i, the principal radii of curvatures in the direction v,
are the eigenvalues (with multiplicity) of $(v{\cdot}h_{ij})$.

Deduce also that the index of the Hessian form of L_p at a critical point
q is the number of focal points (with multiplicities) that lie in the segment
from q to p.

MORSE FUNCTIONS ON COMPACT MANIFOLDS

7. Let M be compact, and regard $C^\infty(M)$ as a Frechet space with the
topology of uniform convergence of functions and their derivatives (make
this precise using local charts). Let $h \in C^\infty(M)$.

a) Show that there is a smooth imbedding $M \hookrightarrow \mathbb{R}^n$ with $h = x_1$.
b) Let c and ε_i be parameters such that $c \longrightarrow \infty$ and the $\varepsilon_i \longrightarrow 0$ such
   that for $p = (-c + \varepsilon_1, \varepsilon_2, \ldots, \varepsilon_n)$, the phase function L_p on M is a
   Morse function. Define

$$g(x) = \frac{L_p(x) - c^2}{2c} \qquad (x \in M).$$

Show that $g \longrightarrow h$ in $C^\infty(M)$.

8. Show that Morse functions on M as well as the Morse functions on M
the critical values of which are all different, form dense open sets in $C^\infty(M)$
(see [GP, p. 47, Exercise 19]).

9. Complete the details in the proofs of Theorem 2.12 and 2.13.

3 Additonal remarks on singularities and oscillatory integrals

3.1 Morse theory

In this section we shall make a few additional remarks on the *general theory* of singularities of smooth functions and maps and its applications to topology and oscillatory integrals. We begin with Morse theory.

We recall that given any symmetric bilinear form B on a real finite dimensional vector space V we can find a basis α_i of the dual space V^* such that, if r is the rank of B,

$$B(v,v) = \sum_{1 \leq i \leq r} \varepsilon_i \alpha_i(v)^2 \qquad (v \in V, \ \varepsilon_i = \pm 1).$$

The number of ε_i that are -1 is called the *index* of B. It is independent of the choice of the basis with the given properties.

The original investigation of Morse involved how the topology of a space was related to the structure of the critical set of a Morse function defined on it. Let M be a smooth *compact* manifold of dimension m, and f a smooth real function on M that is a Morse function. Recall that this means that the only critical points of f are nondegenerate. We have seen that in this case there are only finitely many critical points. At any critical point x of f, we write $I(x)$ or $I(x:f)$ for the index of the Hessian form of f at x. If $m = \dim(M)$, the index satisfies $0 \leq I(x) \leq m$, with $I(x) = 0$ (resp., m) if and only if x is a local minimum (resp., local maximum). For any integer r $(0 \leq r \leq m)$, let us write $M_r = M_r(f)$ for the number of critical points of f with index r; it is natural to call this the rth *Morse number* of M. Let B_r be the Betti numbers of M $(0 \leq r \leq m)$. Then, Morse formulated the relationship between the critical locus structure of f and the topology of M in the following theorem.

Theorem 3.1 (Morse inequalities) *Between the numbers M_r and B_r, the following inequalities are always satisfied:*

$$M_r - M_{r-1} + M_{r-2} - \cdots \pm M_0 \geq B_r - B_{r-1} + B_{r-2} - \cdots \pm B_0 \qquad (0 \leq r \leq m).$$

In particular, we have

$$M_r \geq B_r \qquad (0 \leq r \leq m).$$

Moreover,

$$M_m - M_{m-1} + M_{m-2} - \cdots \pm M_0 = B_m - B_{m-1} + B_{m-2} - \cdots \pm B_0.$$

We may introduce polynomials associated to M, the *Morse polynomial* of f, written $\mathbf{M}(f : M : t)$, and the Poincaré polynomial of M, written $\mathbf{P}(M : t)$, defined by

$$\mathbf{M}(f : M : t) = \sum_{0 \le r \le m} M_r(f)t^r, \qquad \mathbf{P}(M : t) = \sum_{0 \le r \le m} B_r t^r.$$

The Morse inequalities are proved by establishing the following result: *There exists a polynomial $Q(t)$ with nonnegative integer coefficients such that*

$$\mathbf{M}(f : M : t) - \mathbf{P}(M : t) = (1 + t)Q(t). \tag{3.1}$$

Most modern proofs of (3.1) follow essentially the original ideas of Morse introduced in his fundamental paper [Mo]. Let M_a be the set $\{x \in M \mid f(x) \le a\}$. If a is a regular value of f, this is a smooth manifold with boundary, and it is a question of how the topology of M_c changes as c increases from a to a value (regular) b, and an essential role in this analysis is played by Morse's lemma (see [Bo3]).

It turns out that the torsion in the integral homology of M also imposes restrictions on the Morse numbers. Let $g(r)$ be the number of cyclic groups in the canonical decomposition of the torsion subgroup of $H_r(M : \mathbb{Z})$; then ([Sm]),

$$M_r(f) \ge M_r + g(r) + g(r - 1).$$

It is natural to call a Morse function *optimal* if

$$M_r(f) = B_r + g(r) + g(r - 1) \qquad (0 \le r \le m).$$

Optimal Morse functions always exist if $\dim(M) > 5$. A Morse function is called *perfect* if

$$M_r(f) = B_r \qquad (0 \le r \le m).$$

It is clear that the existence of a perfect Morse function implies that there is no torsion and

$$\mathbf{M}(f : M : t) = \mathbf{P}(M : t).$$

As a very simple application, let us compute the Betti numbers of complex projective space $\mathbb{C}P^n$. We regard $\mathbb{C}P^n$ as the space of one-dimensional subspaces of $\mathbb{C}^{n+1}$, which is viewed as a Hilbert space with the usual scalar product $(\,\cdot\,,\,\cdot\,)$. Let H be a Hermitian matrix with distinct eigenvalues $h_0 < h_1 < \cdots < h_n$. The function

$$z \longmapsto \frac{(Hz,z)}{(z,z)} \qquad (z \in \mathbb{C}^{n+1} \setminus \{0\})$$

is scale invariant and so defines a smooth function f_H on $\mathbb{C}P^n$. We will show first that f_H is a Morse function the critical points of which are the points that correspond to the eigenvectors of H, and then we compute the Morse numbers. We may assume that H is diagonal in the standard basis $(e_j)_{0 \le j \le n}$, with $He_j = h_j e_j$. Write $\mathbb{C}P^n$ as the union of the open sets A_j where $z_j \ne 0$. Then $t_k = z_k z_j^{-1}$ $(k \ne j)$ are coordinates on A_j, and

$$f_H(t) = \frac{\sum_{k \ne j} h_k |t_k|^2 + h_j}{\sum_{k \ne j} |t_k|^2 + 1} = h_j + \sum_{k \ne j}(h_k - h_j)|t_k|^2 + \cdots ,$$

showing that $t = 0$, which describes the point p_j defined by e_j, is critical with index $2j$. We now show that there are no other critical points. If $z \in \mathbb{C}^{n+1}$ with $(z,z) = 1$ defines a critical point, differentiation in the direction of v gives

$$\Re(v, Hz) - (Hz, z)\Re(v, z) = 0.$$

As this is true for iv also and for all v, we get

$$Hz = (Hz, z)z,$$

which implies that the point defined by z is some p_j.

Since the index is always even, $M_r = 0$ for odd r, hence $B_r = 0$ for odd r. But then, the relations

$$M_{2k} \ge B_{2k}, \qquad M_0 + M_2 + \cdots + M_{2n} = B_0 + B_2 + \cdots + B_{2m}$$

imply that $M_r = B_r$ for all r. Thus, f_H is a perfect Morse function, $\mathbb{C}P^n$ has no torsion, and

$$\mathbf{P}(\mathbb{C}P^n : t) = 1 + t^2 + \cdots + t^{2n}.$$

Using phase functions on Lie groups we shall extend this result to all flag manifolds.

The argument we have used is perfectly general and is known as the *lacunary principle*; it already goes back to Morse.

Proposition 3.2 *Suppose f is a Morse function such that the Morse numbers satisfy the relations $M_r M_{r+1} = 0$ for all r. Then f is perfect, M has no torsion, and*

$$\mathbf{P}(M : t) = \mathbf{M}(f : M : t).$$

Proof: We shall show by induction on $r \geq 0$ that $M_r t = B_r$ for all r. We have $M_0 \geq B_0 \geq 1$. Hence, $M_1 = 0$, and so $-M_0 \geq B_1 - B_0 \geq -B_0$, which gives $M_0 \leq B_0$. Thus, $M_0 = B_0$. Suppose that $r > 0$ and $M_k = B_k$ for $k < r$. We have $M_r \geq B_r$, and so, if $M_r = 0$, we must have $M_r = B_r = 0$. If $M_r > 0$, then, $M_{r+1} = 0$, so that $-M_r + M_{r-1} - \cdots \geq B_{r+1} - B_r + B_{r-1} - \cdots$, which leads to $-M_r \geq -B_r$ or $M_r \leq B_r$. Hence, $M_r = B_r$. $\square$

The proof of the Morse inequalities is usually obtained by first giving M a Riemannian metric, and deforming M along the flow generated by the gradient of f. Witten [Wi] has generalized this idea and considered *quantum* dynamical systems defined by f, obtaining the Morse inequalities and much more.

3.2 Structure of differentiable maps

The study of smooth maps and their generic structure was begun by H. Whitney in 1955, when he gave a complete description of the generic maps $(\mathbb{R}^2, 0) \longrightarrow (\mathbb{R}^2, 0)$. Since then, the subject has grown enormously and has an immense number of intersections with many areas of modern mathematics. To get a feeling for this development, the reader should consult the two beautiful volumes [AGV] or the very nice introductory book [M]. The central concept is that of stability: An object is stable if it is isomorphic to all nearby objects. The study of stability of smooth maps has led to a classification of generic singularities at least in small dimensions.

It is impossible to give any reasonable idea of the subject in these lectures. We just content ourselves with mentioning two striking results.

Tougeron's theorem   *If* $(\mathbb{R}^n, 0) \longmapsto (\mathbb{R}, 0)$ *is a smooth map with finite multiplicity* μ, *i.e., if*

$$\mu = \dim_{\mathbb{R}} \mathbb{R}[[x_1, \ldots, x_n]]/(\partial f/\partial x_1, \ldots, \partial f/\partial x_n) < \infty$$

(here, $(\ldots)$ refers to the ideal generated), then f is equivalent to its Taylor jet of order $\mu + 1$.

Mather's theorem   *An infinitesimally stable map is stable.*

3.3 Bernstein's theorem

Apart from all the asymptotic expansions on oscillatory integrals that the classification of singularities leads one to, there is a very general theorem proved by Bernstein (and by Atiyah using the theorem of resolution of singularities of Hironaka):

Theorem 3.3   *If f is analytic in a neighborhood of 0 in $\mathbb{R}^n$ and 0 is a critical point, the integral*

$$\int e^{i\tau f(x)} g(x_1, \ldots, x_n) dx_1 \cdots dx_n,$$

where g is localized at 0, has an asymptotic expansion of the form

$$e^{i\tau f(0)} \sum_{k,\alpha} a_{k,\alpha}(g)\tau^\alpha (\log \tau)^k \qquad (\tau \longrightarrow \infty),$$

where α runs through a finite set of decreasing arithmetic progressions of rational numbers.

The reader should view all the special and more explicit results in the background provided by this general theorem.

In addition to the references [AGV] and [M] mentioned already, the reader should refer to the paper [D3], which was very influential in making the ideas of singularities and oscillatory integrals available to the general mathematician.

4   Phase functions on homogeneous spaces

4.1 Distance functions in representation spaces

Let G be a Lie group and V a finite dimensional real representation of G admitting an invariant nondegenerate symmetric bilinear form $\langle \cdot, \cdot \rangle$. We have seen in Section 1.4 that the phase function

$$\varphi_{V,u}(g) = \langle u, g \cdot v \rangle$$

is, up to an additive constant, the distance function from the point u to the manifold $G \cdot v$, the orbit of v; here, the distance may be an indefinite metric if the form $\langle \cdot, \cdot \rangle$ is not positive definite. If G is compact, we can always choose the form to be positive definite and invariant, so that in this case the interpretation as distance function is exact. Note that the orbit $G \cdot v$ is also closed in this case. In this section, we shall take a closer look at these types of phase functions for compact G and discuss briefly some very nice applications to the geometry and topology of compact homogeneous spaces.

Before we discuss these examples, we shall note that this method in principle applies to any homogeneous space G/H where G is a compact Lie group. More precisely, we have the following variant of a well-known theorem of Chevalley proved by him for affine algebraic groups (see also [Mos] and [Pa]).

Proposition 4.1 *Let G be a compact Lie group, and $H \subset G$ a closed subgroup. Then there is a linear representation of G such that H is exactly the stabilizer of a vector in the representation space.*

Proof: By Frobenius reciprocity, we have $L^2(G/H) = \bigoplus_{i \geq 1} F_i$ where each F_i is a finite dimensional sub-G-module with $F_i^H \neq 0$, the superscript H denoting the subspace of vectors fixed by H. Let H_N be the closed subgroup defined by

$$H_N = \{ g \in G \mid gu = u \ \forall u \in F_i^H, \ 1 \leq i \leq N \}.$$

Then $H_1 \supset H_2 \supset \cdots \supset H_N \supset \cdots$, and so, for reasons of dimension and compactness, we have an N_0 such that $H_N = H_{N_0}$ for all $N \geq N_0$. The proposition will follow if we show that $H_{N_0} = H$. Indeed, suppose we have proved this. Let $F = \bigoplus_{i \leq N_0} F_i$, let $(v_j)_{1 \leq j \leq m}$ be a basis of F^H, and let $V = F \oplus F \oplus \cdots \oplus F$ (m copies of F). If $v = (v_1, \ldots, v_m)$, the stabilizer of v in G is $H_{N_0} = H$.

Suppose that $H_{N_0} \neq H$. Since H_{N_0} contains H, we can find an element $h \in H_{N_0} \setminus H$. Let $X = G/H$, and f the natural map $G \longrightarrow X$; write $x_0 = f(1)$, $x = f(h)$. Since H is compact, $H \cdot x_0 = x_0$, and $x_0 \neq x$, we can find an H-invariant continuous function g on X that is 0 on a neighborhood of x_0 and 1 at x. Then $g \in L^2(X)$, and it is fixed by H but not by h. But the projection of g in F_i is in F_i^H and so, as $h \in H_i$, this projection is fixed by h. As this is true for all i, g itself is fixed by h, a contradiction. $\square$

Remark   In Chevalley's theorem where G and H are affine algebraic groups, it is only possible to guarantee that H is the stabilizer of a point in the *projective space* of the representation space. This is because it may happen that G/H is a projective variety (for instance, when H is a Borel subgroup) and so cannot have a realization as an affine orbit. But if we assume that G/H is affine, then the algebraic version of Proposition 4.1 is true.

4.2 Phase functions defined by the adjoint representation

For any Lie group U, we generally write $\mathfrak{u}$ for its Lie algebra and Ad for the adjoint representation of U; for $X \in \mathfrak{u}$, $u \in U$, we write X^u for $\mathrm{Ad}(u)(X)$. For any $X \in \mathfrak{u}$, we write U_X for the subgroup of U fixing X. Let K be a compact connected Lie group with Lie algebra $\mathfrak{k}$. For $X, Y \in \mathfrak{k}$, with $\langle \cdot, \cdot \rangle$ as the Cartan–Killing form of $\mathfrak{k}$, we write

$$h_{X,Y}(k) = \langle X, Y^k \rangle \qquad (k \in K).$$

The function $h_{X,Y}$ is a phase function on the orbit $K \cdot Y \approx K/K_Y$. If $\mathfrak{t}$ is a maximal abelian subalgebra of $\mathfrak{k}$ and $Y \in \mathfrak{t}$ is a regular element of $\mathfrak{t}$, then $K_Y = T$ is the maximal torus defined by $\mathfrak{t}$, namely $T = \exp \mathfrak{t}$; and K/T is the *flag manifold* of K. The use of these phase functions goes back to Bott [Bo2].

We shall consider a slightly more general situation, where K is a maximal compact subgroup of a connected semisimple Lie group G with finite center and Lie algebra $\mathfrak{g}$. Let $\mathfrak{g} = \mathfrak{k} \oplus \mathfrak{s}$ be the Cartan decomposition of $\mathfrak{g}$, where $\mathfrak{s}$ is the orthogonal complement of $\mathfrak{k}$ in $\mathfrak{g}$ with respect to the Cartan–Killing form $\langle \cdot, \cdot \rangle$. Let $\mathfrak{a}$ be a maximal abelian subspace of $\mathfrak{s}$, and $\mathfrak{g}_\alpha$ the root spaces for the decomposition of $\mathfrak{g}$ with respect to $\mathfrak{a}$. We write Δ for the set of roots, Δ^+ for a positive system of roots, and $n(\alpha) = \dim(\mathfrak{g}_\alpha)$. We denote by $\mathbf{w}$ the Weyl group of $(\mathfrak{g}, \mathfrak{a})$, and for any $X \in \mathfrak{a}$ we write $\mathbf{w}_X$ for the centralizer of X in $\mathbf{w}$. If M is the centralizer of $\mathfrak{a}$ in K, then $\mathbf{w}$ is the normalizer of $\mathfrak{a}$ mod M, and so each element $w \in \mathbf{w}$ has a representative x_w in K, even in K^0. For $X, Y \in \mathfrak{s}$, we write

$$f_{X,Y}(k) = \langle X, Y^k \rangle \qquad (k \in K),$$

which defines a phase function on $K \cdot Y \approx K/K_Y$. The study of these phase functions goes back to Hermann [He] and Takeuchi–Kobayashi [TK].

If K is semisimple, then the classical theory of Weyl allows us to view K as a maximal compact subgroup of a complex group G; the Lie algebra $\mathfrak{g}$ of G is then viewed as a real Lie algebra, and the Cartan decomposition is just $\mathfrak{g} = \mathfrak{k} \oplus (-1)^{1/2}\mathfrak{k}$. Thus, the actions of K on $\mathfrak{k}$ and $\mathfrak{s}$ are isomorphic via the map of multiplication by $(-1)^{1/2}$. The natural choice of $\mathfrak{a}$ is now $(-1)^{1/2}\mathfrak{t}$ where $\mathfrak{t}$ is a maximal abelian subalgebra of $\mathfrak{k}$. The complex root spaces are now of dimension 2, when considered as root spaces for $\mathfrak{a}$.

Proposition 4.2 *If $X, H \in \mathfrak{a}$, the critical set of $f_{X,H}$ is* *

$$K_{X,H} = \bigcup_{w \in \mathbf{w}} K_X w K_H = \coprod_{\mathbf{w}_X \backslash \mathbf{w} / \mathbf{w}_H} K_X w K_H$$

In particular, it is smooth.

Proof: First consider $U, V \in \mathfrak{s}$. For $u \in K$ to be critical for $f_{U,V}$, it is necessary and sufficient that for all $Z \in \mathfrak{k}$ we have

$$\left(\frac{d}{dt}\right)_{t=0} \langle U, V^{k \exp tZ} \rangle = 0.$$

The derivative on the left reduces to

$$\langle U, [Z, V^k] \rangle = \langle [V^k, U], Z \rangle,$$

---

* For any subgroup L of K containing M, $x_w L$ does not depend on the representative x_w of w in K and so is written wL.

and this is 0 for all Z if and only if $[V^k, U] = 0$. Thus,

$$k \text{ is critical for } f_{U,V} \iff [V^k, U] = 0. \tag{4.1}$$

We now write $U = X$, $V = H$ and suppose that $X, H \in \mathfrak{a}$. If k is critical, H^k centralizes X and so, as the centralizer of an element of $\mathfrak{a}$ has the same structure as $\mathfrak{g}$, we can find $u \in K_X$ such that $H^{uk} \in \mathfrak{a}$. By a well-known result ([Va, Lemma 16, p. 288]) that asserts that if an element v of K takes an element H_1 of $\mathfrak{a}$ into an element H_2 of $\mathfrak{a}$, then one can find $w \in \mathbf{w}$ taking H_1 to H_2, we can find $w \in \mathbf{w}$ such that $H^{uk} = H^w$. Hence, for some $v \in K_H$, $ukv = x_w$ for some representative x_w of W in K. In other words, $k \in K_X w K_H$. The converse that any such element is critical is immediate, since for such an element we have $[H^k, X] = 0$. To express the union as a disjoint union, first observe that as the representatives of a subgroup $\mathbf{w}_V$ can be chosen in K_V, it follows that $K_X w K_H$ depends on w only through the double coset $\mathbf{w}_X w \mathbf{w}_H$. Moreover, as we can move X and H into the closure of the positive chamber $\mathfrak{a}^+$ of $\mathfrak{a}$, there is no loss of generality in assuming that $X, H \in \mathrm{Cl}(\mathfrak{a}^+)$. Then K_X, K_H are respectively contained in the parabolic subgroups defined by the simple roots vanishing at X, H, respectively, and the disjointness follows from the Bruhat decomposition. $\square$

Remark Let $U \in \mathfrak{s}$ be arbitrary and $V \in \mathfrak{a}$ be regular. Then, as K^0 is compact, there must be at least one critical point (say at a minimum) of $f_{U,V}$. Hence, by (4.1) there is some $k \in K^0$ such that $[V, U^k] = 0$. By the regularity of V, this means that U^k is in $\mathfrak{a}$. From this, it follows in the standard way that all maximal abelian subspaces of $\mathfrak{a}$ are conjugate under K^0. This type of argument goes back to Hunt [Hu] for the conjugacy of maximal tori, and to Helgason [H] in the present context.

We shall now calculate the Hessians of the $f_{X,H}$. Let θ be the Cartan involution of $\mathfrak{g}$ defined by $\mathfrak{k}$, so that θ is the identity on $\mathfrak{k}$ and minus the identity on $\mathfrak{s}$. If we put

$$\mathfrak{k}_\alpha = \mathfrak{k} \cap (\mathfrak{g}_\alpha \oplus \mathfrak{g}_{-\alpha}), \qquad \mathfrak{s}_\alpha = \mathfrak{s} \cap (\mathfrak{g}_\alpha \oplus \mathfrak{g}_{-\alpha}),$$

then it is easy to see that the orthogonal projections from $\mathfrak{g}_{\pm\alpha}$ onto $\mathfrak{k}_\alpha$ and $\mathfrak{s}_\alpha$ are bijective. We write F_α for the orthogonal projection

$$F_\alpha \colon \mathfrak{g}_\alpha \oplus \mathfrak{g}_{-\alpha} \longrightarrow \mathfrak{k}_\alpha.$$

V. S. Varadarajan

It is clear from what we have said that

$$\dim(\mathfrak{k}_\alpha) = n(\alpha).$$

Proposition 4.3   *Let $k = u x_w v$ with $u \in K_X$, $v \in K_H$, and x_w a representative of w in K. Let $\mathcal{H}_k$ be the Hessian form of $f_{X,H}$ at k, viewed as
a symmetric bilinear form on $\mathfrak{k} \times \mathfrak{k}$. Then*

$$\mathcal{H}_k(Z, Z) = - \sum_{\alpha \in \Delta^+} \alpha(H)(w\alpha)(X) \| F_\alpha(Z^v) \|^2 \qquad (Z \in \mathfrak{k}).$$

In particular, $f_{X,H}$ is clean, and its index at k is

$$\sum_{\alpha \in \Delta^+,\ \alpha(H)(w\alpha)(X) > 0} n(\alpha).$$

Proof: Clearly,

$$\mathcal{H}_k(Z, Z) = \left(\frac{d^2}{dt^2} \right)_{t=0} f_{X,H}(k \exp tZ) \qquad (Z \in \mathfrak{k}).$$

But, as $f_{X,H}(k \exp tZ) = \langle X^*, H^{\exp tZ^v} \rangle$ where $X^* = X^{w^{-1}}$, we have

$$\mathcal{H}_k(Z, Z) = \langle X^*, [Z^v, [Z^v, H]] \rangle = -\langle [X^*, Z^v], [H, Z^v] \rangle.$$

We now write, for any $U \in \mathfrak{k}$,

$$U = \sum_{\alpha \in \Delta^+} (U_\alpha + \theta U_\alpha) \bmod \mathfrak{m} \qquad (U_\alpha \in \mathfrak{g}_\alpha),$$

and observe that for any $R \in \mathfrak{a}$, since $\theta \mathfrak{g}_\alpha = \mathfrak{g}_{-\alpha}$,

$$[R, U_\alpha + \theta U_\alpha] = \alpha(R)(U_\alpha - \theta U_\alpha).$$

If we now remember that $\langle \mathfrak{g}_\alpha, \mathfrak{g}_\beta \rangle = 0$ unless $\alpha = -\beta$, and that $\|S\|^2 = -\langle S, S \rangle$ for $S \in \mathfrak{k}$, we get the desired formula. The formula for the index is immediate. $\square$

Proposition 4.4   *Suppose that X is regular. Then $f_{X,H}$ is a Morse function on K/K_H, the critical points of which are at the Weyl group points wK_H. If $X \in \mathfrak{a}^+$ and $H \in \mathrm{Cl}(\mathfrak{a}^+)$, the Hessian of $f_{X,H}$ at wK_H has index*

$$\sum_{\alpha \in \Delta^+,\ w\alpha \in \Delta^+,\ \alpha(H) > 0} n(\alpha).$$

In particular, the phase function has a unique local minimum and local maximum (which are global minima and maxima). If both X and H are in $\mathfrak{a}^+$, then the index of $f_{X,H}$ at the point wM of the flag manifold K/M is

$$\sum_{\alpha \in \Delta^+,\ w\alpha \in \Delta^+} n(\alpha).$$

Proof: Immediate from Proposition 4.3. $\square$

Corollary 4.5   *Suppose that K is semisimple and that G is its complexification viewed as a real Lie group. Then, for $X \in \mathfrak{a}^+$, the Morse indices of $f_{X,H}$ are even; and if both X and H are in $\mathfrak{a}^+$, and if $P(w)$ is the number of positive roots that remain positive after application of w, the index of $f_{X,H}$ at wK_H is $P(\omega)$, which is equal to $2\ell(w_0 w)$ where w_0 is the element of the Weyl group that takes Δ^+ to $-\Delta^+$ and $\ell(\cdot)$ is the length function on the Weyl group (viewed as a Coxeter group).*

Proof: In this case, the numbers $n(\alpha)$ are all 2, and the result is immediate. $\square$

4.3 Applications

Our first application is to the cohomology of the flag manifolds.

Theorem 4.6   *Let K be a compact, connected, semisimple Lie group and $\mathfrak{k}$ its Lie algebra. Then the flag manifolds K/K_H ($H \in \mathfrak{k}$) are all without torsion, and the distance from a generic point is a perfect Morse function. Moreover, for the full flag manifold K/T, where T is a maximal torus of K, the Poincaré polynomial is*

$$P(K/T : t) = \frac{\prod_{1 \le i \le n}(1 - t^{2d_i})}{(1 - t^2)^n},$$

where n is the rank of K, and d_i are the degrees of the homogeneous generators of the Weyl group-invariant polynomials on the Lie algebra t of T.

Proof: We have already seen in Section 3.1 that if all Morse numbers are even, the Morse and Poincaré polynomials are the same and there is no torsion. Now, when K is semisimple, we take G to be the complexification of K. In this case, $\mathfrak{s} = (-1)^{1/2}\mathfrak{k}$, and multiplication by $(-1)^{1/2}$ establishes an isomorphism of the K-actions on $\mathfrak{k}$ and $\mathfrak{s}$. Corollary 4.5 now gives the statements of the proposition except for the computation of the Poincaré polynomial of the full flag manifold $K/T = K/K_H$ where $H \in \mathfrak{a}^+$. But, by Corollary 4.5, the index of $f_{X,H}$ at wT is $2\ell(w_0 w)$, and so the Poincaré polynomial of K/T is

$$\sum_{w \in \mathbf{w}} t^{2\ell(w)}.$$

The desired formula now follows from a well-known result in the theory of finite reflection groups ([St, pp. 130–144]) that asserts

$$\sum_{w \in \mathbf{w}} t^{2\ell(w)} = \frac{\prod_{1 \leq i \leq n}(1 - t^{2d_i})}{(1 - t^2)^n}. \qquad \square$$

Remark   For $K = \mathrm{SU}(n+1)$, the subgroup T is usually taken to be the diagonal subgroup, and so K/T is the manifold of flags $(L_i)_{1 \leq i \leq n+1}$ where L_i is a linear subspace of $\mathbb{C}^{n+1}$ and

$$L_1 \subset L_2 \subset \cdots \subset L_{n+1} = \mathbb{C}^{n+1}, \quad \dim(L_i) = i.$$

Further, $\mathbf{w}$ is the permutation group S_{n+1} of $\{1, 2, \ldots, n+1\}$, and $\ell(w)$ is the number of inversions in $\{w(1), w(2), \ldots, w(n+1)\}$. By Newton's theorem on the symmetric functions, one knows that the numbers d_i are

$$2, 3, \ldots, n+1,$$

and so the result used is

$$\sum_{w \in S_{n+1}} t^{\ell(w)} = \frac{\prod_{1 \leq i \leq n}(1 - t^{d_i})}{(1 - t)^n}.$$

This is trivial to verify when $n = 1$, and the general case follows by induction on n; one has to establish the formula (denoting the left side by $P_n(t)$)

$$P_{n+1}(t) = P_n(t)(1 + t + t^2 + \cdots + t^{n+1}),$$

which can be done by considering the possible values for $w(n+2)$. Indeed, if s_k is the interchange $k \leftrightarrow n + 2$, S_{n+2} is the disjoint union of the cosets $S_{n+1}s_k$ ($1 \leq k \leq n + 2$), and it is easy to check that $\ell(ws_k) = \ell(w) + (n + 2 - k)$ for $w \in S_{n+1}$. Similar but somewhat more involved considerations apply to the other compact classical groups.

For the flag manifold K/K_H, the foregoing procedure gives a very explicit recipe for calculating the Poincaré polynomial. We assume that $H \in \mathrm{Cl}(\mathfrak{a}^+)$ and denote by $Z(H)$ the set of simple roots vanishing at H. The subgroup $\mathbf{w}_H$ is then generated by the reflections corresponding to the elements of $Z(H)$. If $\Delta^{H,+}$ is the set of positive roots taking a value > 0 at H, then for any $w \in \mathbf{w}$ we may define $P^H(w)$ to be the number of positive roots in $\Delta^{H,+}$ that remain positive after application by w; it is clear that $P^H(w) = P^H(ww_H)$, where w_H is any element of $\mathbf{w}_H$; this is due to the obvious fact that $\mathbf{w}_H$ permutes the elements of $\Delta^{H,+}$. From the formula for the index of $f_{X,H}$ at the Weyl group points on K/K_H, it is now clear that

$$\mathbf{P}(K/K_H : t) = \sum_{w \in \mathbf{w}/\mathbf{w}_H} t^{2P^H(w)}.$$

We leave it as an exercise to deduce from this the formula for the Poincaré polynomial of $\mathbb{C}P^n$ computed earlier, as well as the Poincaré polynomials of the manifolds of other types of flags (for example, the Grassmannians). The Morse theoretic computations of the cohomology of the complex flag manifolds go back to Bott [Bo2].

As another application we mention the *convexity theorem of Kostant* [Ko]. The theorem has an infinitesimal and a global version. It is the infinitesimal version that we consider here. We go back to the general setting of $\mathfrak{g}, \mathfrak{a}$. Then, Kostant's theorem states that *if $p_\mathfrak{a}$ is the orthogonal projection $\mathfrak{s} \longrightarrow \mathfrak{a}$, the projection*

$$p_\mathfrak{a}(K \cdot H)$$

is the convex hull of the points of the Weyl group orbit of the point $H \in \mathfrak{a}$. Heckman [Hec] gave a very simple proof of this and the corresponding global version by geometric methods based on the critical point and Hessian calculations of the phase functions $f_{X,H}$ and their global analogs $F_{a,H}$ (which we shall introduce and study in the next subsection). We refer the reader to the work of Heckman for the details.

4.4 The phase functions coming from the Iwasawa decomposition

We work in the context of (G, K) and introduce now a new class of phase functions on K, or rather K/M, the real flag manifold associated to G. They appear to have been studied first in [DKV1]. They arise naturally from the infinite dimensional representation theory of G, and their study was suggested by some questions of spectral estimates on locally homogeneous spaces associated to G.

Recall that we have the Iwasawa decomposition of G:

$$G = KAN \qquad (A = \exp \mathfrak{a}, \ N = \exp(\sum_{\alpha \in \Delta^+} \mathfrak{g}_\alpha)). \qquad (4.2)$$

This means that any element $x \in G$ can be uniquely written as

$$x = k(x) \exp H(x) \, n(x) \qquad (k(x) \in K, \ H(x) \in \mathfrak{a}, \ n(x) \in N). \qquad (4.3)$$

The functions $k(\cdot), H(\cdot), n(\cdot)$ are all analytic, and the Iwasawa decomposition gives an analytic diffeomorphism of G with $K \times A \times N$. Let as write, as usual, M for the centralizer of A in K. Then M normalizes N, and

$$P = MAN$$

is a parabolic subgroup of G, in fact, a minimal one. We have

$$G/AN \simeq K, \qquad G/P \simeq K/M, \qquad (4.4)$$

canonically. When $G = \mathrm{SL}(n, \mathbb{R})$, K/M is the manifold of full flags in $\mathbb{R}^n$, and so one refers to K/M in the general case also as the *flag manifold* of G. The identification (4.4) allows us to have an action of G on K and a compatible action on K/M. Thus, for $x \in G$, $u \in K$,

$$x[u] = k(xu), \qquad x[uM] = k(xu)M. \qquad (4.5)$$

The phase functions we want to investigate are

$$F_{a,\lambda}(k) = \lambda(H(ak)) \qquad (a \in A, \ \lambda \in \mathfrak{a}^*).$$

Since M normalizes N, it follows that these phase functions are defined on K/M. One could have used an arbitrary element x of G instead of $a \in A$; but in view of the polar decomposition

$$x = u_1 a^+ u_2 \qquad (u_1, u_2 \in K, \ a^+ \in \mathrm{Cl}(A^+), \ A^+ = \exp(\mathfrak{a}^+)),$$

it follows that

$$H(xk) = H(a^+ u_2 k);$$

thus, it is enough to restrict oneself to the study of the $F_{a,\lambda}$. These phase functions arise naturally in the representation theory of G. Indeed, the matrix coefficients of the representations of the principal series of G are of the form

$$\int_K e^{(i\lambda - \rho)(H(ak))} g(k)dk,$$

where g is a smooth function on K, and so are oscillatory integrals(!) of the form

$$\int_{K/M} e^{iF_{a,\lambda}(k)} g(a, k)dk,$$

where g is a smooth function on $A \times (K/M)$. The study of these phase functions and their associated oscillatory integrals is therefore very natural from our point of view.

It turns out that the infinitesimal versions of these phase functions are the phase functions $f_{X,H}$ discussed earlier. Indeed, we have

$$\left(\frac{d}{dt}\right)_{t=0} F_{\exp tX, \lambda}(k) = \langle X, H_\lambda^k \rangle, \tag{4.6}$$

where H_λ is the image of λ in $\mathfrak{a}$ under the canonical isomorphism of $\mathfrak{a}^*$ with $\mathfrak{a}$ induced by the (nondegenerate) restriction to $\mathfrak{a} \times \mathfrak{a}$ of the Cartan–Killing form, so that

$$\langle H_\lambda, H' \rangle = \lambda(H') \qquad \forall H' \in \mathfrak{a}.$$

The relation (4.6) is not completely trivial to derive; indeed, its derivation will reveal that some effort is needed in obtaining information about the $F_{a,\lambda}$. Here it is question of showing that

$$\left(\frac{d}{dt}\right)_{t=0} H((\exp tX)k) = p_{\mathfrak{a}}(X^{k^{-1}}) \qquad (X \in \mathfrak{a}), \tag{4.7}$$

$p_\mathfrak{a}$ being the orthogonal projection $\mathfrak{s} \longrightarrow \mathfrak{a}$. Put $X^{k^{-1}} = Z$, $p_\mathfrak{a}(Z) = R$. The orthogonal complement of $\mathfrak{a}$ in $\mathfrak{s}$ is contained in $\mathfrak{k} \oplus \mathfrak{n}$ (where $\mathfrak{n}$ is the Lie algebra of N), and so,

$$Z = R + V + W \qquad (V \in \mathfrak{k}, W \in \mathfrak{n}).$$

Thus,

$$(\exp tX)k = k \exp tX^{k^{-1}} = \exp tV \exp tR \exp tW \exp O(t^2),$$

from which (4.7) follows easily. We shall give presently the main calculation on the derivatives of the Iwasawa projection $H\colon G \longrightarrow \mathfrak{a}$, from which all such differential calculations will follow rather systematically.

For $x, y \in G$, we write $x^y = yxy^{-1}$. Let $U(\mathfrak{g})$ be the universal enveloping algebra of the complexification of $\mathfrak{g}$. The adjoint representation of G on $\mathfrak{g}$ extends to a representation of G on $U(\mathfrak{g})$ by automorphisms. We write u^x for $\mathrm{Ad}(x)(u)$; we have

$$u^{xy} = (u^y)^x, \qquad (uv)^x = u^x v^x \qquad (x, y \in G,\ u, v \in U(\mathfrak{g})).$$

We shall view elements of $U(\mathfrak{g})$ as left invariant differential operators on G. To define this interpretation, we should specify how an element $u = X_1 X_2 \cdots X_r$, where the X_i are in $\mathfrak{g}$, acts as a differential operator. We shall define

$$(\partial(u)f)(x) := f(x; u) = (\partial^r / \partial t_1 \partial t_2 \cdots \partial t_r)_0\, f(x \exp t_1 X_1 \cdots \exp t_r X_r)$$

the subscript 0 denoting the fact that the derivatives are taken at $t_1 = \cdots = t_r = 0$. Now the Iwasawa decomposition $\mathfrak{g} = \mathfrak{k} \oplus \mathfrak{a} \oplus \mathfrak{n}$ gives rise to the decomposition

$$U(\mathfrak{g}) = (\mathfrak{k}U(\mathfrak{g}) + U(\mathfrak{g})\mathfrak{n}) \oplus U(\mathfrak{a}).$$

It therefore makes sense to speak of the projection

$$E_\mathfrak{a}\colon U(\mathfrak{g}) \longrightarrow U(\mathfrak{a}),$$

where the projection is taken mod $(\mathfrak{k}U(\mathfrak{g}) + U(\mathfrak{g})\mathfrak{n})$. It is clear that this projection preserves the degree filtrations on both sides, and that if $u \in$

$U(\mathfrak{g})$ has zero constant term, the same is true of $E_{\mathfrak{a}}(u)$; here, *constant term is a homomorphism*

$$\varepsilon : U(\mathfrak{g}) \longrightarrow \mathbb{C}$$

that sends all elements of $\mathfrak{g}$ to 0. Note that since $\mathfrak{a}$ is abelian, $U(\mathfrak{a})$ is isomorphic canonically to the symmetric algebra over $\mathfrak{a}$. Thus, on $U(\mathfrak{a})$ the degree filtration arises from a gradation. Thus, we may speak in $U(\mathfrak{a})$ of the homogeneous components of any element.

Lemma 4.7   *Let* $y \in G$, $b \in U(\mathfrak{g})$. *Then*

$$H(y; b) = \varepsilon(b) + \left(E_{\mathfrak{a}}(b^{t(y)}) \right)_1. \tag{4.8}$$

Here the subscript 1 *means the homogeneous component of degree* 1, *and* $t(y) = a(y)n(y)$ *is the "triangular part" of* y. *In particular, for* $X \in \mathfrak{g}$,

$$H(y; X) = X^{t(y)}. \tag{4.9}$$

Proof: The fact that H is left invariant under K and right invariant under N implies that

$$H(1; u) = 0 \qquad \forall \, u \in \mathfrak{k} U(\mathfrak{g}) + U(\mathfrak{g})\mathfrak{n}.$$

Now, for $z \in G$,

$$H(yz) = H(t(y)z) = H(z^{t(y)}t(y)) = H(z^{t(y)}) + H(y).$$

So

$$H(y; b) = H(1; b^{t(y)}) = \varepsilon(b) + H(1; E_{\mathfrak{a}}(b^{t(y)})).$$

But as $H(\exp X_1 \cdots \exp X_r) = X_1 + \cdots + X_r$ for $X_i \in \mathfrak{a}$, it is clear that for any $c \in U(\mathfrak{a})$, $H(1; c) = c_1$. $\square$

Remark   It follows from the last assertion of the lemma that for $X \in \mathfrak{a}$, $k \in K$,

$$\left(\frac{d}{dt} \right)_{t=0} H((\exp tX)k) = \left(\frac{d}{dt} \right)_{t=0} H(k \exp tX^{k^{-1}})$$
$$= H(1; X^{k^{-1}})$$
$$= E_{\mathfrak{a}}(X^{k^{-1}})$$
$$= p_{\mathfrak{a}}(X^{k^{-1}}),$$

since, for any $Y \in \mathfrak{s}$, $E_{\mathfrak{a}}(Y) = p_{\mathfrak{a}}(Y)$ because the orthogonal complement of $\mathfrak{a}$ in $\mathfrak{s}$ is contained in $\mathfrak{k} + \mathfrak{n}$. As we noted earlier, this gives (4.7).

We now have the machinery to determine the critical sets of $F_{a,\lambda}$. From (4.9) we see that, for any $Z \in \mathfrak{k}$, $k \in K$,

$$F_{a,\lambda}(k; Z) = \langle Z, H_\lambda^{t(ak)^{-1}} \rangle = \langle Z, H_\lambda^{n(ak)^{-1}} \rangle,$$

so that k is critical for $F_{a,\lambda}$ if and only if $H_\lambda^{n(ak)^{-1}} \in \mathfrak{s}$. But it can be shown that for any $R \in \mathfrak{a}$, $R^N = R + \mathfrak{n}$; so, for any $n \in N$, $R^n \in \mathfrak{s}$ if and only if $R^n - R \in \mathfrak{s} \cap \mathfrak{n} = 0$, i.e., if and only if $n \in N_R$, where N_R is the centralizer of R in N. Thus,

$$k \text{ is critical for } F_{a,\lambda} \iff ak \in KAN_\lambda.$$

Here we have abused the notation and written N_λ instead of N_{H_λ}.

For any $R \in \mathfrak{a}$, we have $G_R = K_R A N_R = K_R \exp \mathfrak{s}_R$, which implies that $KAN_R = KK_R A N_R = KG_R = K \exp \mathfrak{a}_R$, and so the criterion for k to be critical for $F_{\exp X, \lambda}$ is that $\exp Xk \in K \exp \mathfrak{a}_{H_\lambda}$, which reduces to $[X^{k^{-1}}, H_\lambda] = 0$. But this is exactly the condition that k is a critical point for f_{X, H_λ}. We have thus proved

Proposition 4.8 *The critical set of $F_{\exp X, \lambda}$ is the same as that of f_{X, H_λ} and so is*

$$K_{X, H_\lambda} = \coprod_{w \in \mathbf{w}_X \backslash \mathbf{w} / \mathbf{w}_\lambda} K_X w K_\lambda.$$

The formula (4.6) can be generalized by calculating the derivative for all t, not just at $t = 0$. By (4.9) we have, for any $a \in A$,

$$\left(\frac{d}{dt} \right)_{t=0} H((a \exp tX)k) = H(ak; X^{k^{-1}}) = X^{t(ak)k^{-1}}.$$

But if $ak = ubn$ is the Iwasawa decomposition of ak, $t(ak) = bn$, and so $t(ak)k^{-1} = u^{-1}a$. Thus, $X^{t(ak)k^{-1}} = X^{u^{-1}}$. But $u = a[k]$, the transform of k by a under the A-action, and so we get

$$\left(\frac{d}{dt} \right)_{t=0} F_{a \exp tX, \lambda}(k) = \langle X, H_\lambda^{a[k]} \rangle.$$

Taking $a = \exp sX$, we have

$$\frac{d}{dt}F_{\exp tX, \lambda}(k) = \langle X, H_\lambda^{\exp tX[k]} \rangle. \tag{4.10}$$

Integrating this from 0 to 1, we finally obtain

$$F_{\exp X, \lambda}(k) = \int_0^1 f_{X, H_\lambda}(\exp sX[k])ds \qquad (k \in K,\ X \in \mathfrak{a},\ \lambda \in \mathfrak{a}^*).\tag{4.11a}$$

Furthermore, from (4.9) we get

$$F_{a, \lambda}(k; Y) = \langle Y^{t(ak)}, H_\lambda \rangle = \langle Y^{ak}, H_\lambda^{a[k]} \rangle$$

for all $Y \in \mathfrak{k}$. From this it follows easily that, if $a = \exp X$,

$$F_{a, \lambda}(k; Y) = f_{X, H_\lambda}(a[k]; Z), \tag{4.11b}$$

if Y is related to Z by

$$Y = \ \mathrm{Ad}\ k^{-1} \circ (\mathrm{ad}\ X/\sinh\ \mathrm{ad}\ X) \circ\ \mathrm{Ad}\,(a[k])(Z). \tag{4.11c}$$

We now discuss a remarkable right invariance property of the $F_{a, \lambda}$. Recall that the functions $f_{X, H}$ are right K_H-invariant for trivial reasons. However, in general it is not to be expected that this property persists for the $F_{a, \lambda}$. So it is rather surprising that $F_{a, \lambda}$ is right invariant under K_λ provided that λ *is in the closure of the positive chamber, i.e.,* λ *is dominant.* This can be proved by direct analysis; but to illustrate the intimate relationship of the functions $F_{a, \lambda}$ to representation theory, we shall briefly sketch an argument going back to ideas of Harish-Chandra that establishes this right invariance.

Since we want to use finite dimensional representation theory, we extend $\mathfrak{a}$ to a Cartan subalgebra $\mathfrak{h}$ of $\mathfrak{g}$ that is stable under θ (the Cartan involution defined by $\mathfrak{k}$) and choose a positive system of roots of $(\mathfrak{g}_{\mathbb{C}}, \mathfrak{h}_{\mathbb{C}})$ compatible with the positive system Δ^+ chosen already. We may, without changing much, work with a linear group, so that we may assume that G is contained in its simply connected complexification. Then, for any dominant integral linear function Λ on $\mathfrak{h}$, we have a finite dimensional irreducible representation π_Λ that is a representation of G and has the property that elements of K map into unitary operators and elements of A map into positive self-adjoint operators; this is achieved most simply by noting that $\mathfrak{u} = \mathfrak{k} \oplus (-1)^{1/2}\mathfrak{s}$ is a compact form of $\mathfrak{g}_{\mathbb{C}}$ and so we may suppose that $\pi_\Lambda(Z)$ is skew Hermitian for $Z \in \mathfrak{u}$. Let 1_Λ be the highest vector for the representation π_Λ. Write $\overline{N} = \theta(N)$, the unipotent subgroup defined by the system of negative roots. The compatibility of the positive systems on $\mathfrak{h}_{\mathbb{C}}$ and $\mathfrak{a}$ means that 1_Λ is fixed by N, while for any element $\bar{n} \in \overline{N}$ we have $\pi_\Lambda(\bar{n})1_\Lambda - 1_\Lambda$ is orthogonal to 1_Λ. This gives Harish-Chandra's famous formula

$$e^{\Lambda(H(x))} = ||\pi_\Lambda(x)1_\Lambda|| \qquad (x \in G). \tag{4.12}$$

Now, if for a positive root α of $(\mathfrak{g}_{\mathbb{C}}, \mathfrak{h}_{\mathbb{C}})$ we have $\langle \alpha, \Lambda \rangle = 0$, then $\pi_\Lambda(X_{-\alpha})1_\Lambda = 0$, and so $\pi_\Lambda(\bar{n})1_\Lambda = 1_\Lambda$ for $\bar{n} \in \overline{N}_\Lambda$. This leads easily to

Proposition 4.9 *For any λ such that $H_\lambda \in \mathrm{Cl}(\mathfrak{a}^+)$, the function $F_{a,\lambda}$ is right invariant under $K_\lambda = K_{H_\lambda}$.*

To complete this brief treatment of the $F_{a,\lambda}$, it is now necessary to calculate their Hessians at the critical points. This again is done using Lemma 4.7, but the calculations are more difficult than before because we are calculating second order derivatives. We do not give the proof of the following result here, and refer the reader to [DKV1, pp. 343–346]. It can also be obtained from (4.11) and the corresponding formulae for the Hessians of $f_{X,H}$.

Proposition 4.10 *For any $X \in \mathfrak{a}$, $\lambda \in \mathfrak{a}^*$, $w \in \mathbf{w}$, the Hessian form $\mathcal{H}_w$ of $F_{\exp X, \lambda}$ at $k = x_w$ is given by*

$$\mathcal{H}_w(Z,Z) = -\frac{1}{2} \sum_{\alpha \in \Delta^+} \langle \alpha, \lambda \rangle (1 - e^{-2w\alpha(X)}) ||F_\alpha(Z)||^2,$$

where F_α is as in Proposition 4.3. Moreover, throughout the critical manifold $K_X w K_{H_\lambda}$ the value, the signature, and the rank of the Hessian remain constant and are the same as for f_{X,H_λ}. In particular, the Hessian is transversally nondegenerate so that $F_{\exp X, \lambda}$ is clean. Moreover, the value of $F_{\exp X, \lambda}$, and the rank and signature of the Hessian of $F_{\exp X, \lambda}$ at the points of $K_X w K_{H_\lambda}$ are given respectively by

$$\langle w\lambda, X \rangle, \qquad n_w = -\sum_{\alpha \in \Delta_{X,\lambda}^+} n(\alpha),$$

$$\sigma_w = - \sum_{\alpha \in \Delta_{X,\lambda}^+} n(\alpha) \operatorname{sgn}(\langle \alpha, \lambda \rangle (w\alpha)(X)),$$

where

$$\Delta_{X,\lambda}^+ = \{\alpha \in \Delta^+ \mid \langle \alpha, \lambda \rangle (w\alpha)(X) \neq 0\}.$$

Remarks i) Although there is no right invariance under K_λ if λ is not dominant, the second order data are invariant and are the same as for the f_{X,H_λ}.

ii) The global convexity theorem of Kostant [Ko] asserts that the Iwasawa projection maps the K-conjugacy class of any element $\exp X$ ($X \in \mathfrak{a}$) onto the convex hull of the points of the Weyl group orbit of $\exp X$. As we mentioned earlier, Heckman [Hec] obtained relatively easy proofs of this and the infinitesimal convexity theorem of Kostant by using properties of the phase functions $F_{a,\lambda}$ and $f_{X,\lambda}$ and the relationships between them.

iii) The phase functions $f_{X,H}$ and $F_{a,\lambda}$ form *parametrized families*. But it must be noted that they are *highly nongeneric* because of the tremendous *rigidity* in their behavior as the parameters approach the caustic sets. For instance, as long as X and H (or λ) are regular, these are Morse functions, *but their critical data do not change when the parameters vary within the regular domains.* When the parameters enter some root hyperplanes, which are now the caustic sets, the critical points become enlarged in the simplest possible way; some of the critical points are joined by smooth submanifolds that now consist entirely of critical points, and the Hessians are transversally nondegenerate. Again, variation without enlarging the caustic set does not change the critical data until the parameters become more singular, and the phenomenon of enlargement of the critical manifold by linking repeats itself, and so on.

The simplest special case is $G = \mathrm{SL}(2,\mathbb{R})$. If

$$x = \begin{pmatrix} \alpha & \beta \\ \gamma & \delta \end{pmatrix}, \qquad a = \begin{pmatrix} e^t & 0 \\ 0 & e^{-t} \end{pmatrix}, \qquad k = \begin{pmatrix} \cos\theta & \sin\theta \\ -\sin\theta & \cos\theta \end{pmatrix},$$

then

$$H(x) = \frac{1}{2}\log(\alpha^2 + \gamma^2) \begin{pmatrix} 1 & 0 \\ 0 & -1 \end{pmatrix},$$

$$H(ak) = \frac{1}{2}\log(e^{2t}\cos^2\theta + e^{-2t}\sin^2\theta) \begin{pmatrix} 1 & 0 \\ 0 & -1 \end{pmatrix}.$$

For $t \neq 0$, the critical points are $\theta = 0, \pi/2$. When $t = 0$, the phase function becomes constant, and the whole circle is critical. This method of calculation of $H(x)$ can be extended to the classical groups, using explicit representations of G. For instance, for $\mathrm{SL}(n)$ one can use the defining representations and its exterior powers.

4.5 Phase functions on conjugacy classes

An entirely different class of phase functions are obtained if we restrict the same functions to conjugacy classes. For any $X \in \mathfrak{g}$ (resp., $\gamma \in G$), we write

$$C(X) = \{X^g \mid g \in G\}, \quad C(\gamma) = \{\gamma^g = g\gamma g^{-1} \mid g \in G\}.$$

It is known that these are closed if and only if X (resp., γ) is a semisimple element. In this case, they are closed imbedded submanifolds of $\mathfrak{g}$ (resp., G).

The infinitesimal versions are the phase functions $s_{X,Y}$, which are defined on $C(Y)$ by

$$s_{X,Y}(Z) = \langle X, Z \rangle \qquad (Z \in C(Y)), \tag{4.13}$$

where $X, Y \in \mathfrak{g}$. The global versions are defined in complete analogy with what we did before. For $\lambda \in \mathfrak{a}^*$, we define $S_{\gamma,\lambda}$ by

$$S_{\gamma,\lambda}(t) = \lambda(H(t)) \qquad (t \in C(\gamma)). \tag{4.14}$$

It is absolutely remarkable that for these phase functions one can prove the same type of results as before. We state these results without proofs. For the details, one may refer to [D2]. We just introduce one small bit of terminology: closed submanifolds A, B of a smooth manifold C are said to have a *clean intersection* if $A \cap B$ is a submanifold, and at each point c of $A \cap B$ we have

$$T_c(A \cap B) = T_c(A) \cap T_c(B).$$

It is obvious that this is a generalization of transversal intersection, which arises when $T_c(A)$ and $T_c(B)$ are complementary subspaces in $T_c(C)$.

Proposition 4.11 *i) Let $X, Y \in \mathfrak{g}$ be semisimple. Then the critical set of $s_{X,Y}$ is $\mathfrak{g}_X \cap C(Y)$. This is a clean intersection, and it is a disjoint union of finitely many G_X^0-conjugacy classes, all of the same dimension, and all belonging to the same conjugacy class in $(\mathfrak{g}_X)_{\mathbb{C}}$ under the complex adjoint group of $(\mathfrak{g}_X)_{\mathbb{C}}$.*

ii) Let $\lambda \in \mathfrak{a}^*$, and let $\gamma \in G$ be a semisimple element. Then the critical set of $S_{\gamma,\lambda}$ is $G_{H_\lambda} \cap C(\gamma)$. This intersection is clean and is the union of finitely many $G_{H_\lambda}^0$-conjugacy classes, all of the same dimension.

iii) In both of these cases, the Hessians are transversally nondegenerate.

Remark The cleanness and finiteness parts of the proposition follow from well-known results of Richardson [Ri]. In (i), if X is regular and $\mathfrak{h}$ is the Cartan subalgebra containing it, we have $\mathfrak{g}_X = \mathfrak{h}$ and $\mathfrak{h} \cap C(Y)$ is a finite set that is a subset of the *complex* Weyl group orbit of the point $Y_{\mathbb{C}} \in \mathfrak{h}_{\mathbb{C}}$, which is conjugate to Y under the complex adjoint group of $\mathfrak{g}_{\mathbb{C}}$. In this case, $s_{X,Y}$ is therefore a Morse function. It is possible that it has no critical points; this suggests that one has to *compactify* the orbit and consider $s_{X,Y}$ on the compactification.

5 Spherical oscillatory integrals

5.1 Oscillatory integrals with many frequency variables

The study of the phase functions on homogeneous spaces carried out in Section 4 is the foundation on which a theory of oscillatory integrals with these phase functions can be built. These oscillatory integrals arise in many problems of analysis involving semisimple Lie groups, and so it is worthwhile discussing what one can and should say about such integrals.

Whenever one has a family of phase functions (φ_x) depending on a parameter x varying in some manifold X, the phase functions themselves being defined on a manifold M on which there is a volume form ω, we have the oscillatory integrals

$$\int_M e^{i\tau \varphi_x(y)} a(x:y)\omega \qquad (a \in C^\infty(X \times K),\ K \subset M,\ K \text{ compact}), \quad (5.1)$$

and the problems one encounters are of the following type:

i) To investigate the asymptotic behavior of these integrals as $\tau \longrightarrow \infty$ for each $x \in X$.

ii) To get upper bounds for the integrals that are locally uniform in x.

As long as the phase functions are Morse (or at least clean) for generic x, there is no difficulty in treating (i) at the generic points x. But when x enters the *caustic set* of the family, the critical set becomes in general more complicated, and the problem of uniform estimates in the neighborhoods of parameter points that are in the caustic set is usually difficult. We have examined this situation in some detail in Sections 1–3. However, the families of phase functions that we have introduced in Section 4 behave very differently from generic families in two ways, one for the better and one for the worse. On the good side, we have seen in Section 4 that their behavior as the parameters approach the caustic set is very rigid and very simple. One is always in the clean situation, where the critical sets are smooth and the Hessians transversally nondegenerate. But on the bad side, *it is no longer sufficient to estimate the integrals of the form* (5.1). The point is that these phase functions are really *vector valued*, so that if V is the vector space in which they take their values, and λ is an element of V^*, we are really interested in integrals

$$\int_M e^{i\langle \varphi_x(y), \lambda \rangle} a(x : y)\omega \qquad (a \in C^\infty(X \times K), \ K \subset M, \ K \text{ compact}), \quad (5.2)$$

and we want estimates for the behavior of (5.2) when $\lambda \longrightarrow \infty$ in V^*. Clearly, this is a very much more general problem than (5.1). Indeed, if (λ_i) is a basis of V^* and we write

$$\lambda = \sum_i \tau_i \lambda_i, \qquad \varphi_i = \langle \varphi, \lambda_i \rangle,$$

then we may think of (τ_i) as *a collection of frequency variables*, and it is a question of studying what happens to the integral

$$\int_M e^{i \sum_i \tau_i \varphi_i(x,y)} a(x : y)\omega$$

as $|\tau| \longrightarrow \infty$. As we mentioned in Section 1.4, this is a new type of situation and appears to be worth exploring for its own sake. Another way to look at this is to say that in this problem it is a question of the structure of smooth maps of M into V, and so the problem is much more difficult than in the case of scalar phase functions, where Morse's theory describes the generic situation very well.

In the next subsection, we shall investigate the integrals

$$\int_K e^{i\lambda(H(ak))}g(k)dk \qquad (a \in A,\ \lambda \in \mathfrak{a}^*,\ g \in C^\infty(K)) \qquad (5.3)$$

from this "vectorial" point of view. Since these integrals occur naturally in the theory of spherical functions on semisimple Lie groups — indeed, the elementary spherical functions on these groups are represented as such integrals — we shall call the integrals (5.3) *spherical oscillatory integrals*. Here, one can completely solve the problem of vectorial asymptotic estimates when a is in compact subsets of A [DKV1]. We shall give a brief discussion of the ideas behind the solution.

However, in some spectral problems it is also important to understand what happens to the estimates that we obtain for (5.3) when $a \longrightarrow \infty$. This again is a new situation because, at infinity in the parameter space (here it is A), the phase functions do not in general have a limit, and so it is not immediately obvious what should be done. In Section 5.3 we shall state a conjecture about uniform estimates for (5.3) when $a \longrightarrow \infty$. The conjecture, if proved, will imply certain estimates for the elementary spherical functions $\varphi(\lambda : x)$ on a semisimple Lie group when both λ and x go to infinity. The conjecture can be verified in some cases: when the symmetric space G/K has rank 1, and when G is complex. In the last section, we shall discuss an application of the conjectured uniform estimates of the elementary spherical functions to the problem of error estimation in the spectra of spaces of the form $L^2(\Gamma\backslash G/K)$ where Γ is a cocompact discrete subgroup such that $\Gamma\backslash G/K$ is smooth.

The phase functions $s_{X,Y}$ and $S_{\gamma,\lambda}$ on conjugacy classes in $\mathfrak{g}$ and G are clearly important in the theory of characters and orbital integrals. We shall not treat the questions concerning their oscillatory integrals here.

5.2 Uniform estimates for spherical oscillatory integrals: The method of trigonalization

We begin with the asymptotic expansion of (5.3) of Section 5.1 when λ goes to infinity along a ray. This is of course the classical situation and can be handled without any problem in view of the results of Section 4 on the phase functions $F_{a,\lambda}$ and the theory developed in Section 2. We get the following result.

Theorem 5.1 *Fix $a_0 \in A$, $\lambda_0 \in \mathfrak{a}^*$. Then we have the following asymptotic expansion of the function*

$$e^{i\tau F_{a_0,\lambda_0}}$$

as a distribution:

$$e^{i\tau F_{a_0,\lambda_0}} \sim \sum_w e^{i\tau w\lambda_0(\log a_0)} \sum_{r \geq 0} \tau^{-n_w/2} c_{w,r}, \tag{5.4}$$

where $c_{w,r}$ is a distribution of order $\leq 2r$ supported by $K_{a_0} w K_{H_{\lambda_0}}$, the integer n_w is given by Proposition 4.10, and the sum is over a complete set of representatives of $\mathbf{w}_{a_0} \backslash \mathbf{w} / \mathbf{w}_{\lambda_0}$. Moreover, the asymptotic expansion is uniform if a and λ vary around a_0 and λ_0 in an equisingular manner.

If a and λ are regular, this gives the asymptotic expansion of the elementary spherical functions:

$$\varphi(i\lambda : a) \sim \tau^{-\dim(N)/2} \sum_w e^{i\tau w\lambda(\log a)} e^{i\pi/4\sigma_w} \delta_w(a), \tag{5.5}$$

where δ_w is the function on A defined by

$$\delta_w(a) = \gamma \prod_{\alpha \in \Delta^+} |\frac{\langle \alpha, \lambda \rangle}{2\pi} \sinh w\alpha(\log a)|^{-n(\alpha)/2}$$

with γ an absolute constant that depends on the choice of the Haar measures on K and M.

It is not possible to overlook the formal similarity of Theorem 5.1 with the famous Harish-Chandra asymptotics of the elementary spherical functions. There it is a question of $a \longrightarrow \infty$ for each fixed λ, whereas here it is the other way around, with $\lambda \longrightarrow \infty$ as a remains fixed. Actually, if a and λ are regular, the expansion (5.5) can be obtained from the Harish-Chandra expansion of the elementary spherical functions. But Theorem 5.1 is more general; it applies to all amplitudes, hence to all matrix elements of the spherical principal series representations, and also to singular a and λ.

If we desire asymptotic results that are uniform when a and λ vary in full neighborhoods of a_0 and λ_0, then we have to face the problem of caustics, as we have explained repeatedly. Because of the simplicity of the behavior of the phase functions around the caustic sets, one can use Morse lemma with parameters to handle this situation. In fact, if $k_0 \in K$ is any point of $K_0 = K_{a_0, H_{\lambda_0}}$, we can select local coordinates $x_1, \ldots, x_q, y_1, \ldots, y_s$ on K around k_0 in such a way that K_0 is given locally around k_0 by the equations $y_1 = \cdots = y_s = 0$. The phase function $F_{a,\lambda}$ is then a function of the x_i and y_j and has a nondegenerate critical point when $x_1 = \cdots = x_q = 0$, $a = a_0$, $\lambda = \lambda_0$. Hence, treating $x_1, \ldots, x_q, a, \lambda$ as parameters and applying the method of stationary phase as discussed in Section 2, we get uniform estimates for the spherical oscillatory integral. If we now observe that as λ varies around λ_0, the rays $\tau\lambda$ will fill up a conical neighborhood of λ_0 except for a compact set, we obtain the following corollary:

Corollary 5.2   *Fix $a_0 \in A$, $\lambda_0 \in \mathfrak{a}^*$. Then we can find a neighborhood ω of a_0 in A, a conical neighborhood Γ of λ_0 on $\mathfrak{a}^*$, and a continuous seminorm ν on $C_c^\infty(\omega \times K)$ such that for all $a \in \omega$, $\lambda \in \Gamma$, $g \in C_c^\infty(\omega \times K)$, we have*

$$\left| \int_K e^{i\lambda(H(ak))} g(a,k) dk \right| \leq \nu(g_a) \sum_{\mathbf{w}_{a_0} \backslash \mathbf{w} / \mathbf{w}_\lambda} \prod_{\alpha \in \Delta^+_{a_0, H_{\lambda_0}}} (1 + |\langle \alpha, \lambda \rangle|)^{-n(\alpha)/2}.$$

$$(5.6)$$

Unfortunately, as we vary λ_0 over $\mathfrak{a}^*$, this estimate becomes too weak and so cannot be effective in many applications. So there arises the problem of uniform estimates, where we try to keep the same form as for the majorants on the right side of (5.6), but have a larger set of roots over which the product is taken, so that the decay in λ is much greater at infinity in $\mathfrak{a}^*$.

To do this, we have to take a closer look at the structure of the phase functions $F_{a,\lambda}$ as a function of λ. In what follows, we shall forget about a; everything we do will remain valid in a small neighborhood of a. In order to explain the fundamental idea of the uniform estimates theorem, we assume first that λ varies in the closure Λ^+ of the positive chamber, where

$$\Lambda^+ = \{\lambda \mid \langle \alpha, \lambda \rangle \geq 0 \ \ \forall \, \alpha \in \Delta^+\}. \tag{5.7}$$

Then we know that $F_{a,\lambda}$ is right invariant under K_{H_λ}. Let us select an ordering

$$\alpha_1, \alpha_2, \ldots, \alpha_r$$

of the simple roots, and define the dual (ordered) basis (λ_j) in $\mathfrak{a}^*$ by

$$\langle \lambda_j, \alpha_i \rangle = \delta_{ij} \qquad (1 \leq i, j \leq r).$$

Then, we can write

$$F_{a,\lambda} = \sum_{1 \leq j \leq r} \tau_j F_j, \qquad \lambda = \sum_{1 \leq j \leq r} \tau_j \lambda_j, \qquad F_j = F_{a,H_{\lambda_j}}. \tag{5.8}$$

If we put

$$K_j = \text{ centralizer of } \{\lambda_1, \lambda_2, \ldots, \lambda_j\}, \tag{5.9}$$

then we have

$$K = K_0 \supset K_1 \supset K_2 \supset \cdots \supset K_r = M \tag{5.10}$$

and F_j is right K_j-invariant. It may appear that we have lost some information since F_j is right invariant under the much bigger group $K_{H_{\lambda_j}}$; but this is not serious, because we have selected an arbitrary ordering of the simple roots here, and in an ordering where α_j is the first, the information on F_j will be recaptured.

To illustrate how we put the information contained in (5.10) to use, let us consider the identity in K, which is a critical point of all the F_j. The group K_t is the maximal compact subgroup of the reductive group for which the simple (restricted) roots are $\alpha_{t+1}, \alpha_{t+2}, \ldots, \alpha_r$. The theory developed in Section 4 now implies that the restriction to K_{j-1} of F_j has 1 as a critical point, with a Hessian nondegenerate transversally to K_j, so that the rank of the Hessian is the integer n_j, where

$$n_j = \sum_{\alpha \in \Delta_{j,a}^+} n(\alpha),$$

$$\Delta_{j,a}^+ = \{\alpha \in \Delta^+ \mid \alpha(\log a) \neq 0, \ \alpha = \sum_{t \geq j} m_t \alpha_t, m_j > 0\}.$$

We now select a local section X_j through 1 for K_{j-1}/K_j. Then, the right invariance under the K_j satisfying (5.10) allows us to set up an equivalence of functions

$$(F_1, F_2, \ldots, F_r) \simeq (f_1, f_2, \ldots, f_r),$$

where

i) The f_j are defined on $X = X_1 \times X_2 \times \cdots \times X_r$.

ii) f_j depends only on the variables $x^{[j]} = (x_1, x_2, \ldots, x_j)$, $x_j \in X_j$.

iii) The function $f_j(1, 1, \ldots, 1, x_j)$ has $x_j = 1$ as a nondegenerate critical point with Hessian rank n_j.

The properties (i)—(iii) constitute what we mean when we say that the system

$$(F_j)_{1 \leq j \leq r}$$

has the property of *trigonalizability* around the identity 1 of K.

We now consider the oscillatory integral

$$\int_K e^{i \sum_j \tau_j F_j} g(k) dk,$$

where the amplitude is localized at 1 as an integral

$$\int_X e^{i \sum_j \tau_j f_j} g(x_1, \ldots, x_j) \omega, \tag{5.11}$$

where ω is a volume form on X and g is smooth with compact support. The idea is to apply Morse's lemma with parameters and carry out the integrations successively with respect to the variables x_j, *beginning with the last, namely,* x_r.

Lemma 5.3   *The integral (5.11) has the bounds*

$$\left| \int_X e^{i \sum_j \tau_j f_j} g(x_1, \ldots, x_j)) \omega \right| \leq \nu(g) \prod_{1 \leq j \leq r} (1 + \tau_j)^{-n_j/2},$$

where ν is a seminorm on $C_c^\infty(X)$, and the τ are restricted to a domain of the form

$$\tau_1 \geq \gamma \tau_2 \geq \gamma^2 \tau_3 \geq \cdots \geq \gamma^{r-1} \tau_r$$

for some $\gamma > 0$.

Proof: Integration with respect to x_r, with f_r treated as a Morse function with parameters $x_1, \ldots, x_{r-1}$, will produce an estimate that decays in τ_r like $\tau_r^{-n_r/2}$, but the new phase function will now be

$$\sum_{1 \leq j < r-1} \tau_j f_j(x^{[j]}) + \tau_{r-1} f_{r-1}(x^{[r-1]}) + \tau_r g_r(x^{[r-1]}),$$

where $g_r(x^{[r-1]})$ is the critical value of $f_r(x^{[r-1]}, x_r)$ as a function of x_r. We rewrite this as

$$\sum_{1 \le j < r-1} \tau_j f_j(x^{[j]}) + \tau_{r-1}\left(f_{r-1}(x^{[r-1]}) + (\tau_r/\tau_{r-1})g_r(x^{[r-1]})\right),$$

and for the next integration τ_r/τ_{r-1} will be a parameter that has to be small; i.e., for τ_{r-1}, we have to have a domain $\tau_{r-1} \ge \gamma\tau_r$. If we continue this process we arrive at the estimate stated in the lemma with the τ_j restricted to a domain of the form stated. $\square$

The domains that we obtain are only a small part of the positive chamber of the τ_j. But we can modify this analysis in two important ways. First, we can change the ordering of the F_j so that the τ_j get permuted. Second, we can group the variables τ_j into disjoint subsets and then consider the regime where the τ_j in each group go to infinity in a mutually comparable manner, but the τ_j in different groups go to infinity in a domain of the form described in the lemma. It turns out that *the positive chamber of the τ_j can be written as a union of these special domains where we have groups of frequency variables going to infinity each faster than the next, but with the variables in each group going to infinity in a mutually comparable manner.*

At this stage we will have the definitive uniform bound for the spherical oscillatory integral (5.6), but with λ restricted to be in the positive chamber Λ^+. To remove this restriction appears at first to be a formidable technical problem, because we lose the right invariance that played a crucial role in the "trigonalizability" of the system of phase functions with which we are working. But this can be overcome in the following way. Recall that associated to the Fa, λ are the infinitesimal versions $f_{X,H}$ for which, because of the symmetric way in which X and H enter their definition, there is right invariance with respect to the subgroups K_H for all H, not just for the H in the positive chamber. So for them, the property of trigonalizability will persist, and this can be established exactly as before, since the critical sets have the same structure in both cases. But we now use the information contained in the equations (4.11a–c) to transfer the trigonalizability from the $f_{X,H}$ to the $F_{a,\lambda}$, using the flow generated by the A-action on K/M that enters in the relationships expressed by (4.11a–c).

It was in this manner that the following theorem was proved in [DKV1].

It is the definitive uniform estimate for the spherical oscillatory integrals on the flag manifolds, *when the variable a is in a compact subset of A.*

Theorem 5.4 *Let ω be a compact set in A, and for $w \in \mathbf{w}$ let*

$$\Delta_{\omega,w} = \{\alpha \in \Delta^+ \mid w\alpha(\log a) \neq 0 \quad \forall \, a \in \omega\}.$$

Then there is a continuous seminorm ν on $C^\infty(K)$ such that for all $\lambda \in \mathfrak{a}^*$, $a \in \omega$, $g \in C^\infty(K)$, we have

$$\left| \int_K e^{i\lambda(H(ak))}g(k)dk \right| \leq \nu(g) \sum_{w \in \mathbf{w}} \prod_{\alpha \in \Delta^+_{\omega,w}} (1 + |\langle \alpha, \lambda \rangle|)^{-n(\alpha)/2}.$$

5.3 The problem of uniform bounds for the spherical oscillatory integrals at infinity on A

It is a natural question to ask what happens to the estimate in Theorem 5.4 when a is allowed to go to infinity. This question has proved difficult to resolve so far except in some special cases. From the evidence provided by these special cases, it is possible that one should be able to prove that the bounds in Theorem 5.4 remain valid in the cone generated in A by ω, provided we add a factor of the form

$$Ce^{\gamma || \log a ||} \qquad (\gamma \text{ a constant } > 0)$$

to the right side. More precisely, one may make the following conjecture (**C**):

Let $a_0 \in A \setminus \{1\}$, and for any compact set $\omega \subset A$, let

$$C(\omega) = \{\exp tH \mid t \geq 1, \ \exp H \in \omega\}.$$

For $w \in \mathbf{w}$, let

$$\Delta_{a_0,w} = \{\alpha \in \Delta^+ \mid w\alpha(\log a_0) \neq 0\}.$$

Then there is a compact neighborhood ω of a_0 in A, a constant $\gamma > 0$, and a continuous seminorm ν on $C^\infty(K)$ such that for all $\lambda \in \mathfrak{a}^*$, $a \in C(\omega)$, $g \in C^\infty(K)$, we have

$$\left| \int_K e^{i\lambda(H(ak))}g(k)dk \right| \leq e^{\gamma || \log a ||}\nu(g) \sum_{w \in \mathbf{w}} \prod_{\alpha \in \Delta^+_{a_0,w}} (1 + |\langle \alpha, \lambda \rangle|)^{-n(\alpha)/2}.$$

It is possible to prove this when G/K has rank 1 and also in the case when ω is a compact *regular* set in the positive chamber A^+ and λ is restricted to a closed conical region inside the open positive chamber in $\mathfrak{a}^*$. But the general case appears difficult to settle.

If we take the amplitude g defined by

$$g(a, k) = e^{\rho(H(ak))} \qquad (k \in K),$$

then the integral is just the elementary spherical function $\varphi(\lambda : a)$, and the foregoing estimates become

$$|\varphi(\lambda : a)| \leq C e^{\gamma \|\log a\|} \sum_{w \in \mathbf{W}} \prod_{\alpha \in \Delta_{\omega,w}^+} (1 + |\langle \alpha, \lambda \rangle|)^{-n(\alpha)/2}. \qquad (5.12)$$

Actually, I believe that a much stronger estimate is true: There are constants $C > 0$, $q \geq 0$ such that

$$|\varphi(\lambda : a)| \leq C e^{-\rho(\log a)} (1 + \|\log a\|)^q \sum_{w \in \mathbf{W}} \prod_{\alpha \in \Delta_{\omega,w}^+} (1 + |\langle \alpha, \lambda \rangle|)^{-n(\alpha)/2}.$$

$$(5.13)$$

Indeed this can be verified to be true in the following cases.

  i) G is *complex* (because in this case there are explicit formulae for the elementary spherical functions that go back to Harish-Chandra and Gel'fand–Naimark in the early 1950s).
 ii) G/K has rank 1.
iii) G arbitrary, but ω is restricted to be a compact subset in the *interior* of the positive chamber A^+.

5.4 Spectral asymptotics of $\Gamma \backslash G / K$

One reason for the interest in the asymptotics of elementary spherical functions when the frequency variable λ goes to infinity is that the knowledge of such behavior plays a crucial role in the determination of the orders of magnitude of the asymptotics of the spectra of the spaces $\Gamma \backslash G / K$ where Γ is a discrete cocompact subgroup of G. The space $\Gamma \backslash G / K$ need not be smooth, and to make it smooth we have to ensure that Γ contains no elements of finite order. It is a well-known result of Borel that this can always be done by going over to a normal subgroup of Γ of finite index in Γ. So we shall assume that the space $\Gamma\backslash G/K$ is smooth. Then, the representation of G in $L^2(\Gamma\backslash G/K)$ decomposes discretely, and there is a discrete set $S(\Gamma)$ of λ in $\mathfrak{a}_{\mathbb{C}}^*$ with the property that the representations that occur are precisely the spherical ones with parameters λ in $S(\Gamma)$, with *finite multiplicities $m(\lambda)$*. The set $S(\Gamma)$ is called the *spectrum* of $\Gamma\backslash G/K$. It was proved in [DKV2] that the λ that are not in $(-1)^{1/2}\mathfrak{a}^*$ — which form the so-called *exceptional spectrum $S_e(\Gamma)$* — are asymptotically negligible (in a sense to be explained) compared with the λ in $(-1)^{1/2}\mathfrak{a}^*$ — which form the *principal spectrum $S_p(\Gamma)$* — , and that at infinity on $(-1)^{1/2}\mathfrak{a}^*$ the principal spectrum is asymptotic to the Plancherel measure. One way to formulate these results is as follows.

Theorem 5.5   *Let $\Omega(t)_{t\geq 1}$ be a family of bounded open sets with smooth boundary in $(-1)^{1/2}\mathfrak{a}^*$ such that*

i) $\exists$ constants $C_1 > 0$, $C_2 > 0$ such that for all $t \geq 1$ the diameter of the set $\Omega(t)$ is at most $C_1 T$, and its volume is at least $C_2 t^r$ $(r = \dim \mathfrak{a}^*)$.

ii) The measure of the boundary of $\Omega(t)$ is $O(t^{r-1})$.

Then, as $t \longrightarrow \infty$,

$$\sum_{\lambda \in S_e(\Gamma),\ ||\lambda|| \leq t} m(\lambda) = O(t^{r-2}).$$

Furthermore,

$$\sum_{\lambda \in S_p(\Gamma),\ \lambda \in \Omega(t)} m(\lambda) = \int_{\Omega(t)} \beta(\nu)dv + O(t^{n-1}).$$

Here, $\beta(\nu)$ be the density of the Plancherel measure for the decomposition of the natural representation of G on $L^2(G/K)$.

The first statement is the negligibility of the exceptional spectrum, since the main term in the second statement is of the order of t^{n-1}. The error estimate appears not to be very good, and it is an interesting question as to how much sharper one can make it.

This is a question that has attracted some attention. One way to treat it is to observe that the starting point of all the investigations is the *Selberg trace formula*, which consists of a main term and a supplementary part; the retention of the main term only leads to the preceding result. If we now use the full trace formula and use Theorem 5.4 in analyzing the supplementary part of the trace formula, we get the following result, where the O has been replaced by o:

$$\sum_{\lambda \in S_p(\Gamma), \ \lambda \in \Omega(t)} m(\lambda) = \int_{\Omega(t)} \beta(\nu) dv + o(t^{n-1}).$$

To go beyond this, it becomes necessary to analyze the supplementary part of the trace formula more closely, and this depends on accurate estimates on the behavior in λ of the elementary spherical functions $\varphi(\lambda : a)$ when a and λ both go to infinity.

There is a consequence of (5.12) which we shall now formulate that is one of the keys in this type of analysis. In fact, one can prove the following:

Theorem 5.6 *Under the assumption that the estimates (5.12) are true, we have the following. There exist constants $C, c, \eta > 0$ such that for all $\nu \in \mathfrak{a}^*$, $a \in A$,*

$$\beta(\nu)|\varphi(\nu : a)| \leq C e^{c|| \log a||}(1 + ||\nu||)^{\dim(G/K) - r - \eta}. \tag{5.14}$$

In view of our comments earlier, this result is true when G is complex and when $r = 1$. If we now assume that (5.14) is true, we can state a sharper result on the error estimates for the spectrum $S_p(\Gamma)$:

$$\sum_{\lambda \in S_p(\Gamma), \ \lambda \in \Omega(t)} m(\lambda) = \int_{\Omega(t)} \beta(\nu) dv + O\left(\frac{t^{n-1}}{(\log t)^r}\right).$$

It is possible that these are too conservative, and that the true error estimates are much sharper. In any case, it appears almost inevitable that in any treatment of this question, the simultaneous asymptotics of the elementary spherical functions in both the space and the frequency variables will play a crucial role.

References

[AGV] V. I. Arnol'd, S. M. Gusein-Zade, and A. Varchenko, *Singularities of Differentiable Maps*, Birkhäuser, 1988.

[Ai]   G. B. Airy, *On the intensity of light in the neighbourhood of a caustic*, Trans. Camb. Phil. Soc. **6** (1838), 379–403.

[Bo1] R. Bott, *On the iteration of closed geodesics and the Sturm intersection theory*, Comm. Pure Appl. Math. **9** (1956), 171–206.

[Bo2] ———, *On torsion in Lie groups*, Proc. Nat. Acad. Sci. USA **40** (1954), 586–588.

[Bo3] ———, *Selected Papers of Marston Morse*, Springer Verlag, 1981.

[D1]   J. J. Duistermaat, *Fourier Integral Operators*, Lecture notes, Courant Institute, 1973.

[D2]   ———, *The light in the neighborhood of a caustic*, Sem. Bourbaki, n° 490, 19–29 (1976).

[D3]   ———, *Oscillatory integrals, Lagrange immersions, and unfolding of singularities*, Comm. Pure Appl. Math. **27** (1974), 207–281.

[DKV1] J. J. Duistermaat, J. A. C. Kolk, and V. S. Varadarajan, *Functions, flows, and oscillatory integrals on flag manifolds and conjugacy classes in real semisimple Lie groups*, Comp. Math. **49**, (1983), 309–398.

[DKV2] ———, *Spectra of compact, locally symmetric manifolds of negative curvature*, Invent. Math. **52** (1979), 29–93.

[Er]   A. Erdélyi, *Asymptotic Expansions*, Dover, New York, 1956.

[GP]   V. Guillemin and A. Pollack, *Differential Topology*, Prentice-Hall, 1974.

[GS]   V. Guillemin and S. Sternberg, *Geometric Asymptotics*, Mathematical Surveys and Monographs **14**, Amer. Math. Soc., 1977.

[H]   S. Helgason, *Differential Geometry and Symmetric Spaces*, Academic Press, New York, 1962.

[He]   R. Hermann, *Geometric aspects of potential theory in symmetric spaces*, Math. Ann. **153** (1964), 384–394.

[Hec] G. Heckman, *Projections of orbits and asymptotic behaviour of multiplicities of compact Lie groups*, Thesis, University of Utrecht, 1980.

[Hu]   G. A. Hunt, *A theorem of Élie Cartan*, Proc. Amer. Math. Soc. **7** (1956), 307–308.

[Ko]   B. Kostant, *On convexity, the Weyl group, and the Iwasawa decomposition*, Ann. École Norm. Sup. (4) **6** (1973), 413–455.

[M]   J. Martinet, *Singularities of Smooth Functions and Maps*, London Math. Soc. Lecture Notes **58**, 1982.

[Mi]   J. Milnor, *Morse Theory*, Annals of Mathematics Studies **51**, Princeton University Press, Princeton, New Jersey, 1963.

[Mo]   M. Morse, *Relations between the critical points of a real function of n independent variables*, Trans. Amer. Math. Soc. **27** (1925), 345–396.

[Mos] G. D. Mostow, *Equivariant imbeddings in Euclidean space*, Ann. of Math. **65** (1957), 432–446.

[Pa]   R. S. Palais, *Imbeddings of compact diferentiable transformation groups in orthogonal representations*, J. Math. Mech. **6** (1957), 673–678.

[Ri]   R. W. Richardson, Jr., *Conjugacy classes in Lie algebras and algebraic groups*, Ann. of Math. **86** (1967), 1–15.

[Sm]   S. Smale, *A survey of some recent developments in differential topology*, Bull. Amer. Math. Soc. **69** (1963), 131–145.

[St]   R. Steinberg, *Lectures on Chevalley Groups*, Lecture notes, Yale University, 1967.

[TK]   M. Takeuchi and S. Kobayashi, *Minimal imbeddings of R-spaces*, J. Differential Geometry **2** (1968), 203–215.

[Va]   V. S. Varadarajan, *Harmonic Analysis on Real Reductive Groups*, Lecture Notes in Mathematics **576**, Springer-Verlag, 1977.

[Wi]   E. Witten, *Supersymmetry and Morse theory*, J. Differential Geometry **17** (1982), 661–692.

Chapter 5

The Orbit Method and Unitary Representations for Reductive Lie Groups

DAVID A. VOGAN, JR. *

Massachusetts Institute of Technology

CONTENTS

Introduction

Suppose G is a real reductive algebraic group. A unitary representation of G is a Hilbert space endowed with a continuous action of G by unitary operators. It is irreducible if the Hilbert space is nonzero, but it cannot be written as a direct sum in a nontrivial G-invariant way. The orbit method

---

* Supported in part by NSF grant DMS-9402994.

pioneered by Kirillov and Kostant seeks to construct irreducible unitary representations by analogy with quantization procedures in mechanics.

A classical physical system may sometimes be modeled by a configuration space M, a manifold the points of which represent the possible positions of the bodies in the system. The corresponding phase space is the cotangent bundle T^*M, the points of which represent the possible states (positions and momenta) of the bodies in the system. In this setting, a classical observable is a function on T^*M. The quantum mechanical analog of this system is based on the Hilbert space $L^2(M)$ of square-integrable half-densities on M. A state of the quantum system is a unit vector v in $L^2(M)$; $|v|^2$ is a probability density on M, usually thought of as describing the probability of observing the quantum mechanical state in a given classical configuration. A quantum mechanical observable is an operator T on $L^2(M)$; the scalar product $\langle Tv, v \rangle$ is the expected value of the observable on the state v. Some connection between classical and quantum observables is provided by a symbol calculus; an interesting quantum observable is often a differential operator on M, and the corresponding classical observable is related to the symbol of the differential operator (a function on T^*M).

The states of some more complicated classical mechanical systems may be represented as points of a symplectic manifold X. Here, X corresponds to the cotangent bundle T^*M in the previous example, but there is no longer an underlying configuration manifold M. Classical observables are still functions on X. A quantization of this classical system is a Hilbert space $\mathcal{H}(X)$ endowed with an algebra $\mathcal{A}(X)$ of operators. Here, $\mathcal{A}(X)$ is to be some noncommutative analog of the algebra of functions on X. It is not easy to say in general what $\mathcal{H}(X)$ ought to be, except in some special classes of examples like the one in the previous paragraph.

This relationship between classical and quantum mechanics suggests a classical analog of an irreducible unitary representation of G: a homogeneous space for G, endowed with an invariant symplectic structure. Such homogeneous spaces turn out to be very close to coadjoint orbits: orbits of G on the dual of its Lie algebra. The orbit method in representation theory seeks to attach to a coadjoint orbit X (regarded as the phase space of a classical mechanical system admitting G as a group of symmetries) a Hilbert space $\mathcal{H}(X)$ (the state space for the quantized system). If this can be done in sufficiently natural way, then the action of G on X by symplectomorphisms will give rise to an action of G on $\mathcal{H}(X)$ by unitary operators, that is, to a unitary representation $\pi(X)$ of G.

In the case of reductive groups, the orbit method is fairly well understood for semisimple orbits X: That is, one can construct in a natural way a unitary representation $\pi(X)$. The purpose of this chapter is to outline this construction. Following an idea of Dixmier, we will emphasize not so much the Hilbert space $\mathcal{H}(X)$ as the algebra of operators $\mathcal{A}(X)$. There are several advantages to this approach. First, the relationship between X and $\mathcal{A}(X)$ is somewhat more elementary and direct than that between X and $\mathcal{H}(X)$. Second, $\mathcal{A}(X)$ depends only on the complexification of G. Because complex reductive groups are more or less combinatorial in nature, the operator algebras are necessarily uncomplicated.

Here is an outline of the chapter. In Section 1, we recall the definition of real reductive groups and some of their structure theory. Section 2 recalls some general facts about representation theory of Lie groups and enveloping algebras. Section 3 outlines ideas of Dixmier about ideals in enveloping algebras. At the end of Section 3, there is a moderately precise formulation of the problem that the orbit method seeks to solve.

Some of the basic notions from the method of coadjoint orbits are summarized in Section 4. Sections 5 and 6 describe the classification of coadjoint orbits for reductive groups. The orbits are all built from three special classes: hyperbolic, elliptic, and nilpotent. The semisimple orbits are those built only from hyperbolic and elliptic pieces. It turns out (although we will not explain why) that the construction of representations attached to coadjoint orbits can more or less be reduced to these three classes. The first two have been treated completely (as we will explain), but the third is understood only in special cases.

In Section 7, we construct the representations attached to hyperbolic coadjoint orbits. Roughly speaking, these are principal series and "degenerate series" representations induced from one-dimensional characters of parabolic subgroups.

The operator algebras that we construct here are all differential operator algebras. Section 8 is an elementary discussion of such operators, arranged in a form that fits well with the orbit method. The actual construction of some algebras $\mathcal{A}(X)$ appears in Section 9.

Finally, Section 10 contains a (very incomplete) outline of Zuckerman's construction of representations associated to elliptic coadjoint orbits. A complete account will appear in [KV].

Some other expositions of related material include [V4], [G], and [V6].

1   Reductive groups

A real reductive group is an abstract mathematical object of which there are relatively few examples. For this reason, the general theory is informed and guided by the examples to a remarkable extent. This is evident even in one of the simplest definitions of a real reductive group, which is based on some familiar properties of matrices.

Write $G = \mathrm{GL}(n, \mathbb{R})$ for the group of invertible $n \times n$ matrices with real entries, the *general linear group over* $\mathbb{R}$. If $g \in G$, define

$$\theta g = {}^t g^{-1}, \tag{1.1a}$$

the inverse of the transpose of g. Since transpose and inversion are both antiautomorphisms of G (meaning that ${}^t(gh) = {}^t h\, {}^t g$, for example), the map θ is an automorphism. Since transpose and inversion have order 2 and commute with each other, θ has order 2. We call θ the *Cartan involution* of G. Write $K = G^\theta$ for the subgroup of fixed points of θ. This is the group $\mathrm{O}(n)$ of $n \times n$ real orthogonal matrices, the *orthogonal group*. It is compact.

Write $\mathfrak{g}_0 = \mathfrak{gl}(n, \mathbb{R})$ for the Lie algebra of G, consisting of all $n \times n$ matrices with real entries. The differential of θ at the identity is an involutive linear automorphism of $\mathfrak{g}_0$, also written θ. We have

$$\theta X = -{}^t X \tag{1.1b}$$

for $X \in \mathfrak{g}_0$. (Using the same letter for the differential of θ is an abuse of notation, since $G \subset \mathfrak{g}_0$, but no confusion should result.) The $+1$-eigenspace of θ is

$$\mathfrak{k}_0 = n \times n \text{ skew symmetric real matrices;} \tag{1.1c}$$

it is the Lie algebra of K. The -1-eigenspace of θ is

$$\mathfrak{p}_0 = n \times n \text{ symmetric real matrices.} \tag{1.1d}$$

We have

$$\mathfrak{g}_0 = \mathfrak{k}_0 \oplus \mathfrak{p}_0, \tag{1.1e}$$

the *Cartan decomposition of* $\mathfrak{gl}(n, \mathbb{R})$.

There is a corresponding decomposition for G. For that, we need a lemma. Recall that a real matrix g is called *positive definite symmetric* if the bilinear form on $\mathbb{R}^n$ defined by $B_g(v, w) = \langle gv, w \rangle$ is positive definite and symmetric. (Here we have written $\langle \, , \, \rangle$ for the usual inner product on $\mathbb{R}^n$.)

Lemma 1.1   *i) An $n \times n$ real matrix X is symmetric if and only if there is an orthogonal basis of $\mathbb{R}^n$ consisting of eigenvectors of X.*

ii) An $n \times n$ real matrix g is positive definite symmetric if and only if there is an orthogonal basis of $\mathbb{R}^n$ consisting of eigenvectors of g with strictly positive eigenvalues.

iii) The exponential map is an analytic diffeomorphism

$$\exp: \mathfrak{p}_0 \to P$$

from symmetric matrices to positive definite symmetric matrices. The inverse

$$\log: P \to \mathfrak{p}_0$$

is also analytic.

Sketch of proof: Part (i) is standard linear algebra. For (ii), suppose first that $g \in P$. The symmetry of the form B_g makes g symmetric. The existence of an orthonormal basis of eigenvectors follows from (i). If v is a nonzero eigenvector with eigenvalue λ, then the assumed positivity of B_g gives

$$0 < B_g(v, v)/\langle v, v \rangle = (gv, v)/\langle v, v \rangle = (\lambda v, v)/\langle v, v \rangle = \lambda.$$

So the eigenvalues are positive. The converse assertion is elementary. For (iii), it follows from (i) and (ii) that $\exp$ is a one-to-one map of $\mathfrak{p}_0$ onto P. The analyticity of $\log$ is slightly more subtle; it may be proved for example by calculating the Jacobian of $\exp$. $\square$

Proposition 1.2 (Polar or Cartan decomposition for $\mathrm{GL}(n, \mathbb{R})$)   *Suppose $G = \mathrm{GL}(n, \mathbb{R})$, $K = \mathrm{O}(n)$, and $\mathfrak{p}_0$ is the space of $n \times n$ symmetric matrices. Then the map*

$$K \times \mathfrak{p}_0 \to G, \qquad (k, X) \mapsto k \exp(X),$$

is an analytic diffeomorphism of $K \times \mathfrak{p}_0$ *onto* G. *The inverse is given by*

$$g \mapsto (g \exp(-\frac{1}{2} \log((\theta g)^{-1} g)), \frac{1}{2} \log((\theta g)^{-1} g)).$$

Proof: Suppose $g \in \mathrm{GL}(n, \mathbb{R})$. Define $p = (\theta g)^{-1} g = ({}^t g) g$. The bilinear form attached to p is

$$B_p(v, w) = \langle pv, w \rangle = \langle {}^t ggv, w \rangle = \langle gv, gw \rangle,$$

which is symmetric and positive definite. So p is positive definite symmetric. By Lemma 1.1, there is a unique symmetric matrix X with $p = \exp(2X)$. Define $k = g \exp(-X)$. Then,

$${}^t kk = \exp(-X) {}^t gg \exp(-X) = \exp(-X) \exp(2X) \exp(-X) = 1,$$

and so k belongs to $\mathrm{O}(n)$. By construction $g = k \exp(X)$, which proves the surjectivity of the polar decomposition map. The construction also proves the formula for the inverse. $\square$

The polar decomposition allows one to study many structural problems about $\mathrm{GL}(n, \mathbb{R})$ in two steps: one involving the compact group $K = \mathrm{O}(n)$, and one involving the vector space $\mathfrak{p}_0$ (often regarded as a representation of K). Elie Cartan discovered that all real reductive groups share a similar property. It is so fundamental that it may be taken as the definition of the class.

Definition 1.3 A subgroup $G \subset \mathrm{GL}(n, \mathbb{R})$ is called a *linear real reductive group* if

   i) G is closed.
   ii) If X is a symmetric matrix, then $X \in \mathrm{Lie}(G)$ if and only if $\exp(X) \in G$.
   iii) G is preserved by the Cartan involution θ of $\mathrm{GL}(n, \mathbb{R})$. That is, a matrix g belongs to G if and only if ${}^t g$ belongs to G.

Suppose G is such a group. The restriction of θ to G (still denoted θ) is called the *Cartan involution* of G. Write

$$\mathfrak{g}_0 = \mathrm{Lie}(G) \subset \mathfrak{gl}(n, \mathbb{R})$$

for the Lie algebra of G; necessarily, it is preserved by the Cartan involution θ of $\mathfrak{gl}(n, \mathbb{R})$. Accordingly, we can write

$$\mathfrak{g}_0 = \mathfrak{k}_0 \oplus \mathfrak{p}_0$$

for the decomposition into $+1$ and -1 eigenspaces (the skew-symmetric and symmetric matrices in $\mathfrak{g}_0$). Put

$$K = G^\theta = G \cap \mathrm{O}(n),$$

a compact subgroup of G (by condition (i)); its Lie algebra is $\mathfrak{k}_0$.

Proposition 1.4 (Cartan decomposition for linear real reductive groups) *In the setting of Definition 1.3, the map*

$$K \times \mathfrak{p}_0 \to G, \qquad (k, X) \mapsto k \exp(X),$$

is an analytic diffeomorphism of $K \times \mathfrak{p}_0$ onto G.

Proof: Suppose $g \in G$. Write $g = k \exp(X)$ for the polar decomposition of g in $\mathrm{GL}(n, \mathbb{R})$. Then $\exp(2X) = {}^t g g$, which belongs to G by condition (iii) of Definition 1.3. By condition (ii) of the definition, X belongs to $\mathfrak{p}_0$. It follows that $k = g \exp(-X)$ belongs to $G \cap \mathrm{O}(n) = K$. This proves the surjectivity of the Cartan decomposition map. The remaining assertions follow from Proposition 1.2. $\square$

Many of the most interesting real reductive groups are nonlinear; that is, they do not appear as subgroups of $\mathrm{GL}(n, \mathbb{R})$. The following definition is broad enough for us.

Definition 1.5 A *real reductive group* is a Lie group G endowed with a continuous homomorphism $\pi \colon G \to \mathrm{GL}(n, \mathbb{R})$, subject to the following conditions:

i) $\pi(G) = \overline{G}$ is a linear real reductive group (Definition 1.3).
ii) The kernel of π is finite.

Suppose G is such a group. We use the differential of π to identify the Lie algebra $\mathfrak{g}_0$ of G with $\overline{\mathfrak{g}}_0 = \mathrm{Lie}(\overline{G}) \subset \mathfrak{gl}(n, \mathbb{R})$. Define $\mathfrak{k}_0$, $\mathfrak{p}_0$, and $\overline{K}$ as in Definition 1.3, and put

$$K = \pi^{-1}(\overline{K}).$$

By hypothesis (ii), K is a compact subgroup of G with Lie algebra $\mathfrak{k}_0$.

Proposition 1.6 (Cartan decomposition for real reductive groups)   *In the setting of Definition 1.5, the map*

$$K \times \mathfrak{p}_0 \to G, \qquad (k, X) \mapsto k \exp(X),$$

is an analytic diffeomorphism of $K \times \mathfrak{p}_0$ onto G. The map $\theta \colon G \to G$ defined by

$$\theta(k \exp(X)) = k \exp(-X)$$

is an involutory automorphism (the Cartan involution*) with $G^\theta = K$.*

Proof: Given $g \in G$, write

$$\pi(g) = \overline{k} \exp(\overline{X}),$$

with $\overline{X} \in \overline{\mathfrak{p}}_0$. The identification of $\overline{\mathfrak{p}}_0$ with $\mathfrak{p}_0$ provides an element $X \in \mathfrak{p}_0$. The element $k = g \exp(-X)$ satisfies $\pi(k) = \overline{k}$, and so belongs to K. This proves the surjectivity of the decomposition, and the analyticity of the inverse maps. Analyticity of the map to G follows from the analyticity of the exponential map and of multiplication.

For the last assertion, θ is obviously an analytic map of order two with fixed point set K. What must be shown is that $\theta(gh) = \theta(g)\theta(h)$ for all $g, h \in G$. To prove this, consider the function

$$F(g, h) = \theta(gh)\theta(h)^{-1}\theta(g)^{-1}$$

from $G \times G$ to G. We want to show that it carries $G \times G$ to the point 1. Since θ is an automorphism of $\overline{G}$, F takes values in the kernel of π. Since this is a finite set, F is constant on the connected components of $G \times G$. Obviously, F is trivial on $K \times K$; but by the Cartan decomposition, this subgroup meets every connected component of $G \times G$. $\square$

The definition of reductive group has the following "hereditary" property.

Proposition 1.7   *Let $G = K \exp(\mathfrak{p}_0)$ be the Cartan decomposition of a real reductive group with Cartan involution θ. Let H be a subgroup of G, and assume that*

i) H is closed.
ii) If $X \in \mathfrak{p}_0$, then $X \in \mathrm{Lie}(H)$ if and only if $\exp(X) \in H$.
iii) H is preserved by θ.

Then H is a real reductive group with Cartan involution $\theta|_H$.

This is almost immediate from Definition 1.5.

To construct interesting examples of reductive groups, we need a way to verify condition (ii) of Definition 1.3. Here is one.

Lemma 1.8   *Suppose A is an $n \times n$ matrix and X is a symmetric $n \times n$ matrix. Then X commutes with A if and only if $\exp(X)$ commutes with A.*

Proof: The first (resp., second) condition is equivalent to the assertion that the linear transformation A respects the decomposition of $\mathbb{R}^n$ as a direct sum of eigenspaces of X (resp., $\exp(X)$). By Lemma 1.1, these decompositions coincide. $\square$

Lemma 1.9   *Suppose G is a real reductive group and $X \in \mathfrak{p}_0$.*

i) If $g \in G$, then $\mathrm{Ad}(g)X = X$ if and only if g commutes with $\exp X$.
ii) If $Y \in \mathfrak{g}_0$, then $\mathrm{ad}(X)Y = 0$ if and only if $\mathrm{Ad}(\exp X)Y = Y$.

Proof: The two statements are very similar; we consider only the first. Write $\pi \colon G \to \overline{G} \subset \mathrm{GL}(n, \mathbb{R})$ as in Definition 1.5, and $\overline{g} = \pi(g)$, $\overline{X} = d\pi(X)$. "Only if" is trivial, so suppose g commutes with $\exp X$. Applying π, we find that $\overline{g}$ commutes with $\exp \overline{X}$. By Lemma 1.8, $\overline{g}$ commutes with $\overline{X}$; so $\mathrm{Ad}(\overline{g})\overline{X} = \overline{X}$. Consequently, $\overline{\mathrm{Ad}(g)X} = \mathrm{Ad}(\overline{g})\overline{X} = \overline{X}$. Since $d\pi$ is one-to-one, it follows that $\mathrm{Ad}(g)X = X$, as we wished to show. $\square$

Proposition 1.10   *Suppose G is a real reductive group with Cartan involution θ.*

i) If S is a θ-stable subset of G, then

$$H = Z_G(S) = \{g \in G \mid gsg^{-1} = s, \text{ all } s \in S\}$$

is a real reductive group with Cartan involution $\theta|_H$.

ii) If $\mathfrak{s}$ is a θ-stable subset of $\mathfrak{g}_0$, then

$$H = Z_G(\mathfrak{s}) = \{g \in G \mid \mathrm{Ad}(g)Y = Y, \text{ all } Y \in \mathfrak{s}\}$$

is a real reductive group with Cartan involution $\theta|_H$.

Proof: We apply Proposition 1.7. Conditions (a) and (c) are immediate, and (b) is satisfied because of Lemma 1.9. $\square$

A second way to verify condition (ii) of Definition 1.3 is using bilinear forms.

Lemma 1.11   *Suppose B is a bilinear form on $\mathbb{R}^n$, and X is a symmetric $n \times n$ matrix. Then the following two conditions are equivalent:*

1) For all $v, w \in \mathbb{R}^n$, $B(Xv, w) + B(v, Xw) = 0$.

2) For all $v, w \in \mathbb{R}^n$, $B(\exp(X)v, \exp(X)w) = B(v, w)$.

Proof: The conditions may be checked for v and w belonging to eigenspaces of X or $\exp(X)$. Write V_s for the eigenspace of X with eigenvalue $s \in \mathbb{R}$. Then (1) is equivalent to

1') Whenever $s + t \neq 0$, $B(V_s, V_t) = 0$.

Similarly, write W_λ for the eigenspace of $\exp(X)$ with eigenvalue $\lambda > 0$. Then (2) is equivalent to

2') Whenever $\lambda\mu \neq 1$, $B(W_\lambda, W_\mu) = 0$.

Since $V_t = W_{\exp t}$, conditions (1') and (2') are equivalent. $\square$

Any bilinear form B on $\mathbb{R}^n$ may be represented by a unique $n \times n$ matrix A, by the formula

$$B(v, w) = \langle Av, w \rangle; \tag{1.2a}$$

the form on the right is the standard inner product on $\mathbb{R}^n$. The symmetry group of the form is

$$G(B) = \{g \in \mathrm{GL}(n, \mathbb{R}) \mid B(gv, gw) = B(v, w), \text{ all } v, w \in \mathbb{R}^n\}$$
$$= \{g \in \mathrm{GL}(n, \mathbb{R}) \mid {}^t gAg = A\}. \tag{1.2b}$$

We are interested in constructing real reductive groups as symmetry groups of forms. In order to apply Proposition 1.7, we need to know conditions for $G(B)$ to be θ-stable. Here is a simple one.

Proposition 1.12   *Suppose B is a bilinear form on $\mathbb{R}^n$, represented by an $n \times n$ matrix A. If $A^2 = cI$ is a nonzero scalar matrix, then the symmetry group $G(B)$ is preserved by the Cartan involution θ of $\mathrm{GL}(n, \mathbb{R})$. Consequently, $G(B)$ is a real reductive group; the maximal compact subgroup $K(B)$ is the centralizer of A in $\mathrm{O}(n)$.*

Proof: Recall that $\theta g = {}^t g^{-1}$ for $g \in \mathrm{GL}(n, \mathbb{R})$. The condition in (1.2b) for g to belong to $G(B)$ is ${}^t g A g = A$. By inverting both sides, this condition is equivalent to $g^{-1} A^{-1}\, {}^t g^{-1} = A^{-1}$. The hypothesis on A says that $A^{-1} = c^{-1} A$; multiplying our condition on g by c gives $g^{-1} A\, {}^t g^{-1} = A$, or ${}^t(\theta g) A(\theta g) = A$; and this is the condition for θg to belong to $G(B)$. To see that $G(B)$ is reductive, apply Definition 1.3: condition (i) is clear, (ii) is Lemma 1.11, and we have just established (iii). $\square$

Example 1.13   Suppose p and q are nonnegative integers, and $n = p + q$. The standard quadratic form of signature (p, q) on $\mathbb{R}^n$ is the form

$$B_{p,q}(v, w) = v_1 w_1 + \cdots + v_p w_p - v_{p+1} w_{p+1} - \cdots - v_n w_n.$$

The corresponding matrix $A_{p,q}$ is diagonal with p entries equal to 1 and q equal to -1. Obviously $A_{p,q}^2 = I$, and so the symmetry group of $B_{p,q}$ is a linear real reductive group. It is called $\mathrm{O}(p, q)$, the *real orthogonal group of signature* (p, q). The maximal compact subgroup is $\mathrm{O}(p) \times \mathrm{O}(q)$.

For a second example, the standard symplectic form on $\mathbb{R}^{2n}$ is

$$\omega(v, w) = \sum_{i=1}^{n} v_i w_{n+i} - v_{n+i} w_i.$$

The corresponding matrix is often called J; it is

$$J = \begin{pmatrix} 0 & -I_n \\ I_n & 0 \end{pmatrix}.$$

This is precisely the matrix of multiplication by i in a certain identification of $\mathbb{R}^{2n}$ with $\mathbb{C}^n$. A first consequence is that $J^2 = -I$, and so the symmetry group of ω is a linear real reductive group. It is called $\mathrm{Sp}(2n, \mathbb{R})$, the *real symplectic group*. The maximal compact subgroup $K(\omega)$ is the centralizer of J in $\mathrm{O}(2n)$. The centralizer of J consists precisely of the linear transformations that are complex linear when $\mathbb{R}^{2n}$ is identified with $\mathbb{C}^n$. We may therefore identify $K(\omega)$ with the unitary group $\mathrm{U}(n)$.

A third construction of real reductive groups is by changing the base field. Let $\mathbb{F}$ denote one of the three fields $\mathbb{R}$, $\mathbb{C}$, or $\mathbb{H}$, and put $d = \dim_{\mathbb{R}} \mathbb{F}$. The standard basis of $\mathbb{F}$ (namely $\{1\}$, $\{1, i\}$, or $\{1, i, j, k\}$) provides an identification of the right vector space $\mathbb{F}^n$ with $\mathbb{R}^{nd}$. In particular, each element $z \in \mathbb{F}$ defines by right multiplication a linear transformation $\rho(z) \in \text{End}(\mathbb{R}^{nd})$. The $\mathbb{R}$-bilinear form on $\mathbb{F}^n$ defined by

$$\langle v, w \rangle = \text{Re}\left(\sum_{p=1}^{n} v_p \overline{w_p} \right)$$

is just the standard inner product on $\mathbb{R}^{nd}$; here bar denotes the standard antiautomorphism of $\mathbb{F}$ (acting by $+1$ on the first basis element of $\mathbb{F}$ and by -1 on the others). Obviously,

$$\langle \rho(z)v, w \rangle = \langle vz, w \rangle = \text{Re}\left(\sum_{p=1}^{n} v_p z \overline{w_p} \right)$$

$$= \text{Re}\left(\sum_{p=1}^{n} v_p \overline{(w_p \overline{z})} \right) = \langle v, w\overline{z} \rangle = \langle v, \rho(\overline{z})w \rangle. \qquad (1.3a)$$

So

$$\rho(\overline{z}) = {}^t\rho(z), \qquad (1.3b)$$

and

$$\rho(\overline{z}^{-1}) = \theta(\rho(z)) \qquad (1.3c)$$

for $z \in \mathbb{F}$ nonzero.

Now the algebra $\mathfrak{gl}(n, \mathbb{F})$ of $n \times n$ matrices over $\mathbb{F}$ may be identified with the algebra of $\mathbb{F}$-linear transformations of $\mathbb{F}^n$, and so with the $\mathbb{R}$-linear transformations of $\mathbb{R}^{nd}$ commuting with right multiplications by $\mathbb{F}$; that is, with the centralizer of $\rho(\mathbb{F})$ (or $\rho(\mathbb{F}^\times)$) in $\mathfrak{gl}(nd, \mathbb{R})$. In particular,

$$\text{GL}(n, \mathbb{F}) = \text{centralizer of } \rho(\mathbb{F}^\times) \text{ in } \text{GL}(nd, \mathbb{R}); \qquad (1.4)$$

this is a linear real reductive group by (1.3c) and Proposition 1.10. More generally, we have

Proposition 1.14   *Suppose* $\mathbb{F}$ *is a field of dimension* d *over* $\mathbb{R}$, *and* G *is a linear real reductive group in* $\mathrm{GL}(nd, \mathbb{R})$. *Then* $G \cap \mathrm{GL}(n, \mathbb{F})$ *(cf. (1.4)) is a linear real reductive group.*

This follows from Proposition 1.7, Lemma 1.8, and (1.3c).

Example 1.15   Suppose p and q are nonnegative integers, and $n = p + q$. The standard Hermitian form of signature (p, q) on $\mathbb{C}^n$ is the form

$$H_{p,q}(v, w) = v_1 \overline{w_1} + \cdots + v_p \overline{w_p} - v_{p+1} \overline{w_{p+1}} - \cdots - v_n \overline{w_n}.$$

The group of complex-linear transformations preserving this form is called $\mathrm{U}(p, q)$, the *unitary group of signature* (p, q). If we identify $\mathbb{C}^n$ with $\mathbb{R}^{2n}$, then the real part of $H_{p,q}$ is the quadratic form $B_{2p,2q}$ of Example 1.13. It is easy to check that a complex-linear transformation preserving the real part of $H_{p,q}$ must preserve the entire form; so it follows that

$$\mathrm{U}(p, q) = \mathrm{O}(2p, 2q) \cap \mathrm{GL}(n, \mathbb{C}).$$

By Proposition 1.14, this is a linear real reductive group. The proof shows that its maximal compact subgroup is

$$(\mathrm{O}(2p) \times \mathrm{O}(2q)) \cap \mathrm{GL}(n, \mathbb{C}) = \mathrm{U}(p) \times \mathrm{U}(q).$$

2   Representations and operator algebras

Suppose G is a topological group. A *unitary representation* of G is a pair $(\pi, \mathcal{H})$, with $\mathcal{H}$ a complex Hilbert space and $\pi \colon G \to \mathrm{U}(\mathcal{H})$ a homomorphism into the group of unitary operators on $\mathcal{H}$. These are the invertible operators preserving the inner product: the assumption is

$$\langle \pi(g)v, \pi(g)w \rangle = \langle v, w \rangle \qquad (v, w \in \mathcal{H}, \ g \in G). \tag{2.1a}$$

Often, it is convenient to formulate this condition as

$$\langle \pi(g)v, w \rangle = \langle v, \pi(g^{-1})w \rangle \qquad (v, w \in \mathcal{H}, \ g \in G), \tag{2.1b}$$

or simply as $\pi(g)^* = \pi(g^{-1})$. We assume also that π is weakly continuous; that is, that the map

$$G \times \mathcal{H} \to \mathcal{H}, \qquad (g, v) \mapsto \pi(g)v, \qquad (2.1c)$$

is continuous. An *invariant subspace* of $\mathcal{H}$ is a closed subspace $\mathcal{H}_0 \subset \mathcal{H}$ that is preserved by all the operators $\pi(g)$. In this case, the restricted operators define a unitary representation $(\pi_0, \mathcal{H}_0)$ of G. The orthogonal complement $\mathcal{H}_1$ of $\mathcal{H}_0$ is a second invariant subspace, and $\mathcal{H} = \mathcal{H}_0 \oplus \mathcal{H}_1$; we write $\pi = \pi_0 \oplus \pi_1$ accordingly. We say that π is *irreducible* if $\mathcal{H} \neq 0$, and the only invariant subspaces of $\mathcal{H}$ are 0 and $\mathcal{H}$.

Suppose $(\pi_1, \mathcal{H}_1)$ and $(\pi_2, \mathcal{H}_2)$ are unitary representations of G. An *intertwining operator* from π_1 to π_2 is a continuous linear map T from $\mathcal{H}_1$ to $\mathcal{H}_2$ with the property that

$$T\pi_1(g) = \pi_2(g)T \qquad (g \in G).$$

The space of all intertwining operators is written $\mathrm{Hom}_G(\mathcal{H}_1, \mathcal{H}_2)$. Forming the adjoint T^* defines a conjugate-linear isomorphism

$$\mathrm{Hom}_G(\mathcal{H}_1, \mathcal{H}_2) \simeq \mathrm{Hom}_G(\mathcal{H}_2, \mathcal{H}_1).$$

If some intertwining operator is a unitary isomorphism, then we say that π_1 and π_2 are equivalent. We write $\widehat{G}$ for the set of equivalence classes of irreducible unitary representations of G.

The fundamental problem of abstract harmonic analysis is this: to decompose into irreducible representations an arbitrary unitary representation of G. This problem has a fairly good abstract answer for a large class of groups including the real reductive Lie groups. The case of compact groups will be recalled at (2.4). The abstract answer is of little help in finding explicit decompositions for particular interesting representations, however, and this remains an active area of research.

A second basic problem in abstract harmonic analysis is the determination of the set $\widehat{G}$. If this can be accomplished, a question about G may sometimes be analyzed along the following lines. First, the question is made into one about some unitary representation $(\pi, \mathcal{H})$ of G (perhaps on square-integrable functions on a homogeneous space, for example). Next, the representation π is decomposed into irreducible representations. At the same time, one tries to make the original question into a family of questions about the irreducible constituents of π. Finally, this family of questions is answered using our knowledge of all irreducible representations. Of course each step of this program is fraught with peril; but it has been carried out with some success in a variety of cases. A striking example quite close to the topic of this chapter is provided by the book [BW], which analyzes the cohomology of cocompact discrete subgroups of real reductive Lie groups.

To carry out the program just sketched, one needs to understand not *all* irreducible unitary representations, rather just those appearing in whatever representation π one first constructs. With this in mind, we formulate

Problem 2.1   *Suppose G is a real reductive Lie group. Construct a family of irreducible unitary representations of G sufficiently large to decompose many interesting unitary representations into irreducibles.*

Of course this "problem" is not at all well defined, because of the phrase "many interesting." We will make it more precise, and outline what is known about solving it.

To explain the approach we will adopt, it is helpful to begin with the case of a finite group G. The *group algebra* $\mathbb{C}[G]$ is the algebra over $\mathbb{C}$ with basis $\{\delta_g \mid g \in G\}$, and multiplication table

$$\delta_g \delta_h = \delta_{gh} \qquad (g, h \in G). \qquad (2.2a)$$

This algebra has a conjugate-linear antiautomorphism $*$, defined by

$$\left(\sum_{g \in G} a_g \delta_g \right)^* = \sum_{g \in G} \overline{a_g} \delta_{g^{-1}}. \qquad (2.2b)$$

Then, a unitary representation $(\pi, \mathcal{H})$ of G defines an associative algebra homomorphism

$$\pi \colon \mathbb{C}[G] \to \mathrm{End}(\mathcal{H}), \qquad \pi\left(\sum a_g \delta_g \right) = \sum a_g \pi(g), \qquad (2.2c)$$

satisfying

$$\pi(a^*) = \pi(a)^*. \qquad (2.2d)$$

(We say that π is a $*$-*homomorphism.*) Conversely, any $*$-homomorphism π from $\mathbb{C}[G]$ to $\mathrm{End}(\mathcal{H})$ arises from a unitary representation of G on $\mathcal{H}$.

Proposition 2.2   *Suppose G is a finite group. Then the equivalence classes of irreducible unitary representations π of G are in one-to-one correspondence with the maximal two-sided ideals I_π in $\mathbb{C}[G]$. This bijection has the following properties.*

i) If $(\pi, \mathcal{H}_\pi)$ is an irreducible unitary representation, then I_π is the kernel of the algebra homomorphism $\pi\colon \mathbb{C}[G] \to \mathrm{End}(\mathcal{H}_\pi)$ of (2.2c).

ii) Suppose I_π is a maximal two-sided ideal in $\mathbb{C}[G]$. Choose a maximal left ideal $J_\pi \supset I_\pi$, and define $V_\pi = \mathbb{C}[G]/J_\pi$. Then V_π is isomorphic to $\mathcal{H}_\pi$ as a module for $\mathbb{C}[G]$.

This is well known and not very hard to prove. It is not often used directly to determine the irreducible representations of G, because $\mathbb{C}[G]$ is such a complicated algebra. Nevertheless, the proposition suggests a way to approach the representation theory of a group G. First, one should find an associative algebra $A(G)$ the module theory of which is related to the representation theory of G. Next, one should study the ideals in $A(G)$.

For a locally compact group G, a natural candidate for the algebra $A(G)$ is the convolution algebra $L^1(G)$ (with respect to a left Haar measure dg). Given a unitary representation $(\pi, \mathcal{H})$ and $f \in L^1(G)$, one defines $\pi(f) = \int_G f(g)\pi(g)dg$. Then one can make an excellent correspondence between unitary representations of G and appropriate modules for $L^1(G)$. These modules all extend to a certain completion of $L^1(G)$, called $C^*(G)$. For type I groups G (which include the real reductive Lie groups), there is a bijection between irreducible unitary representations of G and primitive ideals in $C^*(G)$.

Now this bijection is a powerful technical tool — for example, it is at the heart of the abstract theory of decomposition into irreducible representations. But as a way to describe irreducible representations explicitly, it is (like Proposition 2.2) of little value. The algebra $C^*(G)$ is too complicated. We would like instead an algebra the module theory of which is perhaps not quite so perfectly related to the unitary representation theory of G, but with an ideal theory we can hope to study directly.

Suppose now that G is a Lie group. Write

$$\mathfrak{g}_0 = \mathrm{Lie}(G), \quad \mathfrak{g} = \mathfrak{g}_0 \otimes_{\mathbb{R}} \mathbb{C}, \quad U(\mathfrak{g}) = \text{universal enveloping algebra of } \mathfrak{g}.$$
$$(2.3)$$

One of the central ideas of Lie theory is that these objects can be used to translate problems about Lie groups into linear algebra. Here is an example, along the lines of Proposition 2.2. Recall that a finite dimensional (nonunitary) representation (π, V_π) of G is just a continuous homomorphism π of G into $\mathrm{GL}(V_\pi)$, the general linear group of some finite dimensional complex vector space V_π.

Proposition 2.3 *Suppose G is a connected, simply connected Lie group. Then the finite dimensional irreducible representations π of G are in one-to-one correspondence with the two-sided maximal ideals I_π of finite codimension in $U(\mathfrak{g})$. This bijection has the following properties.*

i) *Suppose (π, V_π) is a finite dimensional irreducible representation of G. Write $\pi\colon \mathfrak{g}_0 \to \mathfrak{gl}(V_\pi)$ for the differential of π (a Lie algebra representation), and $\pi\colon U(\mathfrak{g}) \to \mathrm{End}(V_\pi)$ for its extension to $U(\mathfrak{g})$ (a homomorphism of associative algebras). Then I_π is the kernel of $\pi\colon U(\mathfrak{g}) \to \mathrm{End}(V_\pi)$.*

ii) *Suppose I_π is a maximal two-sided ideal of finite codimension in $U(\mathfrak{g})$. Choose a maximal left ideal $J_\pi \supset I_\pi$, and define $V_\pi = U(\mathfrak{g})/J_\pi$. Then the action of $\mathfrak{g}_0$ on V_π by left multiplication in $U(\mathfrak{g})$ is a Lie algebra representation; the corresponding group representation of G is isomorphic to π.*

Like Proposition 2.2, this is an elementary result. Because $U(\mathfrak{g})$ is a relatively uncomplicated algebra, Proposition 2.3 is sometimes even directly useful for studying finite dimensional representations.

We turn now to the problem of extending Proposition 2.3 to cover unitary representations. A continuous homomorphism between Lie groups is automatically smooth (and even analytic); this is used in Proposition 2.3 to guarantee the existence of the Lie algebra representation $\pi\colon \mathfrak{g}_0 \to \mathfrak{gl}(V_\pi)$. For infinite dimensional representations the situation is more complicated.

Definition 2.4 Suppose G is a Lie group, and $(\pi, \mathcal{H}_\pi)$ is a unitary representation of G. The vector $v \in \mathcal{H}_\pi$ is called *smooth* (resp., *analytic*) if the map

$$G \to \mathcal{H}_\pi, \qquad g \mapsto \pi(g)v,$$

is smooth (resp., analytic). Write $\mathcal{H}_\pi^\infty$ (resp., $\mathcal{H}_\pi^\omega$) for the set of smooth (resp., analytic) vectors in $\mathcal{H}_\pi$. These are G-invariant linear subspaces of $\mathcal{H}_\pi$, but they are usually not closed.

Suppose $X \in \mathfrak{g}_0$. The *differential of* π *at* X is the linear transformation $\pi(X)$ of $\mathcal{H}_\pi^\infty$ defined by

$$\pi(X)v = \lim_{t \to 0}(1/t)(\pi(\exp tX)v - v).$$

Lemma 2.5 *The differential of* π *is a Lie algebra representation of* $\mathfrak{g}_0$ *on* $\mathcal{H}_\pi^\infty$. *It therefore defines a homomorphism of associative algebras* $\pi \colon U(\mathfrak{g}) \to \mathrm{End}(\mathcal{H}_\pi^\infty)$. *All of the resulting operators preserve the subspace* $\mathcal{H}_\pi^\omega \subset \mathcal{H}_\pi^\infty$.

It is not very difficult to show that $\mathcal{H}_\pi^\infty$ is dense in $\mathcal{H}_\pi$. It is also true that $\mathcal{H}_\pi^\omega$ is dense in $\mathcal{H}_\pi$; this much deeper result, which we will use in the proof of Proposition 2.12, is due to Nelson. In any case, we can now construct one of the maps of Proposition 2.3 for unitary representations.

Definition 2.6 Suppose G is a Lie group, and $(\pi, \mathcal{H}_\pi)$ is a unitary representation of G. The *annihilator of* π *in* $U(\mathfrak{g})$ is the kernel of the homomorphism of Lemma 2.5:

$$\mathrm{Ann}(\pi) = \ker(\pi \colon U(\mathfrak{g}) \to \mathrm{End}(\mathcal{H}_\pi^\infty)).$$

This is a two-sided ideal in $U(\mathfrak{g})$; it is equal to $U(\mathfrak{g})$ if and only if $\mathcal{H}_\pi = 0$.

As some assurance that this definition is well behaved, here is an elementary lemma.

Lemma 2.7 *In the setting of Definition 2.6, suppose* $W \subset \mathcal{H}_\pi^\infty$ *is any subspace that is dense in* $\mathcal{H}_\pi$. *Then*

$$\mathrm{Ann}(\pi) = \mathrm{Ann}(W) = \{u \in U(\mathfrak{g}) \mid \pi(u)w = 0, \text{ all } w \in W\}.$$

So we may compute the annihilator on analytic or (when these are defined) K-finite vectors.

What properties can we expect of the annihilator of an irreducible unitary representation? Fairly simple examples show that it need not be a maximal ideal in general. Ring theory has a natural suggestion to offer.

Definition 2.8   Suppose R is a ring with unit element. A left R-module M is called *simple* if it is not zero, and its only submodules are 0 and M. A two-sided ideal $I \subset R$ is called *(left) primitive* if there is a simple R-module M such that

$$I = \mathrm{Ann}(M) = \{r \in R \,|\, rM = 0\}.$$

Theorem 2.9 (Dixmier [D])   *Suppose G is a connected Lie group and $(\pi, \mathcal{H}_\pi)$ is an irreducible unitary representation of G. Then $\mathrm{Ann}(\pi)$ is a primitive ideal in $U(\mathfrak{g})$.*

The reason this is not obvious is that $\mathcal{H}_\pi^\infty$ is not a simple $U(\mathfrak{g})$ module (unless π is finite dimensional). For reductive G, we will see in Theorem 2.17 how Harish-Chandra constructs a simple $U(\mathfrak{g})$-submodule of $\mathcal{H}_\pi^\infty$ that is dense in $\mathcal{H}_\pi$. In light of Lemma 2.7, this proves Theorem 2.9 in the reductive case. Dixmier's proof in general involves several important ideas, so we will outline the easiest part of it.

Lemma 2.10   *Suppose G is a connected Lie group, $(\pi, \mathcal{H}_\pi)$ is a unitary representation of G, and $V \subset \mathcal{H}_\pi^\omega$ is a $U(\mathfrak{g})$-invariant subspace. Then the closure $\mathcal{H}_0$ of V in $\mathcal{H}_\pi$ is a G-invariant subspace.*

Proof: For any subset S of $\mathcal{H}_\pi$, define

$$S^\perp = \{w \in \mathcal{H}_\pi \,|\, \langle w, s \rangle = 0, \text{ all } s \in S\}.$$

The closure of any subspace S may be characterized as $(S^\perp)^\perp$, and so it suffices to show that $V^\perp$ is G-invariant. For this, fix $w \in V^\perp$ and $v \in V$; we must show that the function $f(g) = \langle \pi(g)w, v \rangle$ is identically zero. Because π is unitary, $f(g) = \langle w, \pi(g^{-1})v \rangle$. Because v is assumed to be an analytic vector, the function f is analytic on G. Because G is connected, it therefore suffices to show that all derivatives of f vanish at the identity on G. A typical derivative of f at the identity is $\langle w, \pi(u)v \rangle$, with $u \in U(\mathfrak{g})$. Since V is assumed to be $U(\mathfrak{g})$-invariant, $\pi(u)v \in V$. Since $w \in V^\perp$, the derivative vanishes, as we wished to show. $\square$

(I am grateful to P. E. Paradan for showing me this elegant argument.)

Lemma 2.11.   *Suppose G is a connected Lie group and $(\pi, \mathcal{H}_\pi)$ is an irreducible unitary representation. Then any nonzero $U(\mathfrak{g})$-invariant subspace $V \subset \mathcal{H}_\pi^\omega$ is dense in $\mathcal{H}_\pi$. Consequently, $\mathrm{Ann}(V) = \mathrm{Ann}(\pi)$.*

(The last assertion uses Lemma 2.7.)

Proposition 2.12.   *Suppose G us a connected Lie group and $(\pi, \mathcal{H}_\pi^\omega)$ is an irreducible unitary representation. Then $\mathrm{Ann}(\pi)$ is a prime ideal in $U(\mathfrak{g})$.*

Proof: Recall that a two-sided ideal I in a (possibly noncommutative) ring R is called *prime* if whenever J_1 and J_2 are two-sided ideals with $J_1 J_2 \subset I$, then either $J_1 \subset I$ or $J_2 \subset I$. So suppose J_1 and J_2 are ideals in $U(\mathfrak{g})$ with $J_1 J_2 \subset \mathrm{Ann}(\pi)$, but $J_2 \not\subset \mathrm{Ann}(\pi)$. Then $J_2 \mathcal{H}_\pi^\omega = V$ is a nonzero $U(\mathfrak{g})$-invariant subspace of $\mathcal{H}_\pi^\omega$. By Lemma 2.11, $\mathrm{Ann}(V) = \mathrm{Ann}(\pi)$. But

$$J_1 V = J_1 J_2 \mathcal{H}_\pi^\omega \subset \mathrm{Ann}(\pi)\mathcal{H}_\pi^\omega = 0;$$

so $J_1 \subset \mathrm{Ann}(\pi)$, as we wished to show. $\square$

Dixmier completes the proof of Theorem 2.9 using

Theorem 2.13 ([D])   *Suppose $I \subset U(\mathfrak{g})$ is a prime ideal. Then I is primitive if and only if the center of the ring of fractions of $U(\mathfrak{g})/I$ is $\mathbb{C}$.*

We omit the details.

We now have a reasonable analog of Proposition 2.3(i): a map from irreducible unitary representations to primitive ideals in $U(\mathfrak{g})$. To get an analog of (ii) (a corresponding parametrization of representations), we need to specialize to real reductive groups. Let us first recall the structure of unitary representations for compact groups.

Suppose K is a compact topological group. Then, every irreducible unitary representation of K is finite dimensional. Fix a model (δ, V_δ) for each equivalence class in $\widehat{K}$. If $(\pi, \mathcal{H}_\pi)$ is an arbitrary unitary representation of K, define

$$\mathcal{H}_\pi^\delta = \mathrm{Hom}_K(V_\delta, \mathcal{H}_\pi). \tag{2.4a}$$

We make $\mathcal{H}_\pi^\delta$ into a Hilbert space as follows. If T and S belong to $\mathcal{H}_\pi^\delta$, then $S^* T$ is a map from V_δ to V_δ commuting with the action of K. By Schur's lemma, it is a scalar operator λI; and we define $\langle T, S \rangle = \lambda$. An equivalent formulation is

$$\langle Tv, Sw \rangle_{\mathcal{H}_\pi} = \langle T, S \rangle_{\mathcal{H}_\pi^\delta} \langle v, w \rangle_{V_\delta} \tag{2.4b}$$

for $v, w \in V_\delta$. Now we can form the Hilbert space tensor product $\mathcal{H}_\pi^\delta \otimes V_\delta$. (Because V_δ is finite dimensional, it coincides with the algebraic tensor product.) There is a natural map

$$\mathcal{H}_\pi^\delta \otimes V_\delta \to \mathcal{H}_\pi, \qquad T \otimes v \mapsto Tv; \tag{2.4c}$$

and (2.4b) guarantees that this map preserves inner products. It is therefore an isomorphism onto its image $\mathcal{H}_\pi(\delta)$, the δ-*isotypic subspace of* $\mathcal{H}_\pi$. This is the largest subspace of $\mathcal{H}_\pi$ on which K acts by a sum of copies of δ. By Schur's lemma again, $\mathcal{H}_\pi(\delta)$ and $\mathcal{H}_\pi(\delta')$ are orthogonal whenever δ and δ' are inequivalent. Consequently,

$$\mathcal{H}_\pi \simeq \widehat{\bigoplus_{\delta \in \widehat{K}}} \mathcal{H}_\pi(\delta) \simeq \widehat{\bigoplus_{\delta \in \widehat{K}}} \mathcal{H}_\pi^\delta \otimes V_\delta, \tag{2.4d}$$

the direct sums being Hilbert space direct sums.

A vector $v \in \mathcal{H}_\pi$ is called K-*finite* if it is contained in a finite dimensional K-invariant subspace. We write $\mathcal{H}_\pi^K$ for the space of K-finite vectors. Using (2.4d), we find that

$$\mathcal{H}_\pi^K \simeq \bigoplus_{\delta \in \widehat{K}} \mathcal{H}_\pi(\delta) \simeq \bigoplus_{\delta \in \widehat{K}} \mathcal{H}_\pi^\delta \otimes V_\delta, \tag{2.4e}$$

the direct sums now being algebraic rather than Hilbert space. The representation π is called *admissible for* K if all of the spaces $\mathcal{H}_\pi(\delta)$ are finite dimensional; that is, if every irreducible representation of K has finite multiplicity in π.

Suppose now that G is a real reductive Lie group with maximal compact subgroup K. A unitary representation $(\pi, \mathcal{H}_\pi)$ is called *admissible* if it is admissible for K.

Theorem 2.14 (Harish-Chandra [HC1])   *Every irreducible unitary representation of a real reductive Lie group G is admissible.*

This is a rather difficult result, relying on a deep study of the adjoint action of K on $U(\mathfrak{g})$.

Admissible unitary representations have an excellent algebraic description.

Theorem 2.15 (Harish-Chandra [HC1]) *Suppose $(\pi, \mathcal{H}_\pi)$ is an admissible unitary representation of a real reductive Lie group G. Write $\mathcal{H}_\pi^K$ for the space of K-finite vectors in $\mathcal{H}_\pi$.*

i) $\mathcal{H}_\pi^K$ is a $U(\mathfrak{g})$-invariant subspace of the analytic vectors $\mathcal{H}_\pi^\omega$. In particular, $\mathcal{H}_\pi^K$ carries representations of the group K and the Lie algebra $\mathfrak{g}$.

ii) There is a bijection between closed G-invariant subspaces of $\mathcal{H}_\pi$ and arbitrary $(\mathfrak{g}, K)$-invariant subspaces of $\mathcal{H}_\pi^K$. The correspondence from left to right sends a closed subspace W to W^K (the K-finite vectors in W); from right to left it sends V to $\overline{V}$ (the closure of V in $\mathcal{H}_\pi$).

To complete this circle of ideas, we will describe formally the algebraic objects arising in Theorem 2.15. (The definition is taken from [L].)

Definition 2.16 Suppose G is a real reductive Lie group with maximal compact subgroup K. A $(\mathfrak{g}, K)$-*module* is a complex vector space V endowed with representations of the Lie algebra $\mathfrak{g}$ and the group K, subject to the following conditions.

  i) The action of K is locally finite and smooth. That is, every $v \in V$ belongs to a finite dimensional K-invariant subspace $F \subset V$, and the action of K on F is smooth.

 ii) The differential of the action of K (which makes sense by (i)) is equal to the action of $\mathfrak{k}_0 = \operatorname{Lie}(K) \subset \mathfrak{g}$.

iii) For $k \in K$, $v \in V$, and $X \in \mathfrak{g}$, we have

$$k \cdot (X \cdot v) = \operatorname{Ad}(k)(X) \cdot (k \cdot v).$$

For k in the identity component K_0 of K, condition (iii) is a consequence of (i) and (ii). We may therefore omit condition (iii) when K (or, equivalently, G) is connected.

A $(\mathfrak{g}, K)$-submodule of V is a complex subspace W invariant under the representations of $\mathfrak{g}$ and K. (By conditions (i) and (ii), invariance under K_0 follows from invariance under $\mathfrak{g}$. If K is connected, a $(\mathfrak{g}, K)$-submodule is therefore just a $\mathfrak{g}$-submodule.) We say that V is *irreducible* if it is not zero, and the only submodules are 0 and V. Finally, we say that V is *unitary* if it is endowed with a positive definite Hermitian form $\langle\,,\,\rangle$ satisfying

$$\langle k \cdot v, k \cdot w \rangle = \langle v, w \rangle, \qquad \langle X \cdot v, w \rangle + \langle v, X \cdot w \rangle = 0,$$

for $v, w \in V$, $k \in K$, and $X \in \mathfrak{g}_0$.

Theorem 2.15(i) guarantees that the space $\mathcal{H}_\pi^K$ of K-finite vectors in an admissible unitary representation is a unitary $(\mathfrak{g}, K)$-module; it is called the *Harish-Chandra module of* π. Theorems 2.14 and 2.15(ii) say that $\mathcal{H}_\pi^K$ is irreducible as a $(\mathfrak{g}, K)$-module whenever π is irreducible. Harish-Chandra's last basic result is a converse.

Theorem 2.17 (Harish-Chandra [HC1])   *Suppose G is a real reductive Lie group. The map $\pi \mapsto \mathcal{H}_\pi^K$ is a bijection from (equivalence classes of) irreducible unitary representations of G onto (equivalence classes of) irreducible unitary $(\mathfrak{g}, K)$-modules.*

When G is connected, we have seen that an irreducible $(\mathfrak{g}, K)$-module is irreducible as a representation of $\mathfrak{g}$. This theorem may therefore be viewed as an infinite dimensional analog of Proposition 2.3(ii). It provides a construction of irreducible unitary representations of G from certain (very special) irreducible $U(\mathfrak{g})$-modules.

We can now refine slightly Problem 2.1.

Problem 2.18   *Suppose G is a real reductive Lie group; assume for simplicity that G is connected.*

 i) *Construct a family of interesting primitive ideals $I \subset U(\mathfrak{g})$.*

 ii) *For each primitive ideal I as in (i), construct a finite set of irreducible unitary representations π of G, satisfying $\mathrm{Ann}(\pi) = I$ (Definition 2.6).*

This formulation is still imperfect, but it begins to reflect what we will actually do. The constructions in (ii) will generally take place in three steps. First, we will construct some $U(\mathfrak{g})$-modules W with annihilator equal to I. This step is usually fairly easy; some possibilities for W will often be suggested by the construction of I in (i). Next, we will construct from each W a $(\mathfrak{g}, K)$-module V, still annihilated by I. The principle of this construction (due to Zuckerman) is simple and elegant, but analyzing it in

detail can be quite difficult. (One minor point is that the annihilator of
V may be strictly larger than I.) Finally, we will apply Harish-Chandra's
Theorem 2.17 to get a unitary representation of G.

3   Primitive ideals and Dixmier algebras

In this section we will consider more carefully part (i) of Problem
2.18: the construction of a family of primitive ideals related to unitary
representations. A more detailed account may be found in [V3] and [V7].

The first point is that not every primitive ideal can be the annihilator
of a unitary representation. Suppose for a moment that G is a connected
noncompact simple Lie group. Then G has a large family of irreducible
finite dimensional representations (parametrized by a cone in a lattice of
dimension equal to the rank of G). By Proposition 2.3, it follows that
$U(\mathfrak{g})$ has a large family of maximal ideals of finite codimension. On the
other hand, a *unitary* finite dimensional representation is a homomorphism
$\pi\colon G \to \mathrm{U}(n)$. Because G is noncompact and simple, such a homomorphism
must be trivial. This proves

Lemma 3.1   *Suppose G is a connected noncompact simple Lie group,
and suppose $I \subset U(\mathfrak{g})$ is a maximal ideal of finite codimension. Then I is
the annihilator of a unitary representation if and only if $I = \mathfrak{g}U(\mathfrak{g})$ (the
augmentation ideal).*

What distinguishes the augmentation ideal among all maximal ideals
of finite codimension? If I is such an ideal, then the Wedderburn theorem
guarantees that

$$U(\mathfrak{g})/I \simeq M_n(\mathbb{C}), \tag{3.1}$$

the algebra of $n \times n$ matrices. Since $\mathfrak{g}$ is semisimple, $n = 1$ occurs only for
the augmentation ideal. So we may ask what distinguishes 1×1 matrices
from larger ones. One answer is the absence of zero divisors.

Definition 3.2   Suppose I is a two-sided ideal in a ring R. We say that
the ideal I (or the quotient ring R/I) is *completely prime* if whenever a
and b are elements of R with $ab \in I$, then either $a \in I$ or $b \in I$.

It is easy to check that a completely prime ideal is prime. (The definition of prime was included in the proof of Proposition 2.12.) Here is

some further evidence of the connection between completely prime ideals and unitary representations.

Proposition 3.3 ([V3, Proposition 7.12]) *Suppose G is a connected complex reductive Lie group, and $\pi \in \widehat{G}$ is an irreducible unitary representation. Then the annihilator $I_\pi \subset U(\mathfrak{g})$ (Definition 2.6) is completely prime.*

The proof is very easy, requiring no structural information about π. Exactly the same result is true for $G = \mathrm{GL}(n, \mathbb{R})$; but in this case the proof requires a complete and detailed knowledge of $\widehat{G}$. Any hopes of further generalization are dashed on the rocks of $G = \mathrm{SU}(2)$. This group has an irreducible unitary representation of each dimension $n > 0$; and (3.1) guarantees that the corresponding primitive ideal is completely prime only for $n = 1$. (More subtle examples are available for noncompact simple groups as well. Suppose π is a holomorphic discrete series representation of $\mathrm{Sp}(4, \mathbb{R})$ with Harish-Chandra parameter $\lambda = (\lambda_1, \lambda_2)$. This means that $\lambda_1 > \lambda_2 > 0$ are positive integers. It turns out that $\mathrm{Ann}(\pi)$ is completely prime if and only if $\lambda_1 - \lambda_2 = 1$.)

We can find a way out of this disappointment by looking carefully at the example of $G = \mathrm{SU}(2)$. This group acts holomorphically on the Riemann sphere $\mathbb{C}P^1$. The n-dimensional irreducible representation π_n arises as the space of holomorphic sections of a certain holomorphic line bundle $\mathcal{L}_n \to \mathbb{C}P^1$. Define $D(\mathbb{C}P^1)_n$ to be the algebra of holomorphic differential operators on sections of $\mathcal{L}_n$. This is an algebra of "twisted differential operators" on $\mathbb{C}P^1$. (We will return to a more detailed and general discussion of twisted differential operator algebras in Section 9.) The action of G on $\mathcal{L}_n$ defines a homomorphism of associative algebras

$$\phi_n \colon U(\mathfrak{g}) \to D(\mathbb{C}P^1)_n. \tag{3.2}$$

Write I_n for the kernel of ϕ_n.

Proposition 3.4   *The ideal I_n is a completely prime primitive ideal in $U(\mathfrak{g})$, contained in the annihilator $\mathrm{Ann}(\pi_n)$.*

Sketch of proof: That I_n is completely prime follows from the absence of zero divisors in the twisted differential operator algebra $D(\mathbb{C}P^1)_n$. That it is contained in $\mathrm{Ann}(\pi_n)$ follows from the realization of π_n as holomorphic sections of $\mathcal{L}_n$. $\square$

The lesson to be drawn from this example is that an interesting unitary representation π may appear naturally as a module for some completely prime quotient $U(\mathfrak{g})/I$, even if $\mathrm{Ann}(\pi)$ properly contains I. It is easy to modify Problem 2.18 in accordance with this lesson; we simply weaken the condition in (ii) to $\mathrm{Ann}(\pi) \supset I$.

There is a hint here of a second lesson as well. The realization of π_n on holomorphic sections of $\mathcal{L}_n$ exhibits π_n as a module not only for $U(\mathfrak{g})/I_n$, but also for the full differential operator algebra $D(\mathbb{C}P^1)_n$. In this example the homomorphisms ϕ_n are all surjective, so that there is no difference between $U(\mathfrak{g})/I_n$ and $D(\mathbb{C}P^1)_n$. When we treat general reductive groups, however, we will encounter homomorphisms

$$\phi_\lambda \colon U(\mathfrak{g}) \to D(X)_\lambda. \tag{3.3}$$

Here, X is a "partial flag variety" for $\mathfrak{g}$ (a quotient of a complex reductive group $G_\mathbb{C}$ by a parabolic subgroup $Q_\mathbb{C}$); λ is a character of the Lie algebra $\mathfrak{q}$; and $D(X)_\lambda$ is a twisted differential operator algebra on X. In this setting, the homomorphism ϕ_λ is usually but not always surjective. Perhaps the simplest example when ϕ_λ is not surjective has $\mathfrak{g} = \mathfrak{sp}(4,\mathbb{C})$, $X = \mathbb{C}P^3$ (the variety of lines in the natural four-dimensional representation of $\mathfrak{g}$), and $D(X)_\lambda$ the algebra of differential operators on "half forms" on X. (The top exterior power of the cotangent bundle of X has a well defined square root $\mathcal{L} \to X$ in this example; $D(X)_\lambda$ is the algebra of holomorphic differential operators on sections of $\mathcal{L}$.) In any case, we will construct modules for $D(X)_\lambda$, and not just for $U(\mathfrak{g})/\ker \phi_\lambda$. In order to accommodate this extra structure in something like Problem 2.18, we need a definition.

Definition 3.5 Suppose $G_\mathbb{C}$ is a complex reductive algebraic group with Lie algebra $\mathfrak{g}$. A *Dixmier algebra* for $G_\mathbb{C}$ is a pair (A, ϕ) satisfying the following conditions.

   i) A is an algebra over $\mathbb{C}$, equipped with an algebraic action of $G_\mathbb{C}$ by algebra automorphisms $\mathrm{Ad}(g)$.
   ii) The map $\phi \colon U(\mathfrak{g}) \to A$ is an algebra homomorphism, respecting the adjoint actions of $G_\mathbb{C}$ on $U(\mathfrak{g})$ and A. The differential ad of the adjoint action of $G_\mathbb{C}$ on A is the difference of the left and right actions of $\mathfrak{g}$ defined by ϕ:

$$\mathrm{ad}(X)(a) = \phi(X)a - a\phi(X) \qquad (X \in \mathfrak{g},\ a \in A).$$

iii) A is finitely generated as a $U(\mathfrak{g})$-module.

iv) Each irreducible $G_{\mathbb{C}}$-module occurs at most finitely often in the adjoint action of $G_{\mathbb{C}}$ on A.

We say that the Dixmier algebra is *completely prime* if A is a completely prime algebra. This immediately implies that the kernel I of ϕ is a completely prime ideal in $U(\mathfrak{g})$, and one can show (using condition (iv)) that I must also be primitive.

If I is a primitive ideal in $U(\mathfrak{g})$, then $U(\mathfrak{g})/I$ is a Dixmier algebra for any connected $G_{\mathbb{C}}$. In the setting of (3.3), $D(X)_\lambda$ is a completely prime Dixmier algebra for $G_{\mathbb{C}}$. The adjoint action arises from the action of $G_{\mathbb{C}}$ on X, by change of variable in the differential operators.

It is an idea of Dixmier that for a connected complex algebraic group $G_{\mathbb{C}}$ there should be a close connection between completely prime primitive ideals in $U(\mathfrak{g})$ and orbits of $G_{\mathbb{C}}$ on $\mathfrak{g}^*$. Borho, Joseph, and others found that for reductive $G_{\mathbb{C}}$, this "close connection" cannot be a reasonable bijection. The goal of [V3] was to find a geometric description of all completely prime primitive ideals. That attempt failed, as was shown by work of Mc-Govern. Here is a weaker statement (taken from [V7, Conjecture 2.3]) that appears to be consistent with everything we now understand about Dixmier algebras.

Conjecture 3.6 *Suppose $G_{\mathbb{C}}$ is a connected complex reductive algebraic group. Let $\mathcal{O} \subset \mathfrak{g}^*$ be a coadjoint orbit for $G_{\mathbb{C}}$, and let $\tilde{\mathcal{O}} \to \mathcal{O}$ be a connected covering on which $G_{\mathbb{C}}$ acts compatibly. Write $R(\tilde{\mathcal{O}})$ for the algebra of regular functions on $\tilde{\mathcal{O}}$; this algebra carries a natural algebraic action of $G_{\mathbb{C}}$ by algebra automorphisms. Attached to $\tilde{\mathcal{O}}$ there should be a completely prime Dixmier algebra $(A(\tilde{\mathcal{O}}), \phi(\tilde{\mathcal{O}}))$, with the property that $A(\tilde{\mathcal{O}})$ is isomorphic to $R(\tilde{\mathcal{O}})$ as algebraic representations of $G_{\mathbb{C}}$. The "Dixmier correspondence" $\tilde{\mathcal{O}} \mapsto (A(\tilde{\mathcal{O}}), \phi(\tilde{\mathcal{O}}))$ should be injective.*

The requirements of the conjecture are very far from specifying $A(\tilde{\mathcal{O}})$ completely; but, fortunately, they are most restrictive precisely in those cases when we know least about how to construct $A(\tilde{\mathcal{O}})$.

The following result allows us to relate Conjecture 3.6 to primitive ideals.

Lemma 3.7. *Suppose $G_{\mathbb{C}}$ is a complex reductive algebraic group, and (A, ϕ) is a completely prime Dixmier algebra for $G_{\mathbb{C}}$. Then $\ker \phi = I(A, \phi)$ is a completely prime primitive ideal in $U(\mathfrak{g})$.*

Proof: Since A is completely prime, so is its subalgebra $U(\mathfrak{g})/I(A, \phi)$. So $I(A, \phi)$ is a completely prime ideal, and therefore prime; we need only show it is primitive. Write $\mathcal{Z}(\mathfrak{g})$ for the center of $U(\mathfrak{g})$. As a consequence of Theorem 2.13, $I(A, \phi)$ will be primitive if and only if $I(A, \phi) \cap \mathcal{Z}(\mathfrak{g})$ is a maximal ideal in $\mathcal{Z}(\mathfrak{g})$; that is, if and only if

$$\phi(\mathcal{Z}(\mathfrak{g})) \subset \mathbb{C}. \tag{3.4a}$$

Now,

$$\mathcal{Z}(\mathfrak{g}) = \{z \in U(\mathfrak{g}) \mid \mathrm{Ad}(g)(z) = z, \text{ all } z \in (G_{\mathbb{C}})_0\}. \tag{3.4b}$$

Let A_0 be the subalgebra of A on which $(G_{\mathbb{C}})_0$ acts trivially. By Definition 3.5(iv), A_0 is finite dimensional. Since it is also a completely prime algebra over $\mathbb{C}$, we must have $A_0 = \mathbb{C}$. Since (3.4b) guarantees that $\phi(\mathcal{Z}(\mathfrak{g})) \subset A_0$, (3.4a) follows. $\square$

In the setting of Conjecture 3.6, we write

$$I(\tilde{\mathcal{O}}) = \ker \phi(\tilde{\mathcal{O}}) \subset U(\mathfrak{g}) \tag{3.5}$$

for the completely prime primitive ideal provided by the conjecture and Lemma 3.7. The correspondence sending $\mathcal{O}$ to $\{I(\tilde{\mathcal{O}}) \mid \tilde{\mathcal{O}} \text{ a cover of } \mathcal{O}\}$ is (conjecturally) a kind of multivalued Dixmier correspondence from coadjoint orbits to completely prime primitive ideals.

From the point of view of primitive ideal theory, the most serious problem with Conjecture 3.6 is that this correspondence is not surjective: Not every completely prime Dixmier algebra is of the form $A(\tilde{\mathcal{O}})$ for a coadjoint orbit cover $\tilde{\mathcal{O}}$. Even the underlying correspondence (3.5) to completely prime primitive ideals is not surjective. To understand why this is not entirely bad, we recall an example from [J] and [V3].

Suppose $G_{\mathbb{C}}$ is of type G_2. There is exactly one coadjoint orbit $\mathcal{O}_8$ in $\mathfrak{g}^*$ of dimension 8, and it is simply connected. Joseph found a (unique) completely prime primitive ideal $I(\mathcal{O}_8)$ with the property that $U(\mathfrak{g})/I(\mathcal{O}_8)$ is isomorphic to $R(\mathcal{O}_8)$ as representations of $G_{\mathbb{C}}$. It is therefore reasonable to define $A(\mathcal{O}_8) = U(\mathfrak{g})/I(\mathcal{O}_8)$ as the Dixmier algebra predicted by Conjecture 3.6.

Let G be a simply connected split real reductive Lie group of type G_2. It turns out that G has exactly one irreducible $(\mathfrak{g}, K)$-module V with $\mathrm{Ann}(V) = I(\mathcal{O}_8)$. This $(\mathfrak{g}, K)$-module corresponds to an isolated unitary representation π; in the classification of $\widehat{G}$ given in [V8], π is the unique isolated point among the Langlands quotients of the principal series for the nonlinear group. It is constructed in [V8] as the restriction to G of a ladder representation of $\widetilde{\mathrm{SO}}(4,3)$. Certainly π is an interesting unitary representation of G, and evidence that the approach of Problem 2.18 will find it is welcome.

Joseph found a second completely prime primitive ideal $I'(\mathcal{O}_8)$ closely related to $\mathcal{O}_8$. (Under the adjoint action of $G_\mathbb{C}$, $U(\mathfrak{g})/I(\mathcal{O}_8)$ is slightly smaller than $R(\mathcal{O}_8)$.) The Dixmier correspondence of Conjecture 3.6 (or even (3.5)) has no room for $I'(\mathcal{O}_8)$, however; this ideal is simply omitted. There is exactly one irreducible $(\mathfrak{g}, K)$-module V' with $\mathrm{Ann}(V') = I'(\mathcal{O}_8)$. But it turns out that V' is not unitary; so from the point of view of finding unitary representations, the omission of $I'(\mathcal{O}_8)$ is harmless (or even desirable).

Encouraged by this example, we are going to refine Problem 2.18 to accommodate Dixmier algebras. Here is the setting.

Definition 3.8   Suppose G is a real reductive Lie group. Let $G_\mathbb{C}$ be a connected complex reductive algebraic group with Lie algebra $\mathfrak{g} = \mathrm{Lie}(G)_\mathbb{C}$. We assume that the adjoint action of G on $\mathfrak{g}$ factors through a homomorphism

$$j\colon G \to G_\mathbb{C}$$

inducing the identity map on $\mathfrak{g}$. (This can always be arranged by an appropriate choice of $G_\mathbb{C}$ if G is of "inner type": that is, if every automorphism $\mathrm{Ad}(g)$ (for $g \in G$) of $\mathfrak{g}$ is inner.) In this setting, we say that G is of *inner type* $G_\mathbb{C}$.

Suppose now that (A, ϕ) is a Dixmier algebra for $G_\mathbb{C}$. An (A, K)-*module* is a complex vector space endowed with a module structure for the algebra A and a representation of the group K, subject to the following conditions. (Compare Definition 2.16.)

i) The action of K is locally finite and smooth.

ii) The differential of the action of K (which makes sense by (i)) is equal to the action of $\phi(\mathfrak{k}_0) \subset A$.

iii) For $k \in K$, $v \in V$, and $a \in A$, we have

$$k \cdot (a \cdot v) = [\text{Ad}(j(k))(a)] \cdot (k \cdot v).$$

(Just as in Definition 2.16, condition (iii) for $k \in K_0$ is a consequence of (i) and (ii).)

A *Hermitian transpose* on A is a conjugate-linear antiautomorphism $*$ of A of order 2:

$$(ab)^* = b^* a^*, \quad (za)^* = \bar{z} a^* \qquad (a, b \in A, \ z \in \mathbb{C}).$$

We assume in addition that $*$ is compatible with the usual Hermitian transpose on $U(\mathfrak{g})$ defined by the real form $\mathfrak{g}_0$:

$$[\phi(X + iY)]^* = \phi(-X + iY) \qquad (X, Y \in \mathfrak{g}_0).$$

Suppose finally that V is an (A, K)-module and that $*$ is a Hermitian transpose on A. We say that V is *unitary* if it is endowed with a positive definite Hermitian form $\langle \, , \, \rangle$ satisfying

$$\langle k \cdot v, k \cdot w \rangle = \langle v, w \rangle, \qquad \langle a \cdot v, w \rangle = \langle v, a^* \cdot w \rangle,$$

for $v, w \in V$, $k \in K$, and $a \in A$.

The map ϕ provides a forgetful functor that makes any (unitary) (A, K)-module into a (unitary) $(\mathfrak{g}, K)$-module. This functor sends (A, K)-modules of finite length to $(\mathfrak{g}, K)$-modules of finite length, but it need not send irreducibles to irreducibles.

Theorem 3.9. *In the setting of Definition 3.8, suppose V is a unitary (A, K)-module. Then the Hilbert space completion $\mathcal{H}(V)$ carries a unitary representation $\pi(V)$ of G. The space $\mathcal{H}(V)^\infty$ carries a natural action of A. This action preserves the space $\mathcal{H}(V)^K$ of K-finite vectors (which are automatically smooth), making $\mathcal{H}(V)^K$ an (A, K)-module. If V has finite length, then $\mathcal{H}(V)^K$ is equal to V as an (A, K)-module.*

This is an immediate consequence of Harish-Chandra's Theorem 2.17.

Here is a refinement of Problem 2.18.

Problem 3.10   *Suppose G is real reductive Lie group of inner type $G_{\mathbb{C}}$ (Definition 3.8). Let $\mathcal{O} \subset \mathfrak{g}^*$ be a coadjoint orbit for $G_{\mathbb{C}}$, and let $\tilde{\mathcal{O}} \to \mathcal{O}$ be a connected covering on which $G_{\mathbb{C}}$ acts compatibly.*

   i) Construct a completely prime Dixmier algebra $(A(\tilde{\mathcal{O}}), \phi(\tilde{\mathcal{O}}))$ as in Conjecture 3.6.

   ii) Construct a finite collection $\{W_i(\tilde{\mathcal{O}}) \mid i = 1, \ldots, r\}$ of modules for $A(\tilde{\mathcal{O}})$.

   iii) For each i, construct from $W_i(\tilde{\mathcal{O}})$ an $(A(\tilde{\mathcal{O}}), K)$-module $V_i(\tilde{\mathcal{O}})$ (Definition 3.8).

   iv) For each i, construct a Hermitian transpose $*$ on $A(\tilde{\mathcal{O}})$ (Definition 3.8) and a unitary structure $\langle \, , \, \rangle$ on $V_i(\tilde{\mathcal{O}})$ (Definition 3.8).

If all these steps are completed, then Theorem 3.9 makes the Hilbert space completion $\mathcal{H}_i(\tilde{\mathcal{O}})$ of $V_i(\tilde{\mathcal{O}})$ into a unitary representation of G.

The method of coadjoint orbits suggests roughly how the modules $W_i(\tilde{\mathcal{O}})$ should be parametrized. Write $\mathfrak{g}_0^*$ for the real dual of the real Lie algebra of G. It is contained in $\mathfrak{g}^*$ as a real form.

Lemma 3.11   *In the setting of Problem 3.10, the intersection $\mathcal{O}(\mathbb{R}) = \mathcal{O} \cap \mathfrak{g}_0^*$ is a finite union of orbits of G:*

$$\mathcal{O}(\mathbb{R}) = \mathcal{O}_1(\mathbb{R}) \cup \cdots \cup \mathcal{O}_s(\mathbb{R}).$$

The covering map $\tilde{\mathcal{O}} \to \mathcal{O}$ induces (possibly disconnected) G-equivariant coverings

$$\tilde{\mathcal{O}}_i(\mathbb{R}) \to \mathcal{O}_i(\mathbb{R})$$

having the same degree as $\tilde{\mathcal{O}}$.

This is elementary. What the orbit method suggests is that each of the modules in Problem 3.10 should be parametrized by one of the real orbit covers $\tilde{\mathcal{O}}_i(\mathbb{R})$, together with some additional data. (This idea can be made precise and correct for semisimple orbits, but it requires further refinement in general.) In Section 4 we will begin to study real coadjoint orbits, and to see how one might attach representations to them.

4   Structure of coadjoint orbits

In this section we recall some general structure theory for coadjoint orbits. We work at first with an arbitrary real Lie group G, writing

$$\mathfrak{g}_0 = \text{Lie}(G), \qquad \mathfrak{g}_0^* = \text{Hom}_{\mathbb{R}}(\mathfrak{g}_0, \mathbb{R}). \tag{4.1a}$$

We write G_0 for the identity component of G. Often we write elements of G as lower case Roman letters, elements of $\mathfrak{g}_0$ as upper case Roman letters, and elements of $\mathfrak{g}_0^*$ as lower case Greek letters. The *coadjoint action* of G on $\mathfrak{g}_0^*$ is just the transpose of the adjoint action:

$$[\text{Ad}^*(g)\xi](Y) = \xi(\text{Ad}(g^{-1})Y) \qquad (\xi \in \mathfrak{g}_0^*,\ g \in G,\ Y \in \mathfrak{g}_0). \tag{4.1b}$$

The differential of this action is written

$$\text{ad}^*\colon \mathfrak{g}_0 \to \text{End}(\mathfrak{g}_0^*), \qquad [\text{ad}^*(X)\xi](Y) = \xi(\text{ad}(-X)Y) = \xi([Y, X]). \tag{4.1c}$$

The isotropy group for Ad^* at ξ is written G_ξ:

$$G_\xi = \{g \in G \mid \text{Ad}^*(g)\xi = \xi\}. \tag{4.1d}$$

The Lie algebra of G_ξ is

$$\begin{aligned}
\mathfrak{g}_{\xi,0} &= \{X \in \mathfrak{g}_0 \mid \text{ad}^*(X)\xi = 0\} \\
&= \{X \in \mathfrak{g}_0 \mid \xi([Y, X]) = 0, \text{all } Y \in \mathfrak{g}_0\}. \tag{4.1e}
\end{aligned}$$

To each $\xi \in \mathfrak{g}_0^*$ we attach a skew-symmetric bilinear form ω_ξ on $\mathfrak{g}_0$, defined by

$$\omega_\xi(X, Y) = \xi([X, Y]) = [\text{ad}^*(Y)\xi](X) = [-\text{ad}^*(X)\xi](Y). \tag{4.1f}$$

Lemma 4.1 *With notation as in (4.1), the radical of ω_ξ is equal to $\mathfrak{g}_{\xi,0}$. Consequently, ω_ξ descends to a nondegenerate symplectic form (still denoted ω_ξ) on*

$$\mathfrak{g}_0/\mathfrak{g}_{\xi,0} \simeq T_\xi(G \cdot \xi)$$

(the tangent space at ξ to the coadjoint orbit through ξ). As ξ' varies over the orbit $W = G \cdot \xi$, the family of $\omega_{\xi'}$ defines a closed two-form ω_W and, therefore, a G-invariant symplectic structure on W.

Proof: That the radical of ω_ξ is $\mathfrak{g}_{\xi,0}$ is clear from (4.1e) and (4.1f). It follows that ω_W is a two-form on W. Each element $X \in \mathfrak{g}_0$ defines a vector field X_W on W. These span the tangent space TW at each point; so to prove that ω_W is closed, it suffices to show that $d\omega_W(X_W, Y_W, Z_W) = 0$ for all $X, Y, Z \in \mathfrak{g}_0$. We compute

$$d\omega_W(X_W, Y_W, Z_W)$$
$$= X_W \cdot \omega_W(Y_W, Z_W) - Y_W \cdot \omega_W(X_W, Z_W) + Z_W \cdot \omega_W(X_W, Y_W)$$
$$- \omega_W([X_W, Y_W], Z_W) + \omega_W([X_W, Z_W], Y_W) - \omega_W([Y_W, Z_W], X_W).$$

Evaluating at the point $\xi' \in W$, we get

$$= X_W \cdot \xi'([Y, Z]) - Y_W \cdot \xi'([X, Z]) + Z_W \cdot \xi'([X, Y])$$
$$- \xi'([[X, Y], Z]) + \xi'([[X, Z], Y]) - \xi'([[Y, Z], X]).$$

In the first three terms, we are differentiating the function on W obtained by applying the variable linear functional ξ' to a fixed element of $\mathfrak{g}_0$. This amounts to applying $-\mathrm{ad}^*$ to ξ'. We get

$$= -(\mathrm{ad}^*(X)\xi')([Y, Z]) + (\mathrm{ad}^*(Y)\xi')([X, Z]) - (\mathrm{ad}^*(Z)\xi')([X, Y])$$
$$- \xi'([[X, Y], Z] - [[X, Z], Y] + [[Y, Z], X])$$
$$= -\xi'([[Y, Z], X]) + \xi'([[X, Z], Y]) - \xi'([[X, Y], Z])$$
$$- \xi'([[X, Y], Z] - [[X, Z], Y] + [[Y, Z], X])$$
$$= -2\xi'([[X, Y], Z] - [[X, Z], Y] + [[Y, Z], X]).$$

The argument of ξ' vanishes by the Jacobi identity, and so ω_W is closed. The G-invariance is clear from the definition. $\square$

Lemma 4.1 says that any coadjoint orbit is in a natural way a symplectic homogeneous space. The converse (that any symplectic homogeneous space is a coadjoint orbit) is not quite true, for two reasons. First, a coadjoint orbit has a slightly stronger structure (which we will describe in a moment). Second, this additional structure lifts to covering spaces.

Definition 4.2. Suppose (W, ω) is a symplectic manifold. The symplectic form provides a smooth identification of the tangent bundle of W with the cotangent bundle: To the tangent vector $X \in T_w(W)$, we associate the cotangent vector $\tau(X)$ defined by

$$\tau(X)(Y) = \omega_w(Y, X) \qquad (Y \in T_w(W));$$

this is the contraction of $-\omega_w$ with X. If f is a smooth function on W, then df is a one-form (a smooth section of the cotangent bundle). We may therefore define

$$X_f = \tau^{-1}(df),$$

a smooth vector field on W, called the *Hamiltonian vector field of* f. Using the action of vector fields on functions, we now define the *Poisson bracket* of the smooth functions f and g by

$$\{f, g\} = X_f \cdot g = dg(X_f) = \omega(X_f, X_g) = -X_g \cdot f.$$

Proposition 4.3   *The Poisson bracket defines a Lie algebra structure on $C^\infty(W)$. The map $f \mapsto X_f$ is a Lie algebra homomorphism from $C^\infty(W)$ to the Lie algebra of vector fields on W. Its kernel consists of the locally constant functions on W.*

  Suppose W is a coadjoint orbit in $\mathfrak{g}_0^*$, and $Y \in \mathfrak{g}_0$. Write $f(Y)$ for the smooth function on W obtained by restricting to W the linear function Y on $\mathfrak{g}_0^*$. Then the corresponding Hamiltonian vector field $X_{f(Y)}$ is equal to the vector field Y_W induced by the action of G on W. The map $Y \mapsto f(Y)$ is a Lie algebra homomorphism.

Proof: The assertions about general symplectic manifolds are standard (see, for example, [Ar, Chapter 8] or [AM, Chapter 3]; both sources use slightly different sign conventions from ours). For the rest, we compute (for $Y, Z \in \mathfrak{g}_0$)

$$\begin{aligned}
\omega_W(Z_W, Y_W) &= f([Z, Y]) & \text{(definition of } \omega_W) \\
&= Z_W \cdot f(Y) & \text{(as at the end of the proof of Lemma 4.1)} \\
&= df(Y)(Z_W) \\
&= \tau(X_{f(Y)})(Z_W) & \text{(definition of } X_{f(Y)}) \\
&= \omega_W(Z_W, X_{f(Y)}) & \text{(definition of } \tau).
\end{aligned}$$

Because ω_W is nondegenerate, and the vector fields Z_W span each tangent space, it follows that $Y_W = X_{f(Y)}$. At the same time, we have shown that

$$\{f(Z), f(Y)\} = X_{f(Z)} \cdot f(Y) = Z_W \cdot f(Y) = f([Z, Y]),$$

proving the last assertion. □

Definition 4.4 (see [Ko, Section 5])   A *Hamiltonian G-space* is a symplectic manifold W equipped with a symplectic action of G and a linear map $f: \mathfrak{g}_0 \to C^\infty(W, \mathbb{R})$, with the following properties.

 i) The map f is a Lie algebra homomorphism (for the Poisson bracket Lie algebra structure).

 ii) The Hamiltonian vector field $X_{f(Y)}$ on W associated to $Y \in \mathfrak{g}_0$ is equal to the vector field Y_W obtained by differentiating the action of G in the direction Y.

iii) The map f is G-equivariant:

$$f(\mathrm{Ad}(g)Y)(w) = f(Y)(g^{-1} \cdot w).$$

Condition (i) is actually a consequence of (ii) and (iii), but we include it because of its appealing simplicity. For g in the identity component G_0 of G, condition (iii) is a consequence of (i) and (ii). Condition (ii) also guarantees that the action of G_0 is symplectic; so the entire definition may be phrased more succinctly for connected G.

Suppose (W, f) is a Hamiltonian G-space. The *moment map* for W is the G-equivariant smooth map

$$\mu: W \to \mathfrak{g}_0^*, \qquad \mu(w)(Y) = f(Y)(w).$$

Proposition 4.3 implies that each coadjoint orbit is a Hamiltonian G-space (the requirement in (iii) being easy to verify). Its moment map is the identity. It is also easy to see that a G-equivariant covering of a Hamiltonian G-space is again a Hamiltonian G-space. The following result is a partial converse.

Proposition 4.5 ([Ko, Theorem 5.4.1]) *Suppose W is a homogeneous Hamiltonian G-space. Then the moment map is a covering of a coadjoint orbit $G \cdot \xi$. Consequently, $W \simeq G/G_{\xi,1}$, with $G_{\xi,1}$ an open subgroup of the isotropy group G_ξ.*

Proof: Since μ is G-equivariant, its image must be a single orbit $G \cdot \xi$. Fix a point $w \in W$ with $\mu(w) = \xi$, and define $G_{\xi,1}$ to be the isotropy group at w. Obviously this is a subgroup of G_ξ; we need only show it is open. This amounts to showing that the differential of μ is one-to-one. Because W is a homogeneous space for G, the tangent space T_w is spanned by the vector fields $Y_W = X_{f(Y)}$. We must show that Y_W vanishes at w if and only if $Y_{G\cdot\xi}$ vanishes at ξ. By the nondegeneracy of the symplectic form on W, $X_{f(Y)}$ vanishes at w if and only if $\omega_w(X_{f(Y)}, X_{f(Z)}) = 0$ for all $Z \in \mathfrak{g}_0$. By the definition of the Poisson bracket on W, this is the same as the vanishing of all $\{f(Z), f(Y)\}(w)$. By assumption (i) in Definition 4.4 and the definition of μ, this is the same as the vanishing of all $[Z, Y]$ (for varying Z) at ξ. By (4.1e), this last condition is the same as $Y \in \mathfrak{g}_{\xi,0}$; that is, $Y_{G\cdot\xi}$ vanishes at ξ. $\square$

According to Proposition 4.5, the coverings $\tilde{\mathcal{O}}$ appearing in Problem 3.10 are precisely the complex homogeneous Hamiltonian $G_{\mathbb{C}}$-spaces. This abstract characterization may lend a little respectability to what appears to be an ad hoc setting.

We turn now to a preliminary examination of how representations are attached to coadjoint orbits.

Definition 4.6 Suppose G is a Lie group, and $\xi \in \mathfrak{g}_0^*$. An *integral orbit datum at ξ* is an irreducible unitary representation $(\tau, \mathcal{H}_\tau)$ of the isotropy group G_ξ, subject to the condition

$$\tau(\exp X) = e^{i\xi(X)} \cdot \mathrm{Id}_{\mathcal{H}_\tau} \qquad (X \in \mathfrak{g}_{\xi,0}). \tag{4.2}$$

The orbit datum $(\tau, \mathcal{H}_\tau)$ at ξ is *equivalent* to $(\tau', \mathcal{H}_{\tau'})$ at ξ' if there is a $g \in G$ so that $\mathrm{Ad}^*(g)\xi = \xi'$, and τ is equivalent to the representation $h \mapsto \tau'(ghg^{-1})$ of G_ξ. (The second requirement makes sense because $G_{\xi'} = gG_\xi g^{-1}$.) The orbit $G \cdot \xi$ is called *integral* if it admits an integral orbit datum.

To understand this definition, notice first that the linear functional

$$i\xi \colon \mathfrak{g}_{\xi,0} \to i\mathbb{R} \tag{4.3a}$$

is automatically a Lie algebra homomorphism by (4.1e). If $G_{\xi,0}$ is simply connected, we therefore get automatically a unique group homomorphism

$$\tau_0 \colon G_{\xi,0} \to \mathrm{U}(1), \qquad \tau_0(\exp X) = e^{i\xi(X)}, \tag{4.3b}$$

on the identity component $G_{\xi,0}$ of G_ξ. In general (when $G_{\xi,0}$ need not be simply connected), the requirement in (4.3b) still specifies at most one τ_0; the problem is existence. Define

$$L_{\xi,0} = \{X \in \mathfrak{g}_{\xi,0} \mid \exp(X) = e\}. \tag{4.3c}$$

Then, τ_0 exists if and only if

$$\xi(L_{\xi,0}) \subset 2\pi\mathbb{Z}. \tag{4.3d}$$

(The necessity of this condition is clear, and sufficiency is not too difficult.)

Finally, it is not difficult to show that an integral orbit datum exists if and only if τ_0 exists; so (4.3d) is precisely the condition for $G \cdot \xi$ to be integral. When G_ξ is disconnected, τ is usually not unique.

There is an obvious way to construct a unitary representation of G from an integral orbit datum: by unitary induction from G_ξ to G. That is, we consider continuous functions

$$C(G/G_\xi, \mathcal{H}_\tau) = \{f \colon G \to \mathcal{H}_\tau \mid f(gh) = \tau(h)^{-1}f(g) \quad (g \in G,\ h \in G_\xi)\}. \tag{4.4a}$$

Such a function is said to be of *compact support modulo* G_ξ if there is a compact subset K of G so that f vanishes outside KG_ξ. We write $C_c(G/G_\xi, \mathcal{H}_\tau)$ for such compactly supported functions. Suppose f_1 and f_2

belong to $C(G/G_\xi, \mathcal{H}_\tau)$. Then we can define a complex-valued function on G by

$$\langle f_1, f_2 \rangle_{\text{loc}}(g) = \langle f_1(g), f_2(g) \rangle_{\mathcal{H}_\tau}. \qquad (4.4b)$$

Because τ is unitary, (4.4a) implies that $\langle f_1, f_2 \rangle_{\text{loc}}$ is actually a function on G/G_ξ. If one of the f_i belongs to $C_c(G/G_\xi, \mathcal{H}_\tau)$, then $\langle f_1, f_2 \rangle_{\text{loc}}$ is compactly supported on G/G_ξ. In that case we may define

$$\langle f_1, f_2 \rangle = \int_{G/G_\xi} \langle f_1, f_2 \rangle_{\text{loc}}(x) dx. \qquad (4.4c)$$

Here the G-invariant measure dx on $G/G_\xi \simeq G \cdot \xi$ arises naturally from the symplectic structure; the volume form may be taken to be the top exterior power of the symplectic form. In this way, we get a G-invariant positive definite Hermitian form on $C_c(G/G_\xi, \mathcal{H}_\tau)$. Its Hilbert space completion is called $L^2(G/G_\xi, \mathcal{H}_\tau)$. This space carries a unitary representation $\text{Ind}_{G_\xi}^G(\tau)$ of G, given on $C_c(G/G_\xi, \mathcal{H}_\tau)$ by left translation.

The difficulty is that this induced representation is almost never irreducible. (An interesting exception occurs when G_ξ is open in G, that is, when the orbit $G \cdot \xi$ is discrete.) To get an irreducible representation, we would like to make a similar construction on a smaller space; that is, we wish to impose additional conditions on the functions f in (4.4a). One natural idea is to extend the representation τ to a larger subgroup $H \supset G_\xi$. It turns out to be good to restrict attention to extensions τ_H still satisfying the analog of (4.2):

$$\tau_H(\exp X) = e^{i\xi(X)} \cdot \text{Id}_{\mathcal{H}_\tau} \qquad (X \in \mathfrak{h}_0 \supset \mathfrak{g}_{\xi,0}). \qquad (4.5a)$$

Such an extension of τ can exist only if $i\xi$ is a Lie algebra homomorphism from $\mathfrak{h}_0$ to $i\mathbb{R}$, that is, only if

$$\xi([X, Y]) = 0 \qquad (X, Y \in \mathfrak{h}_0). \qquad (4.5b)$$

According to (4.1f), this requirement is equivalent to the requirement that the symplectic form ω_ξ vanish on $\mathfrak{h}_0/\mathfrak{g}_{\xi,0} \subset \mathfrak{g}_0/\mathfrak{g}_{\xi,0} \simeq T_\xi(G \cdot \xi)$. We digress for a moment to recall some linear algebra related to this last condition.

Suppose (V, ω) is a real symplectic vector space, and $Z \subset V$ is any subspace. Define

$$Z^\perp = \{ v \in V \mid \omega(v, z) = 0, \text{ all } z \in Z \}. \qquad (4.6a)$$

Then,

$$\dim Z + \dim Z^\perp = \dim V. \tag{4.6b}$$

We say that Z is *isotropic* if $\omega|_Z = 0$, equivalently, if $Z \subset Z^\perp$. By (4.6b), this implies that

$$\dim V = \dim Z + \dim Z^\perp \geq 2 \dim Z \qquad (Z \text{ isotropic}). \tag{4.6c}$$

Dually, Z is *coisotropic* if $Z \supset Z^\perp$. This implies that

$$\dim V = \dim Z + \dim Z^\perp \leq 2 \dim Z \qquad (Z \text{ coisotropic}). \tag{4.6d}$$

We say that L is *Lagrangian* if it is both isotropic and coisotropic. In this case

$$\dim L = \frac{1}{2} \dim V \qquad (L \text{ Lagrangian}). \tag{4.6e}$$

Evidently a subspace L is Lagrangian if and only if it is isotropic and $\dim L = \frac{1}{2} \dim V$. Lagrangian subspaces always exist; all are conjugate under the action of the group $\mathrm{Sp}(\omega)$.

Next, suppose (M, ω) is a symplectic manifold. A submanifold $\Lambda \subset M$ is *Lagrangian* (resp., *isotropic* or *coisotropic*) if $T_\lambda \Lambda$ is a Lagrangian (resp., isotropic or coisotropic) subspace of $T_\lambda M$ for every $\lambda \in \Lambda$.

We want to shrink the space (4.4a) as much as possible, to have a good chance of getting an irreducible representation. In the setting of (4.5), this means taking the subgroup H as large as possible. We know already that $\mathfrak{h}_0/\mathfrak{g}_{\xi,0}$ must be isotropic; so it is natural to impose the requirement that $\mathfrak{h}_0/\mathfrak{g}_{\xi,0}$ be Lagrangian.

Definition 4.7   Suppose $(\tau, \mathcal{H}_\tau)$ is an integral orbit datum at ξ. An *invariant real polarization of* τ is a closed subgroup $H \supset G_\xi$, and an extension τ_H of τ to H, satisfying

i) $\tau_H(\exp X) = e^{i\xi(X)} \cdot \mathrm{Id}_{\mathcal{H}_\tau} \quad (X \in \mathfrak{h}_0)$.
ii) H is generated by G_ξ and H_0.
iii) $\dim H/G_\xi = \frac{1}{2} \dim G/G_\xi$.

We will show in Section 7 how to construct invariant real polarizations in one large class of examples; but let us consider briefly how one might look for them in general. According to (4.5), the existence of τ_H forces $\mathfrak{h}_0/\mathfrak{g}_{\xi,0}$ to be an isotropic subspace of $\mathfrak{g}_0/\mathfrak{g}_{\xi,0} = T_\xi(G \cdot \xi)$. Then (iii) makes it a Lagrangian subspace. To construct a polarization, we therefore need first of all a Lagrangian subspace of the symplectic vector space $T_\xi(G \cdot \xi)$. There will be many such subspaces. Each is of the form $\mathfrak{h}_0/\mathfrak{g}_{\xi,0}$, with $\mathfrak{h}_0$ a subspace of $\mathfrak{g}_0$ containing $\mathfrak{g}_{\xi,0}$. We will show that each Lagrangian subspace gives rise to at most one invariant real polarization.

To correspond to a polarization, $\mathfrak{h}_0$ must first be invariant under the adjoint action of G_ξ, and it must be a Lie algebra. When these two conditions are satisfied, there is a Lie subgroup H of G with Lie algebra $\mathfrak{h}_0$, satisfying (ii) and (iii) of Definition 4.7. The requirement that H be closed is not automatically satisfied and is an additional restriction on $\mathfrak{h}_0$. Because of (ii), condition (i) (together with the requirement that τ_H extend τ) determines τ_H uniquely; but there is a simple topological obstruction to its existence, which further constrains $\mathfrak{h}_0$.

Because of these requirements, it may easily happen that no invariant real polarizations exist. When they do exist, there are often many. The only natural notion of "equivalence" of polarizations is conjugation under the group G_ξ. But we have seen that each polarization is determined by a G_ξ-invariant Lagrangian subspace; so each equivalence class consists of a single element. For this reason, the problem of relating different polarizations is a difficult one, and we will ignore it entirely. (In the special setting of Section 7, we will find one distinguished polarization; others exist, however.)

In the setting of Definition 4.7, we have seen that the tangent space $T_\xi(H \cdot \xi)$ is a Lagrangian subspace of $T_\xi(G \cdot \xi)$. Since the symplectic structure is G-invariant, it follows that $H \cdot \xi \simeq H/G_\xi$ is a Lagrangian submanifold of $G \cdot \xi$. A similar argument proves

Lemma 4.8   *In the setting of Definition 4.7, consider the natural projection*

$$\pi: G \cdot \xi \to G/H, \qquad \pi(g \cdot \xi) = gH.$$

Then the fibers of π are connected Lagrangian submanifolds of $G \cdot \xi$.

Let us consider now the construction of a unitary representation of G from an invariant polarization (H, τ_H) at ξ. Certainly, we can define a space of continuous functions

$$C(G/H, \mathcal{H}_\tau) = \{ f: G \to \mathcal{H}_\tau \mid f(gh) = \tau_H(h)^{-1} f(g) \quad (g \in G, \ h \in H) \}.$$
$$(4.7a)$$

For future reference, notice that we can rewrite at least the smooth functions $C^\infty(G/H, \mathcal{H}_\tau)$ as follows. The Lie algebra $\mathfrak{g}_0$ acts on smooth functions by differentiation on the right:

$$(\rho(X)f)(g) = \frac{d}{dt} f(g \exp(tX))|_{t=0}. \qquad (4.7b)$$

This action makes equally good sense on smooth functions with values in $\mathcal{H}_\tau$. Then

$$
\begin{aligned}
C^\infty(&G/H, \mathcal{H}_\tau) \\
&= \{f \in C^\infty(G/G_\xi, \mathcal{H}_\tau) \mid \rho(X)f = -i\xi(X)f \quad (X \in \mathfrak{h}_0)\} \\
&= \{f \in C^\infty(G, \mathcal{H}_\tau) \mid f(gh) = \tau(h)^{-1}f(g),\ \rho(X)f = -i\xi(X)f \\
&\qquad\qquad (g \in G,\ h \in G_\xi,\ X \in \mathfrak{h}_0)\}. \qquad (4.7c)
\end{aligned}
$$

Write $C_c(G/H, \mathcal{H}_\tau)$ for the subspace of functions of compact support modulo H. Just as in (4.4b), we can define

$$\langle f_1, f_2 \rangle_{\mathrm{loc}}(g) = \langle f_1(g), f_2(g) \rangle_{\mathcal{H}_\tau}, \qquad (4.7d)$$

a complex-valued function on G/H. However we cannot imitate (4.4c) with an integral over G/H: Even in very simple examples, this space may not admit a G-invariant measure. To circumvent this problem, we recall the notion of half-density.

Definition 4.9 Suppose V is a finite dimensional real vector space and t is a real number. A *t-density on V* is a symbol $c|dx|^t$, with $c \in \mathbb{R}$ and dx a Lebesgue measure on V. We identify $c|dx|^t$ with $c'|dx'|^t$ if $dx' = j\,dx$ and $c = c'j^t$. The t-densities on V form a one-dimensional real vector space $D_t(V)$. For $t = 1$, they are just the multiples of Lebesgue measure on V. We have natural isomorphisms

$$D_t(V) \otimes D_s(V) \simeq D_{t+s}(V). \qquad (4.8a)$$

Suppose that M is a smooth manifold. Define a real line bundle $\mathcal{D}_t$ on M by $\mathcal{D}_t(m) = D_t(T_m M)$. We call $\mathcal{D}_t$ the *t-density bundle on M*; a section of $\mathcal{D}_t$ is called a *t-density on M*. If $t = 1$, sections of $\mathcal{D}_1$ may be identified with densities on M. In particular, if δ is a compactly supported continuous section of $\mathcal{D}_1$, there is a natural integral

$$\int_M \delta(m) \in \mathbb{R} \qquad (\delta \in C_c(M, \mathcal{D}_1)). \tag{4.8b}$$

There are also natural isomorphisms

$$\mathcal{D}_t \otimes \mathcal{D}_s \simeq \mathcal{D}_{t+s} \tag{4.8c}$$

as line bundles on M.

We return now to the setting of Definition 4.7, and the problem of construction of a unitary representation from τ_H. Write $D_t = D_t(\mathfrak{g}_0/\mathfrak{h}_0)$. The adjoint action of H gives rise to a representation ϕ of H on $\mathfrak{g}_0/\mathfrak{h}_0$, and so to an action χ_t of H on D_t. (It is easy to check that $\chi_t(h) = |\det \phi(h)|^{-t}$.) Consider the space of continuous functions

$$C(G/H, \mathcal{H}_\tau \otimes D_{1/2}) = \{f : G \to \mathcal{H}_\tau \otimes D_{1/2} \mid f(gh) = (\tau_H \otimes \chi_{1/2})(h^{-1})f(g)\} \tag{4.9a}$$

(What we are doing is twisting the bundle on G/H defined by τ_H by the half-density bundle.) The inner product on $\mathcal{H}_\tau$ and (4.8a) provide a sesquilinear pairing

$$\begin{aligned} &\langle\,,\,\rangle_{1/2} : \mathcal{H}_\tau \otimes D_{1/2} \times \mathcal{H}_\tau \otimes D_{1/2} \to D_1^{\mathbb{C}}, \\ &\langle v \otimes \delta, v' \otimes \delta'\rangle_{1/2} = \langle v, v'\rangle \cdot \delta \otimes \delta'. \end{aligned} \tag{4.9b}$$

Given f_1 and f_2 in $C(G/H, \mathcal{H}_\tau \otimes D_{1/2})$, we therefore get

$$\langle f_1, f_2\rangle_{\mathrm{loc}} = \langle f_1(\cdot), f_2(\cdot)\rangle_{1/2} \in C(G/H, D_1). \tag{4.9c}$$

This is a density on G/H, compactly supported if f_1 or f_2 is. By integrating densities, we therefore get a positive definite G-invariant quadratic form on $C_c(G/H, \mathcal{H}_\tau \otimes D_{1/2})$. Its Hilbert space completion $L^2(G/H, \mathcal{H}_\tau \otimes D_{1/2})$ carries a unitary representation of G, called $\mathrm{Ind}_H^G(\tau_H)$. This is often a reasonable unitary representation to attach to the orbit datum (ξ, τ). The most serious shortcoming of the construction is that invariant real polarizations often do not exist.

Example 4.10   Suppose $G = SL(2, \mathbb{R})$, so that $\mathfrak{g}_0$ consists of 2×2 real matrices of trace zero. We consider some examples of elements $\xi \in \mathfrak{g}_0^*$.

i) $\xi \begin{pmatrix} a & b \\ c & -a \end{pmatrix} = \nu a$, $0 \neq \nu \in \mathbb{R}$. The element ξ is hyperbolic (see Lemma 5.5). Its isotropy group is

$$G_\xi = \left\{ \begin{pmatrix} x & 0 \\ 0 & x^{-1} \end{pmatrix} \mid x \in \mathbb{R}^\times \right\}.$$

There are exactly two orbit data at ξ: both are one-dimensional unitary characters, and

$$\tau_+ \begin{pmatrix} x & 0 \\ 0 & x^{-1} \end{pmatrix} = |x|^{i\nu}, \qquad \tau_- \begin{pmatrix} x & 0 \\ 0 & x^{-1} \end{pmatrix} = |x|^{i\nu} \cdot \mathrm{sgn}(x).$$

Each has an invariant real polarization by the subgroup

$$H = \left\{ \begin{pmatrix} x & y \\ 0 & x^{-1} \end{pmatrix} \mid x \in \mathbb{R}^\times, y \in \mathbb{R} \right\};$$

the characters $\tau_{\pm,H}$ are trivial on elements $\begin{pmatrix} 1 & y \\ 0 & 1 \end{pmatrix}$. The unitarily induced representations of (4.9) are always irreducible.

ii) $\xi \begin{pmatrix} a & b \\ c & -a \end{pmatrix} = c$. The element ξ is nilpotent (see Lemma 5.3). The corresponding symplectic form is

$$\omega_\xi \left(\begin{pmatrix} a & b \\ c & -a \end{pmatrix}, \begin{pmatrix} a' & b' \\ c' & -a' \end{pmatrix} \right) = 2(ca' - ac'),$$

and the isotropy group is

$$G_\xi = \left\{ \begin{pmatrix} \epsilon & t \\ 0 & \epsilon \end{pmatrix} \mid \epsilon = \pm 1 \right\}.$$

Again the orbit data are two one-dimensional unitary characters $\tau_\pm$, defined by

$$\tau_+ \begin{pmatrix} \epsilon & t \\ 0 & \epsilon \end{pmatrix} = 1, \qquad \tau_- \begin{pmatrix} \epsilon & t \\ 0 & \epsilon \end{pmatrix} = \epsilon.$$

Each has an invariant real polarization by the subgroup H of (i). In this case, $\mathrm{Ind}_H^G(\tau_{+,H})$ is irreducible, but $\mathrm{Ind}_H^G(\tau_{-,H})$ is a direct sum of two irreducible components.

iii) $\xi \begin{pmatrix} a & b \\ c & -a \end{pmatrix} = t/2(b-c)$, $0 \neq t \in \mathbb{R}$. This element is elliptic (see Lemma 5.6). Its isotropy group is

$$G_\xi = \left\{ \begin{pmatrix} \cos\theta & \sin\theta \\ -\sin\theta & \cos\theta \end{pmatrix} \mid \theta \in \mathbb{R} \right\}.$$

Write Z for the matrix $\begin{pmatrix} 0 & 1 \\ -1 & 0 \end{pmatrix}$, so that the general element of G_ξ is $\exp\theta Z$. Since $\xi(Z) = t$, the requirement (4.2) for an integral orbit datum is

$$\tau_t(\exp\theta Z) = e^{it\theta}.$$

This is a well defined character of G_ξ if and only if $t \in \mathbb{Z}$; so we conclude that ξ is integral if and only if $t \in \mathbb{Z}$. The subgroup G_ξ is maximal in G, and so there is no invariant real polarization of τ_t.

5   (Co)adjoint orbits for reductive groups

We turn now to a more detailed study of coadjoint orbits in the case of real reductive groups. We single out three special classes, called hyperbolic, elliptic, and nilpotent. A hyperbolic coadjoint orbit is isomorphic to an affine bundle over a real flag variety. We will eventually attach representations to such orbits by real analysis on the real flag variety. An elliptic coadjoint orbit carries an invariant (indefinite) Kähler structure; it is isomorphic to an open orbit in a complex flag variety. We will attach representations to elliptic orbits by complex analysis techniques. A nilpotent coadjoint orbit has (in general) neither real nor complex structure of these kinds; so we do not know how to attach representations to it.

In the remainder of this section, we will define the three special classes of orbits and show how to realize a general coadjoint orbit as a combination of them.

So suppose G is a real reductive Lie group, with Cartan involution θ, maximal compact subgroup K, and Lie algebra $\mathfrak{g}_0$. Definition 1.5 allows us to identify $\mathfrak{g}_0$ with a Lie subalgebra of $\mathfrak{gl}(n, \mathbb{R})$ closed under transpose. On $\mathfrak{gl}(n, \mathbb{R})$ we can define the *trace form*

$$\langle X, Y \rangle = \operatorname{tr} XY. \tag{5.1a}$$

This is a symmetric bilinear form, invariant by the adjoint action of $\mathrm{GL}(n,\mathbb{R})$:

$$\langle \mathrm{Ad}(g)X, \mathrm{Ad}(g)Y \rangle = \langle X, Y \rangle \tag{5.1b}$$

for $X, Y \in \mathfrak{gl}(n,\mathbb{R})$ and $g \in \mathrm{GL}(n,\mathbb{R})$. By direct computation one finds that $\langle\,,\,\rangle$ is positive definite on symmetric matrices and negative definite on skew symmetric matrices, and that it makes these two subspaces orthogonal to each other. One immediate consequence is that $\langle\,,\,\rangle$ is nondegenerate. We also deduce

Lemma 5.1 *Suppose G is a real reductive Lie group with a homomorphism $\pi \colon G \to \mathrm{GL}(n,\mathbb{R})$ as in Definition 1.5. Then the restriction to $\mathfrak{g}_0$ of the trace form on $\mathfrak{gl}(n,\mathbb{R})$ (cf. (5.1)) has the following properties:*

i) $\langle \mathrm{Ad}(g)X, \mathrm{Ad}(g)Y \rangle = \langle X, Y \rangle, \quad X, Y \in \mathfrak{g}_0, g \in G.$

ii) $\langle\,,\,\rangle$ *is positive definite on $\mathfrak{p}_0$, negative definite on $\mathfrak{k}_0$, and has $\langle \mathfrak{p}_0, \mathfrak{k}_0 \rangle = 0$ (notation as in Definition 1.3).*

iii) $\langle\,,\,\rangle$ *is nondegenerate on $\mathfrak{g}_0$.*

Consequently $\langle\,,\,\rangle$ defines an isomorphism $\mathfrak{g}_0 \simeq \mathfrak{g}_0^*$ carrying the adjoint action of G on $\mathfrak{g}_0$ to the coadjoint action of G on $\mathfrak{g}_0^*$.

Because of this lemma, coadjoint orbits for reductive groups are closely related to adjoint orbits; and adjoint orbits in turn are closely related to conjugacy classes of matrices. We recall some facts from linear algebra.

Lemma 5.2 *Suppose X is an $n \times n$ real matrix. The following properties of X are equivalent.*

a) *The linear transformation of $\mathbb{C}^n$ defined by X is diagonalizable.*

b) *The minimal polynomial of X has no repeated factors.*

c) *The conjugacy class*

$$\mathrm{GL}(n,\mathbb{R}) \cdot X = \{gXg^{-1} \mid g \in \mathrm{GL}(n,\mathbb{R})\}$$

is closed in $\mathfrak{gl}(n,\mathbb{R})$.

d) *There is an $X' = gXg^{-1}$ in the conjugacy class of X so that X' commutes with ${}^tX'$.*

When these conditions are satisfied, we say that X is *semisimple*.

Sketch of proof: We will sketch the proof that (b) implies (c), which is perhaps one of the least familiar parts of the argument. Let $p = p_1 \cdots p_r$ be a factorization of the minimal polynomial of X, with p_i an irreducible real polynomial. Each p_i is either linear or quadratic with no real roots. Let d_i be the dimension of $\ker p_i(X)$. We have

$$\mathbb{R}^n = \bigoplus_{i=1}^{r} \ker p_i(X), \qquad (5.2a)$$

so $\sum d_i = n$. Also,

$$\mathrm{GL}(n, \mathbb{R}) \cdot X = \{ Y \in \mathfrak{gl}(n, \mathbb{R}) \mid \dim \ker p_i(Y) \geq d_i, \text{ all } i \}. \qquad (5.2b)$$

(Obviously, $\mathrm{GL}(n, \mathbb{R}) \cdot X$ is contained in the right side of (5.2b). For the other containment, notice that for any matrix Z, the sum $\sum_{i=1}^{r} \ker p_i(Z)$ must be direct, as the p_i are distinct irreducible polynomials. For Y in the right side of (5.2b), it follows by dimension counting that

$$\mathbb{R}^n = \bigoplus_{i=1}^{r} \ker p_i(Y), \qquad (5.2c)$$

and that $\dim \ker p_i(Y) = d_i$. This implies easily that Y is conjugate to X.) To complete the proof of (c), notice that the right side of (5.2b) is closed in $\mathfrak{gl}(n, \mathbb{R})$; this follows from the compactness of the Grassmannian manifolds of d-dimensional subspaces of $\mathbb{R}^n$. $\square$

Lemma 5.3 *Suppose X is an $n \times n$ real matrix. The following properties of X are equivalent.*

a) *X is nilpotent.*

b) *The characteristic polynomial of X is t^n.*

c) *The closure of the conjugacy class of X contains 0.*

d) *The conjugacy class of X contains a multiple rX of X, with $r > 0$ and $r \neq 1$.*

e) *The conjugacy class of X contains every multiple rX of X, with $r > 0$.*

f) *There is an element $A \in \mathfrak{gl}(n, \mathbb{R})$ with $[A, X] = X$.*

When these conditions are satisfied, we say that X is *nilpotent*.

Proof: We first prove the ascending implications, beginning with (f). Then $\exp(sA)(X)\exp(-sA) = \exp(\mathrm{ad}(sA))(X) = e^s X$, which proves (e). Trivially (e) implies (d). Assume (d); that is, that $gXg^{-1} = rX$ for some positive $r \neq 1$. Possibly replacing g by g^{-1}, we may assume $r < 1$. Then, $g^n X g^{-n} = r^n X \to 0$, proving (c). Assume (c). The characteristic polynomial $\det(tI - Z)$ is constant on conjugacy classes and depends continuously on Z; so the characteristic polynomial of X must coincide with that of 0, which is t^n. Assume (b). Then $X^n = 0$, since every matrix satisfies its characteristic polynomial; so X is nilpotent.

To finish, we prove that (a) implies (f). Define

$$H = \{ r > 0 \mid gXg^{-1} = rX, \text{ some } g \in \mathrm{GL}(n,\mathbb{R}) \}.$$

This is a subgroup of the multiplicative group of positive real numbers. If s and s' are any two positive numbers, then sX is conjugate to $s'X$ if and only if s and s' belong to the same coset of H. Now every multiple of X is nilpotent, and there are only finitely many conjugacy classes of nilpotent $n \times n$ matrices. It follows that H has finite index m in $\mathbb{R}^{>0}$. Consequently every mth power belongs to H, and so $H = \mathbb{R}^{>0}$. Therefore, every positive multiple of X is conjugate to X. It follows that the tangent space at X to the conjugacy class of X contains X. Now the tangent space at Z to the conjugacy class of Z is $[\mathfrak{gl}(n,\mathbb{R}), Z]$; so (f) follows. $\square$

Lemma 5.4 (Jordan decomposition) *Suppose X is an $n \times n$ real matrix. Then there are a semisimple matrix X_s and a nilpotent matrix X_n uniquely characterized by the following two properties:*

i) $X = X_s + X_n$.
ii) $[X_s, X_n] = 0$.

In addition, we have

iii) X_s and X_n may be expressed as polynomials without constant term in X.
iv) Any matrix commuting with X commutes with X_s and X_n.
v) Any subspace of $\mathbb{R}^n$ preserved by X is also preserved by X_s and X_n.

We turn now to an analogous decomposition of a semisimple matrix.

Lemma 5.5 *Suppose X is an $n \times n$ real matrix. The following properties of X are equivalent.*

i) X is diagonalizable.

ii) The minimal polynomial of X is a product of distinct linear factors.

iii) The conjugacy class of X contains a symmetric matrix.

When these conditions are satisfied, we say that X is *hyperbolic*. The proof of the lemma is easy, and we omit it.

Lemma 5.6 *Suppose X is an $n \times n$ real matrix. The following properties of X are equivalent.*

i) The linear transformation of $\mathbb{C}^n$ defined by X is diagonalizable with purely imaginary eigenvalues.

ii) The minimal polynomial of X is a product of distinct factors of the form $t^2 + a^2$ and t.

iii) The conjugacy class of X contains a skew-symmetric matrix.

When these conditions are satisfied, we say that X is *elliptic*. Again we omit the proof of the lemma.

Here is a complement to the Jordan decomposition of Lemma 5.4.

Lemma 5.7 *Suppose X is a semisimple $n \times n$ real matrix. Then there are a hyperbolic matrix X_h and an elliptic matrix X_e uniquely characterized by the following two properties:*

i) $X = X_h + X_e$.

ii) $[X_h, X_e] = 0$.

In addition, we have

iii) X_h and X_e may be expressed as polynomials in X.

iv) Any matrix commuting with X commutes with X_h and X_e.

v) Any subspace of $\mathbb{R}^n$ preserved by X is also preserved by X_h and X_e.

vi) If X commutes with ^tX, then $X_h = \frac{1}{2}(X + {}^tX)$ and $X_e = \frac{1}{2}(X - {}^tX)$.

vii) Suppose that X is a derivation of an algebra structure on $\mathbb{R}^n$. Then X_e and X_h are derivations as well.

Proof: We first establish the existence of the decomposition. Because the definitions of semisimple, hyperbolic, and elliptic are all invariant under conjugation (see Lemmas 5.2, 5.5, and 5.6), we may replace X by a conjugate matrix. By Lemma 5.2(d), we may therefore assume that X commutes with $^t X$. In this case the matrices X_h and X_e defined in (vi) obviously commute and have sum X. Also X_h is symmetric, and therefore hyperbolic (Lemma 5.5(c)), and X_e is skew-symmetric, and therefore elliptic (Lemma 5.6(c)). This proves the existence, as well as (vi). For the uniqueness, regard X, X_h, and X_e as linear transformations of $\mathbb{C}^n$. They commute with each other, and each is diagonalizable; so they are simultaneously diagonalizable (with respect to some basis $v_1, \ldots, v_n$ of $\mathbb{C}^n$). Write z_i for the diagonal entries of X as a matrix in this basis, and a_i and b_i for those of X_e and X_h. Then $z_i = a_i + b_i$, a_i is real (Lemma 5.5(a)) and b_i is purely imaginary (Lemma 5.6(b)). Therefore, $a_i = \operatorname{Re} z_i$, $b_i = \sqrt{-1}\operatorname{Im} z_i$. This means that X_h acts on each eigenspace of X by the real part of the corresponding eigenvalue. Similarly, X_e acts on each eigenspace of X by the imaginary part of the eigenvalue. These descriptions establish the uniqueness of X_h and X_e, and properties (iv) and (v) in the lemma follow easily. To prove (iii), we need to know the existence of real polynomials p_h and p_e with the properties that

$$ p_h(z_i) = \operatorname{Re} z_i, \qquad p_e(z_i) = \sqrt{-1}\operatorname{Im} z_i. $$

Because the nonreal z_i occur in complex conjugate pairs, this is elementary.

Finally, assume that X is a derivation of an algebra structure $\circ$. After complexification, this means exactly that if v and v' are eigenvectors of eigenvalues z and z', then $v \circ v'$ is an eigenvector of eigenvalue $z + z'$. The foregoing description of X_h and X_e shows that they inherit this property from X. $\square$

With these facts about matrices in hand, we turn to their generalizations for reductive groups.

Definition 5.8   Suppose G is a real reductive Lie group with $\pi\colon G \to \mathrm{GL}(n, \mathbb{R})$ as in Definition 1.5. Recall that the Lie algebra $\mathfrak{g}_0$ of G is identified with a subalgebra of $\mathfrak{gl}(n, \mathbb{R})$. An element $X \in \mathfrak{g}_0$ is called *semisimple* (resp., *nilpotent, hyperbolic,* or *elliptic*) if X has the corresponding property as a matrix (Lemma 5.2, 5.3, 5.5, or 5.6).

To give more intrinsic characterizations of these properties, we will consider not the action of X on $\mathbb{R}^n$, but rather the adjoint action on $\mathfrak{g}_0$:

$$ \operatorname{ad}(X)(Y) = [X, Y]. \tag{5.3a} $$

The kernel of the adjoint action is the center $\mathfrak{z}(\mathfrak{g}_0)$:

$$\mathfrak{z}(\mathfrak{g}_0) = \{Z \in \mathfrak{g}_0 \mid [Z, Y] = 0, \text{ all } Y \in \mathfrak{g}_0\}. \tag{5.3b}$$

Obviously, this is preserved by θ, and so it is the direct sum of its intersections with $\mathfrak{k}_0$ and $\mathfrak{p}_0$:

$$\mathfrak{z}(\mathfrak{g}_0) = \mathfrak{z}_{\mathfrak{k}}(\mathfrak{g}_0) \oplus \mathfrak{z}_{\mathfrak{p}}(\mathfrak{g}_0). \tag{5.3c}$$

The "image" of the adjoint action is the derived algebra

$$\mathfrak{g}_0' = [\mathfrak{g}_0, \mathfrak{g}_0]. \tag{5.3d}$$

Because of the invariance under ad of the trace form (5.1), the derived algebra is precisely the orthogonal complement of the center for the trace form. Lemma 5.1 therefore implies

$$\mathfrak{g}_0 = \mathfrak{g}_0' \oplus \mathfrak{z}(\mathfrak{g}_0), \tag{5.3e}$$

a direct sum of θ-stable ideals. (Notice that although we used structure from matrices to prove (5.3e), the summands are defined in terms of the Lie algebra structure of $\mathfrak{g}_0$.) It is not difficult to show that the Lie algebra $\mathfrak{g}_0'$ is semisimple, and we will apply to it some standard structural results.

Theorem 5.9   *In the setting of Definition 5.8 and (5.3), suppose* $\rho: \mathfrak{g}_0' \to$ *End V is a finite dimensional representation of the semisimple Lie algebra* $\mathfrak{g}_0'$, *and* $X' \in \mathfrak{g}_0'$.

   i) If $\mathrm{ad}(X')$ *is semisimple, then* $\rho(X')$ *is semisimple.*
   ii) If $\mathrm{ad}(X')$ *is nilpotent, then* $\rho(X')$ *is nilpotent.*
   iii) If T *is a derivation of* $\mathfrak{g}_0'$, *then there is a unique element* $X_T \in \mathfrak{g}_0'$ *with* $T = \mathrm{ad}(X_T)$.
   iv) If $X' = X_s' + X_n'$ *is the Jordan decomposition of* X' *as an* $n \times n$ *matrix (Lemma 5.4), then* X_s' *and* X_n' *belong to* $\mathfrak{g}_0'$.
   v) If X' *is semisimple, then there is a* $g \in G$ *so that* $\mathrm{Ad}(g)(X') = gX'g^{-1}$ *commutes with* $\theta(\mathrm{Ad}(g)(X'))$.

   Parts (i) and (ii) are proved in [Hu, Theorem 6.5]; part (iii) is [Hu, Theorem 5.3]; part (iv) is [Hu, Theorem 6.4]; and (v) is essentially [W, Lemma 2.3.3].

Lemma 5.10   *In the setting of Definition 5.8, write $X = X' + X_{\mathfrak{z}}$ for the decomposition according to (5.3e).*

i) X is semisimple if and only if $\mathrm{ad}(X)$ is semisimple.

ii) X is nilpotent if and only if $X_{\mathfrak{z}} = 0$ and $\mathrm{ad}(X)$ is nilpotent.

iii) X is hyperbolic if and only if $X_{\mathfrak{z}} \in \mathfrak{p}_0$ and $\mathrm{ad}(X)$ is hyperbolic.

iv) X is elliptic if and only if $X_{\mathfrak{z}} \in \mathfrak{k}_0$ and $\mathrm{ad}(X)$ is elliptic.

Proof: The matrices X' and $X_{\mathfrak{z}}$ commute with each other; and $\mathrm{ad}(X) = \mathrm{ad}(X')$. The matrix $X_{\mathfrak{z}}$ is a sum of commuting symmetric and skew symmetric matrices (cf. (5.3c)), and it is therefore automatically semisimple.

We will analyze the bracket action $\mathrm{ad}_{\mathfrak{gl}(n,\mathbb{R})}(X)$ of X on all $n \times n$ matrices. This space is naturally identified with $\mathbb{R}^n \otimes (\mathbb{R}^n)^*$, and we compute

$$\mathrm{ad}_{\mathfrak{gl}(n,\mathbb{R})}(X)(v \otimes w) = Xv \otimes w - v \otimes {}^t X w. \tag{5.4}$$

If X is semisimple with eigenvalues $\{z_i\}$, it follows immediately that $\mathrm{ad}_{\mathfrak{gl}(n,\mathbb{R})}(X)$ is semisimple with eigenvalues $\{z_i - z_j\}$. The implication "only if" in (i) is immediate. For the converse, assume that $\mathrm{ad}(X) = \mathrm{ad}(X')$ is semisimple. By Theorem 5.9(i), it follows that X' is semisimple. Therefore, $X = X' + X_{\mathfrak{z}}$ is a commuting sum of semisimple matrices, and so X is semisimple.

For (ii), suppose first that X is nilpotent; say $X^p = 0$. By (5.4), we see that $(\mathrm{ad}(X))^{2p} = 0$; so $\mathrm{ad}(X)$ is nilpotent. By Theorem 5.9(ii), X' is nilpotent. The expression $X = X' + X_{\mathfrak{z}}$ is a commuting sum of a nilpotent and a semisimple matrix, and so it must be the Jordan decomposition of X. Since X is nilpotent, it follows that $X_{\mathfrak{z}} = 0$. Conversely, assume that $X_{\mathfrak{z}} = 0$ and that $\mathrm{ad}(X) = \mathrm{ad}(X')$ is nilpotent. By Theorem 5.9(ii), X' is nilpotent; so since $X_{\mathfrak{z}} = 0$, X is nilpotent.

For (iii) and (iv), assume first that X is hyperbolic; that is, that X is diagonalizable with real eigenvalues. Then (5.4) implies that $\mathrm{ad}(X) = \mathrm{ad}(X')$ is as well; so $\mathrm{ad}(X)$ is hyperbolic. Similarly, X elliptic implies $\mathrm{ad}(X)$ elliptic.

To complete the proofs of (iii) and (iv), we will use a lemma. Here is the setting. Suppose $X' \in \mathfrak{g}_0'$ is a semisimple element. According to Lemma 5.7, the hyperbolic and elliptic parts T_h and T_e of the semisimple derivation $\mathrm{ad}(X')$ of $\mathfrak{g}_0'$ are also (semisimple) derivations. By Theorem 5.9(iii), they are given by the adjoint action of unique elements $X'_{[h]}$ and

$X'_{[e]}$ of $\mathfrak{g}'_0$. On the other hand, the semisimple matrix X' has hyperbolic and elliptic parts X'_h and X'_e.

Lemma 5.11 *In the setting just described, $X'_{[h]} = X'_h$ and $X'_{[e]} = X'_e$. In particular, the hyperbolic and elliptic parts of X' belong to $\mathfrak{g}'_0$.*

Proof: Everything here behaves nicely with respect to conjugation by elements of G. According to Theorem 5.9(iv), we may therefore assume that X' commutes with $\theta X'$. In this case, we have

$$X'_h = \frac{1}{2}(X' - \theta X'), \qquad X'_e = \frac{1}{2}(X' + \theta X')$$

(Lemma 5.7(vi)). Evidently, these matrices belong to $\mathfrak{g}'_0$. We therefore have

$$\mathrm{ad}(X') = \mathrm{ad}(X'_h) + \mathrm{ad}(X'_e),$$

a commuting sum of derivations of $\mathfrak{g}'_0$. By what we have already proved, the first term is hyperbolic and the second elliptic as endomorphisms of $\mathfrak{g}'_0$. By the uniqueness in Lemma 5.7, they are the hyperbolic and elliptic parts of $\mathrm{ad}(X')$. □

We return now to finish the proofs of (iii) and (iv). In general, we will write

$$X_{\mathfrak{z}} = X_{\mathfrak{z}\mathfrak{k}} + X_{\mathfrak{z}\mathfrak{p}}$$

for the decomposition of a central element according to (5.3c). Suppose X is hyperbolic. We have already shown that $\mathrm{ad}(X) = \mathrm{ad}(X')$ is hyperbolic, and so Lemma 5.11 implies that X' is hyperbolic. The three matrices X', $X_{\mathfrak{z}\mathfrak{k}}$, and $X_{\mathfrak{z}\mathfrak{p}}$ are therefore hyperbolic, elliptic, and hyperbolic, respectively; and they commute with each other. It follows that $X = (X' + X_{\mathfrak{z}\mathfrak{p}}) + (X_{\mathfrak{z}\mathfrak{k}})$ exhibits X as a commuting sum of hyperbolic and elliptic matrices. By the uniqueness in Lemma 5.7, the second term is zero; that is, $X_{\mathfrak{z}}$ belongs to $\mathfrak{p}_0$, as we wished to show. Conversely, assume that $\mathrm{ad}(X)$ is hyperbolic and that $X_{\mathfrak{z}} \in \mathfrak{p}_0$. Lemma 5.11 shows that X' is hyperbolic; so X is a sum of two commuting hyperbolic matrices and is therefore hyperbolic. The proof of (iv) is identical. □

Theorem 5.12   *Suppose G is a real reductive Lie group as in Definition 1.5, and $X \in \mathfrak{g}_0$. Then there are hyperbolic, elliptic, and nilpotent elements X_h, X_e, and X_n of $\mathfrak{g}_0$ (Definition 5.8) uniquely characterized by the following two properties:*

i) $X = X_h + X_e + X_n$.

ii) $[X_h, X_e] = [X_h, X_n] = [X_e, X_n] = 0$.

This decomposition has the following additional properties.

iii) Any element $g \in G$ fixing X (that is, $\mathrm{Ad}(g)(X) = X$) also fixes X_h, X_e, and X_n.

iv) The adjoint representation of $\mathfrak{g}_0$ preserves the decomposition in (i): $\mathrm{ad}(X_h)$ is hyperbolic, $\mathrm{ad}(X_e)$ is elliptic, and $\mathrm{ad}(X_n)$ is nilpotent.

This is an easy consequence of Theorem 5.9, Lemma 5.10, and Lemma 5.11; we leave the details to the reader.

Corollary 5.13   *Suppose G is a real reductive Lie group as in Definition 1.5, and $X \in \mathfrak{g}_0$ is a semisimple element.*

i) There is a $g \in G$ so that $\mathrm{Ad}(g)(X)$ commutes with $\theta(\mathrm{Ad}(g)(X))$.

ii) Suppose $[X, \theta X] = 0$. Then the centralizer

$$G_X = \{ g \in G \mid \mathrm{Ad}(g)(X) = X \}$$

is a real reductive Lie group, with Cartan involution the restriction to G_X of θ. In addition,

$$G_X = (G_{X_h})_{X_e} = (G_{X_e})_{X_h}.$$

Proof: Part (i) is an easy consequence of Theorem 5.9(v) (and (5.3c)). For (ii), the descriptions of G_X as iterated centralizers follow from Theorem 5.12(iii). That G_X is reductive then follows from Proposition 1.10(ii). □

6   Interlude on the classification of (co)adjoint orbits

Corollary 5.13 leads to a systematic procedure for describing all the coadjoint orbits for a real reductive Lie group. Although we will not really need it, the classification is simple and attractive. The method of coadjoint orbits suggests as well that it should bear a family resemblance to the classification of unitary representations. For these reasons, we will outline the classification here. Most of the proofs are easy; the more difficult results may be found in many places (including [W, Chapter 2]).

Definition 6.1   Suppose G is a real reductive Lie group. A *Cartan subspace* for G is a maximal abelian subalgebra $\mathfrak{a}_0$ of $\mathfrak{p}_0$. Given such a subspace, we define

$$M = \{k \in K \mid \mathrm{Ad}(k)(X) = X, \text{all } X \in \mathfrak{a}_0\},$$

$$M' = \{k \in K \mid \mathrm{Ad}(k)(\mathfrak{a}_0) = \mathfrak{a}_0\}.$$

The groups M' and M are compact, and M is an open normal subgroup of M'. The quotient is therefore a finite group

$$W(G, \mathfrak{a}_0) = M'/M \subset \mathrm{Aut}(\mathfrak{a}_0),$$

the *Weyl group of* $\mathfrak{a}_0$ *in* G. (The action on $\mathfrak{a}_0$ is by Ad.)

Theorem 6.2   *Suppose G is a real reductive group, and $\mathfrak{a}_0$ is a Cartan subspace of $\mathfrak{p}_0$. Then the inclusions*

$$\mathfrak{a}_0 \subset \mathfrak{p}_0 \subset \mathfrak{g}_0$$

induce bijections among the following three sets:

i) orbits of $W = W(G, \mathfrak{a}_0)$ on $\mathfrak{a}_0$

ii) orbits of K on $\mathfrak{p}_0$

iii) hyperbolic orbits of G on $\mathfrak{g}_0$

Suppose X and Y belong to $\mathfrak{a}_0$. Then the group centralizers G_X and G_Y are equal if and only if the Weyl group centralizers W_X and W_Y are equal. In particular, there are only finitely many possibilities (up to conjugation in G) for the centralizer of a hyperbolic element.

Example 6.3   Suppose $G = O(p, q)$ (Example 1.13) with $p \leq q$. After a simple change of basis, the matrix $A_{p,q}$ of the quadratic form $B_{p,q}$ may be replaced by the matrix

$$A^1_{p,q} = \begin{pmatrix} 0 & I_p & 0 \\ I_p & 0 & 0 \\ 0 & 0 & -I_{p-q} \end{pmatrix}.$$

We write $B^1(p, q)$ for this new form and $O^1(p, q)$ for the corresponding group. If $g \in GL(p, \mathbb{R})$, then it is easy to check that $G^1 = O^1(p, q)$ contains the matrix

$$\lambda^1(g) = \begin{pmatrix} g & 0 & 0 \\ 0 & {}^t g^{-1} & 0 \\ 0 & 0 & I_{p-q} \end{pmatrix}.$$

The map λ^1 provides an embedding

$$GL(p, \mathbb{R}) \subset O^1(p, q) \simeq O(p, q),$$

and all maps respect the Cartan involutions.

Now the space of diagonal matrices is a Cartan subspace for $GL(p, \mathbb{R})$, naturally isomorphic to $\mathbb{R}^p$. It is not difficult to check that the image $\mathfrak{a}^1_0$ of the diagonal matrices under $d\lambda^1$ is a Cartan subspace of $\mathfrak{g}^1_0$; it consists of diagonal matrices with the first p entries equal to the negatives of the next p, and the last $p - q$ entries equal to zero. The Weyl group of this Cartan subspace acts by permuting and changing the signs of the coordinates of $\mathbb{R}^p$; it is isomorphic to the hyperoctahedral group

$$W(G, \mathfrak{a}^1_0) = S_p \times (\mathbb{Z}/2\mathbb{Z})^p,$$

a semidirect product with the second factor normal.

By Theorem 6.2, each hyperbolic orbit has a unique representative

$$X = (X_1, \ldots, X_p) \in \mathbb{R}^p, \qquad X_1 \geq \cdots \geq X_p \geq 0.$$

We can compute the centralizer G^1_X as follows. Write $p = p_r + \cdots + p_1 + p_0$, in such a way that

$$X_1 = \cdots = X_{p_r} > X_{p_r + 1} = \cdots = X_{p_r + \cdots + p_1} > X_{p_r + \cdots + p_1 + 1} = 0.$$

Then,

$$G_X \simeq \mathrm{GL}(p_r, \mathbb{R}) \times \cdots \times \mathrm{GL}(p_1, \mathbb{R}) \times \mathrm{O}^1(p_0, p_0 + q - p).$$

(There is a natural embedding of this group in G^1, which is easy to construct using λ^1.)

There is a parallel result for elliptic elements.

Definition 6.4 Suppose G is a real reductive Lie group. A *Cartan subalgebra* for K is a maximal abelian subalgebra $\mathfrak{t}_0$ of $\mathfrak{k}_0$. Given such a subspace, we define

$$T = \{k \in K \mid \mathrm{Ad}(t)(X) = X, \text{all } X \in \mathfrak{t}_0\},$$
$$T' = \{k \in K \mid \mathrm{Ad}(k)(\mathfrak{t}_0) = \mathfrak{t}_0\}.$$

The group T is called a *small Cartan subgroup of K*. If K is connected, it is a torus. The group T' is compact, and T is an open normal subgroup of T'. The quotient is therefore a finite group

$$W(G, \mathfrak{t}_0) = T'/T \subset \mathrm{Aut}(\mathfrak{t}_0),$$

the *Weyl group of $\mathfrak{t}_0$ in G*. (The action on $\mathfrak{t}_0$ is by Ad.)

Theorem 6.5 *Suppose G is a real reductive group, and $\mathfrak{t}_0$ is a Cartan subalgebra of $\mathfrak{k}_0$. Then the inclusions*

$$\mathfrak{t}_0 \subset \mathfrak{k}_0 \subset \mathfrak{g}_0$$

induce bijections among the following three sets:

  i) orbits of $W = W(G, \mathfrak{t}_0)$ on $\mathfrak{t}_0$
  ii) orbits of K on $\mathfrak{k}_0$
  iii) elliptic orbits of G on $\mathfrak{g}_0$

Suppose X and Y belong to $\mathfrak{t}_0$. Then the group centralizers G_X and G_Y are equal if and only if

  1) the Weyl group centralizers W_X and W_Y are equal; and
  2) the elements X and Y annihilate exactly the same weights of $\mathfrak{t}_0$ on $\mathfrak{p}_0$.

In particular, there are only finitely many possibilities (up to conjugation in G) for the centralizer of a elliptic element.

Example 6.6   Suppose $G = \mathrm{O}(p, q)$ (Example 1.13). Recall that $K \simeq \mathrm{O}(p) \times \mathrm{O}(q)$. Write $p' = [p/2]$ (the greatest integer in $p/2$) and $q' = [q/2]$. Then there are obvious maps

$$\mathrm{SO}(2)^{p'+q'} \simeq \mathrm{SO}(2)^{p'} \times \mathrm{SO}(2)^{q'} \subset \mathrm{O}(p) \times \mathrm{O}(q) \subset \mathrm{O}(p, q).$$

The Lie algebra of $\mathrm{SO}(2)$ consists of skew symmetric 2×2 matrices, and so may be identified naturally with $\mathbb{R}$. The differential of the first inclusion therefore gives

$$\tau \colon \mathbb{R}^{p'} \times \mathbb{R}^{q'} \to \mathfrak{k}_0.$$

It is easy to check that the image of τ is a Cartan subalgebra $\mathfrak{t}_0$ of $\mathfrak{k}_0$. The Weyl group acts by permutation and sign changes of the first p' and last q' coordinates separately:

$$W(G, \mathfrak{t}_0) = (S_{p'} \times (\mathbb{Z}/2\mathbb{Z})^{p'}) \times (S_{q'} \times (\mathbb{Z}/2\mathbb{Z})^{q'}),$$

a product of two hyperoctahedral groups. According to Theorem 6.5, every elliptic orbit in $\mathfrak{g}_0$ has a unique representative

$$Z = [(Z_1^1, \dots, Z_{p'}^1), (Z_1^2, \dots, Z_{q'}^2)] \in \mathbb{R}^{p'+q'},$$
$$Z_1^1 \geq \cdots \geq Z_{p'}^1 \geq 0, \quad Z_1^2 \geq \cdots \geq Z_{q'}^2 \geq 0.$$

We can compute the centralizer G_Z as follows. Write $p' = p_r' + \cdots + p_1' + p_0'$ and $q' = q_r' + \cdots q_1' + q_0'$, in such a way that

$$Z_1^1 = \cdots = Z_{p_r'}^1 > Z_{p_r'+1}^1 = \cdots = Z_{p_r+\cdots+p_1'}^1 > Z_{p_r'+\cdots+p_1'+1}^1 = 0,$$
$$Z_2^1 = \cdots = Z_{q_r'}^2 > Z_{q_r'+1}^2 = \cdots = Z_{q_r+\cdots+q_1'}^2 > Z_{q_r'+\cdots+q_1'+1}^2 = 0,$$
$$Z_{p_r'+\cdots+p_j'+1}^1 = Z_{q_r'+\cdots+q_j'+1}^2 \quad (j = r+1, r-1, \dots, 1).$$

Then,

$$G_Z \simeq \mathrm{U}(p_r', q_r') \times \cdots \times \mathrm{U}(p_1', q_1') \times \mathrm{O}^1(p - 2(p_r' + \cdots + p_1'), q - 2(q_r' + \cdots + q_1')).$$

(The idea is that the element Z acts as $Z_{p_r'+\cdots+p_{j+1}'+1}^1$ times a complex structure on a subspace of the form $\mathbb{R}^{2p_j'+2q_j'}$. The centralizer of Z therefore acquires a factor consisting of the complex-linear transformations of this subspace that preserve the quadratic form. This is the indefinite unitary group $U(p'_j, q'_j)$ of Example 1.15.)

Here is the procedure for constructing all adjoint orbits for a reductive group G. First, choose a Cartan subspace $\mathfrak{a}_0 \subset \mathfrak{p}_0$, and compute its Weyl group. For each element $X_h \in \mathfrak{a}_0$, consider the reductive group G_{X_h}; Theorem 6.2 guarantees that there are only finitely many such groups. (They are described for $G = O(p,q)$ in Example 6.3.) For each such group, choose a Cartan subalgebra $\mathfrak{t}_{X_h,0}$ of $\mathfrak{k}_{X_h,0}$, and compute its Weyl group. For each element $X_e \in \mathfrak{t}_{X_h,0}$, consider the reductive group G_{X_h,X_e}; Theorem 6.5 says that there are still only finitely many such groups. (When G_{X_h} is a product, it suffices to treat each factor separately. The case of $O(p,q)$ is handled in Example 6.6; to treat all of the hyperbolic centralizers for $O(p,q)$, we would need to consider also $GL(n, \mathbb{R})$. The conclusion is that for $O(p,q)$, each group G_{X_h,X_e} is a product of factors of the form $GL(a, \mathbb{R})$, $GL(b, \mathbb{C})$, $U(c,d)$, and $O(e,f)$; there is just one factor of this last form.)

To make a list of all semisimple adjoint orbits, we choose one representative X_h in each Weyl group orbit on $\mathfrak{a}_0$; and then (for each X_h) one representative X_e in each Weyl group orbit on $\mathfrak{t}_{X_h,0}$. The resulting elements

$$X_s = X_h + X_e$$

provide exactly one representative of each semisimple orbit on $\mathfrak{g}_0$.

This procedure can be continued to give all adjoint orbits: We just need a way to give a representative for each of the (finitely many) nilpotent adjoint orbits for each of the (finitely many) reductive groups G_{X_h,X_e}. For the classical groups this problem is not too hard, and for the exceptional groups it has been solved. We refer to [CM] for more information.

7   Hyperbolic elements, real polarizations, and parabolic subgroups

We turn now to the problem of attaching representations to orbits for a reductive group G. Recall from Lemma 5.1 that we can identify $\mathfrak{g}_0$ with $\mathfrak{g}_0^*$ by an isomorphism $X \mapsto \xi_X$:

$$\xi_X(Y) = \langle X, Y \rangle = \operatorname{tr} XY. \tag{7.1a}$$

Lemma 5.1 allows us to identify the isotropy algebra and group for the coadjoint action at ξ_X (see (4.1)) as

$$G_{\xi_X} = \{g \in G \mid \mathrm{Ad}(g)X = X\}, \qquad \mathfrak{g}_{\xi_X,0} = \{Y \in \mathfrak{g}_0 \mid [Y, X] = 0\}. \quad (7.1b)$$

We may sometimes write G_X or $\mathfrak{g}_{X,0}$ accordingly. The symplectic form of (4.1f) is

$$\omega_{\xi_X}(Y, Z) = \langle X, [Y, Z] \rangle = -\langle Z, [Y, X] \rangle; \quad (7.1c)$$

notice that this formula is skew symmetric in all three variables X, Y, and Z. We sometimes write simply ω_X. We call ξ_X semisimple (or nilpotent, elliptic, or hyperbolic) if X is.

Proposition 7.1 *In the setting of (7.1), suppose that* $\mathrm{ad}(X)$ *is diagonalizable with real eigenvalues; this happens in particular if X is hyperbolic (Lemma 5.10). Write* $\mathfrak{g}_{t,0}$ *for the t-eigenspace of* $\mathrm{ad}(X)$, *so that*

$$\mathfrak{g}_0 = \sum_{t \in \mathbb{R}} \mathfrak{g}_{t,0}, \qquad \mathfrak{g}_{X,0} = \mathfrak{g}_{0,0}. \quad (7.2a)$$

Define

$$\mathfrak{p}_X = \sum_{t \geq 0} \mathfrak{g}_{t,0}, \qquad \mathfrak{n}_{X,0} = \sum_{t > 0} \mathfrak{g}_{t,0}, \qquad N_X = \exp \mathfrak{n}_{X,0}. \quad (7.2b)$$

i) The decomposition (7.2a) makes $\mathfrak{g}_0$ *an $\mathbb{R}$-graded Lie algebra:*

$$[\mathfrak{g}_{s,0}, \mathfrak{g}_{t,0}] \subset \mathfrak{g}_{s+t,0}.$$

ii) The subspace $\mathfrak{g}_{t,0}$ *is orthogonal to* $\mathfrak{g}_{s,0}$ *with respect to* ω_X *unless* $s = -t$.
iii) The adjoint action of N_X on X defines a diffeomorphism

$$\gamma \colon N_X \to X + \mathfrak{n}_{X,0}, \qquad \gamma(n) = \mathrm{Ad}(n)(X).$$

iv) The coadjoint action of N_X on ξ_X defines a diffeomorphism

$$\gamma^* \colon N_X \to \{\lambda \in \mathfrak{g}_0^* \mid \lambda|_{\mathfrak{p}_X} = \xi_X|_{\mathfrak{p}_X}\}, \qquad \gamma^*(n) = \mathrm{Ad}^*(n)(\xi_X).$$

v) *The group N_X is connected, simply connected, and nilpotent. It is normalized by G_X, and it meets G_X exactly in the identity element. The semidirect product*

$$P_X = G_X N_X$$

is a closed subgroup of G; more precisely, this is the Levi decomposition of a parabolic subgroup.

vi) *Suppose τ is an integral orbit datum at ξ_X (Definition 4.6). Then τ has a unique extension τ_{P_X} to P_X. This extension may be characterized by*

$$\tau_{P_X}(gn) = \tau(g) \qquad (g \in G_X, \ n \in N_X).$$

The pair (τ_{P_X}, P_X) is an invariant real polarization of τ.

Proof: Part (i) is elementary. For (ii), suppose $Y \in \mathfrak{g}_{s,0}$ and $Z \in \mathfrak{g}_{t,0}$. Then (7.1c) implies that

$$\omega_X(Y, Z) = s\langle Y, Z \rangle = -t\langle Y, Z \rangle,$$

and (ii) follows.

For (iii), list the positive eigenvalues of $\mathrm{ad}(X)$ as $t_1 < t_2 < \cdots < t_l$. Suppose Y_i and Z_i are elements of $\mathfrak{g}_{t_i,0}$. Then we calculate

$$\mathrm{Ad}(\exp Y_i)(X + \sum_j Z_j) = \exp(\mathrm{ad}(Y_i))(X + \sum_j Z_j). \tag{7.3a}$$

On the right side we are exponentiating the linear transformation $\mathrm{ad}(Y_i)$ of $\mathfrak{g}_0$. Now $\mathrm{ad}(Y_i)(X) = -t_i Y_i$, and $\mathrm{ad}(Y_i)$ carries $\mathfrak{g}_{t,0}$ into $\mathfrak{g}_{t+t_i,0}$. Consequently,

$$\mathrm{ad}(Y_i)(\mathfrak{g}_{t_j,0}) \subset \sum_{k>i} \mathfrak{g}_{t_k,0}. \tag{7.3b}$$

Inserting this information in (7.3a) gives

$$\mathrm{Ad}(\exp Y_i)(X + \sum_j Z_j) = X + \sum_{j=1}^{i-1} Z_j + (Z_i - t_i Y_i) + \sum_{j=i+1}^{l} p_j(Y_i, Z_1, \ldots, Z_j). \tag{7.3c}$$

Here $p_j(Y_i, Z_1, \ldots, Z_j) \in \mathfrak{g}_{t_j,0}$ depends in a polynomial way on Y_i and linearly on the various Z_k (with $k \leq j$).

We now consider the map from $\mathfrak{n}_{X,0}$ to $X + \mathfrak{n}_{X,0}$ defined by

$$\pi(Y_1 + \cdots + Y_l) = \mathrm{Ad}(\exp(Y_l) \cdots \exp(Y_1))(X). \qquad (7.3d)$$

We calculate the adjoint action here one factor at a time, using (7.3c). The conclusion is

$$\pi(Y_1 + \cdots + Y_l) = X + \sum_{j=1}^{l}(-t_j Y_j + q_j(Y_1, \ldots Y_{j-1})). \qquad (7.3e)$$

Here, $q_j(Y_1, \ldots Y_{j-1}) \in \mathfrak{g}_{t_j,0}$ depends in a polynomial way on the various Y_k. This description shows that π is a diffeomorphism.

The map π is a composition of

$$\tau: \mathfrak{n}_{X,0} \to N_X, \qquad \tau(Y_1 + \cdots + Y_l) = \exp(Y_1) \cdots \exp(Y_l),$$

and the map γ of (iii) in the proposition. An argument along similar lines to the one just given shows that τ is surjective. Whenever γ is smooth, τ is smooth and surjective, and $\gamma \circ \tau$ is a diffeomorphism, it follows that γ and τ are diffeomorphisms. This proves (iii). Part (iv) is just a reformulation of (iii) using the identification of Lemma 5.1.

For (v), N_X is connected by definition. The proof of (iii) provided a diffeomorphism τ from $\mathfrak{n}_{X,0}$ to N_X, and so N_X is simply connected. Nilpotence is immediate from (i). By (iii), only the identity element of N_X fixes X, so that G_X meets N_X exactly in the identity element. The adjoint action of G_X preserves each eigenspace $\mathfrak{g}_{t,0}$ of $\mathrm{ad}(X)$, and so it preserves $\mathfrak{n}_{X,0}$. It follows that G_X normalizes N_X. It follows that P_X is a subgroup of G. As a consequence of (iii), we have

$$P_X = \{g \in G \mid \mathrm{Ad}(g)(X + \mathfrak{n}_{X,0}) = (X + \mathfrak{n}_{X,0})\}.$$

This shows that P_X is closed in G. That P_X is parabolic means that it contains a minimal parabolic subgroup of G. This is more or less obvious from standard constructions of minimal parabolic subgoups; we omit the details. (Because we do not require G to be in the Harish-Chandra class, there is some question about exactly how a minimal parabolic subgroup should be defined. We define it so that a Levi subgroup is the centralizer of a maximal abelian subalgebra consisting of hyperbolic elements. Such a subspace is just a G-conjugate of a Cartan subspace in $\mathfrak{p}_0$ (Definition 6.1).)

For (vi), it follows from (v) that τ_{P_X} is well defined. Because distinct eigenspaces of $\mathrm{ad}(X)$ are orthogonal with respect to $\langle\,,\,\rangle$, the linear functional ξ_X is zero on $\mathfrak{g}_{t,0}$ for $t \neq 0$. In particular, ξ_X vanishes on $\mathfrak{n}_X$; so τ_{P_X} satisfies condition (i) of Definition 4.7. Condition (ii) of the definition follows from (v) of the proposition. By (ii) of the proposition, the nondegeneracy of ω_X forces the pairing it defines between $\mathfrak{g}_{t,0}$ and $\mathfrak{g}_{-t,0}$ to be nondegenerate for $t \neq 0$; so, in particular, these two spaces have the same dimension. Consequently,

$$\dim \mathfrak{p}_{X,0}/\mathfrak{g}_{X,0} = \dim \mathfrak{n}_{X,0} = (\dim \mathfrak{g}_0/\mathfrak{g}_{X,0})/2.$$

This is condition (iii) of Definition 4.7, proving that (τ_{P_X}, P_X) is an invariant real polarization. That τ_{P_X} is the only extension of τ is elementary; we omit the details. $\square$

Although this construction of polarizations was the main goal of this section, we may as well discuss orbit data in this setting.

Proposition 7.2 *In the setting of (7.1), suppose that ξ is a hyperbolic element of $\mathfrak{g}_0^*$. Then the set of integral orbit data at ξ is naturally in one-to-one correspondence with the irreducible representations of the (finite) group of connected components of G_ξ. In particular, this set is nonempty, and so the orbit $G \cdot \xi$ is integral.*

Proof: By Lemma 5.1, we may write $\xi = \xi_X$ with X a hyperbolic element of $\mathfrak{g}_0$. By Theorem 6.2, we may (after conjugating by an element of G) assume that $X \in \mathfrak{p}_0$. By Proposition 1.10(ii), it follows that G_ξ is a real reductive group with Cartan involution $\theta|_{G_\xi}$. Obviously, the linear functional on $\mathfrak{g}_{\xi,0}$ defined by X is just $\xi|_{\mathfrak{g}_{X,0}}$. What this means is that we have reduced Proposition 7.2 to the case $G = G_\xi$, which we now assume. This assumption means precisely that X belongs to

$$\mathfrak{a}_{1,0} = \{Z \in \mathfrak{p}_0 \mid G_Z = G\}$$
$$= \{Z \in \mathfrak{p}_0 \mid \mathrm{ad}(\mathfrak{g})(Z) = 0, \quad \mathrm{Ad}(K)(Z) = Z\}. \quad (7.5a)$$

Now define

$$\mathfrak{p}_{1,0} = \text{orthogonal complement of } \mathfrak{a}_{1,0} \text{ in } \mathfrak{p}_0. \quad (7.5b)$$

It is easy to check that $\mathfrak{g}_{1,0} = \mathfrak{k}_0 + \mathfrak{p}_{1,0}$ is a Lie subalgebra of $\mathfrak{g}_0$. It is the orthogonal complement of $\mathfrak{a}_{1,0}$ in $\mathfrak{g}_0$. Define

$$G_1 = K \cdot \exp \mathfrak{p}_{1,0}, \qquad A_1 = \exp \mathfrak{a}_{1,0}. \qquad (7.5c)$$

If $Y \in \mathfrak{p}_{1,0}$ and $Z \in \mathfrak{a}_{1,0}$, then Y and Z commute; so $\exp(Y + Z) = \exp(Y)\exp(Z)$. Now, it follows from the Cartan decomposition (Proposition 1.4) that G_1 is a real reductive group, and that

$$G = G_1 \times A_1, \qquad (7.5d)$$

a direct product. Furthermore, A_1 is a vector group, isomorphic to its Lie algebra under the exponential map. The group G_1 has compact center.

Now the linear functional ξ_X is given by inner product with $X \in \mathfrak{a}_{1,0}$, and so it is trivial on $\mathfrak{g}_{1,0}$. According to Definition 4.6, an integral orbit datum at ξ is an irreducible unitary representation $(\tau, \mathcal{H}_\tau)$ of G, such that τ is trivial on the identity component $G_{1,0}$, and

$$\tau(\exp Z) = e^{i\langle Z, X \rangle} \cdot \mathrm{Id}_{\mathcal{H}_\tau}$$

for $Z \in \mathfrak{a}_{1,0}$. Evidently such representations correspond precisely to the representations of $G_1/G_{1,0} \simeq G/G_0$, as we wished to show. $\square$

Propositions 7.1 and 7.2, together with the construction of (4.9), provide a finite set of unitary representations of G attached to each hyperbolic coadjoint orbit. It can be shown that these representations are all irreducible. We are left with two (closely related) problems: to interpret this construction as part of a solution of Problem 3.10; and then to extend it to a wider class of coadjoint orbits. In Section 9 we will treat the first of these problems. The Dixmier algebras we need will be algebras of differential operators. To clarify the formal algebraic construction we will use for these algebras, we examine separately the familiar case of $\mathbb{R}^n$.

8   Taylor series and differential operators on $\mathbb{R}^n$

Suppose $f \in C^\infty(\mathbb{R}^n)$. The Taylor series of f at 0 is a formal power series

$$\sum_{\alpha \in \mathbb{N}^n} c_\alpha(f) x^\alpha, \qquad c_\alpha(f) = \frac{1}{\alpha!} \frac{\partial^\alpha f}{\partial x^\alpha}(0). \qquad (8.1a)$$

(Here, we use standard conventions for multi-indices: $x^\alpha = x_1^{\alpha_1} \cdots x_n^{\alpha_n}$, $\alpha! = \alpha_1! \cdots \alpha_n!$, and so on.) Every formal power series arises as the Taylor series of some smooth function f. We write $\mathbb{C}[[x_1, \ldots, x_n]]$ for the space of all formal power series. Then formation of Taylor series provides a surjective linear map

$$T \colon C^\infty(\mathbb{R}^n) \to \mathbb{C}[[x_1, \ldots, x_n]], \qquad T(f) = \sum_{\alpha \in \mathbb{N}^n} c_\alpha(f) x^\alpha. \qquad (8.1b)$$

What is essential about the Taylor series is the collection of complex numbers c_α; or, what amounts to the same thing, the values of all the derivatives of f at 0. To formalize this point of view, define

$$U = \mathbb{C}[\partial/\partial x_1, \ldots, \partial/\partial x_n]$$
$$= \text{algebra of constant coefficent differential operators on } \mathbb{R}^n. (8.1c)$$

This algebra has as a basis the operators $\partial^\alpha/\partial x^\alpha$. To specify a linear functional on U, we must therefore specify its value at each of these basis elements; and these values can be arbitrary. That is,

$$\operatorname{Hom}_{\mathbb{C}}(U, \mathbb{C}) \simeq \text{families of complex numbers } \{t_\alpha \mid \alpha \in \mathbb{N}^n\}. \qquad (8.1d)$$

We may therefore regard Taylor series as a surjective map

$$\tau \colon C^\infty(\mathbb{R}^n) \to \operatorname{Hom}_{\mathbb{C}}(U, \mathbb{C}), \qquad \tau(f)(u) = (u \cdot f)(0). \qquad (8.1e)$$

The connection with the formal power series in (8.1a) is

$$c_\alpha(f) = \frac{1}{\alpha!} \tau(f)\left(\frac{\partial^\alpha}{\partial x^\alpha}\right). \qquad (8.1f)$$

We could also describe this as an isomorphism

$$\operatorname{Hom}_{\mathbb{C}}(U, \mathbb{C}) \simeq \mathbb{C}[[x_1, \ldots, x_n]], \qquad \mu \mapsto \sum_\alpha \frac{1}{\alpha!} \mu\left(\frac{\partial^\alpha}{\partial x^\alpha}\right) x^\alpha. \qquad (8.1g)$$

We want to understand the algebra $D^{\mathrm{an}}(\mathbb{R}^n)$ of differential operators on $\mathbb{R}^n$ with analytic coefficients. Such operators — indeed, even the algebra $D^{\mathrm{form}}(\mathbb{R}^n)$ of differential operators with formal power series coefficients — act on $\mathbb{C}[[x_1,\ldots,x_n]]$, and therefore on Taylor series. These actions are faithful. (For D^{an}, the reason is that a nonzero differential operator cannot annihilate all analytic functions.) Consequently,

$$U \subset D^{\mathrm{an}}(\mathbb{R}^n) \subset D^{\mathrm{form}}(\mathbb{R}^n) \subset \mathrm{End}(\mathbb{C}[[x_1,\ldots,x_n]]) \simeq \mathrm{End}(\mathrm{Hom}_\mathbb{C}(U,\mathbb{C})).$$
(8.2)

The problem is to identify the differential operators among all the linear transformations of formal power series. One often thinks of differential operators as characterized by the property of not increasing support. By considering only the action on Taylor series, we have in some sense already taken advantage of that property. What we *can* study on the level of Taylor series is order of vanishing. For this purpose it is helpful to write

$$|\alpha| = \alpha_1 + \cdots + \alpha_n \qquad (\alpha \in \mathbb{N}^n).$$
(8.3a)

Define

$$U_p = \text{span of monomials } \partial^\alpha/\partial x^\alpha \text{ with } |\alpha| \le p,$$
(8.3b)

the constant coefficient operators of order at most p. (We will sometimes be careless about this terminology, saying that any element of U_p is of order p even though it might belong to U_{p-1}.) These subspaces form an increasing filtration of U. Similarly, we can define

$$D_p^{\mathrm{an}}(\mathbb{R}^n) = \{\sum_{|\alpha|\le p} f_\alpha(x)\frac{\partial^\alpha}{\partial x^\alpha} \quad (f_\alpha \in C^\omega(\mathbb{R}^n))\},$$
(8.3c)

$$D_p^{\mathrm{form}}(\mathbb{R}^n) = \{\sum_{|\alpha|\le p} f_\alpha(x)\frac{\partial^\alpha}{\partial x^\alpha} \quad (f_\alpha \in \mathbb{C}[[x_1,\ldots,x_n]])\},$$
(8.3d)

the differential operators of order at most p with analytic or formal power series coefficients. By the *leading coefficients* of an operator in D_p^{form} written as in (8.3d), we will mean the formal power series f_α with $|\alpha| = p$. This means that before we can speak of leading coefficients, we must specify the order p we have in mind. If the operator happens to belong to D_{p-1}^{form}, then its leading coefficients are all zero. Now define

$$\mathrm{Hom}_\mathbb{C}(U,\mathbb{C})_p = \{\mu \in \mathrm{Hom}_\mathbb{C}(U,\mathbb{C}) \mid \mu(U_p) = 0\}.$$
(8.3e)

Under the isomorphism (8.1g), this corresponds to

$$\mathbb{C}[[x_1, \ldots x_n]]_p = \big\{ \sum c_\alpha x^\alpha \mid c_\alpha = 0, \text{all } |\alpha| \le p \big\}. \qquad (8.3f)$$

Proposition 8.1 *Suppose $f \in C^\infty(\mathbb{R}^n)$, and $p \ge 0$. The following conditions on f are equivalent.*

  i) *For every $|\alpha| \le p$, $(\partial^\alpha f / \partial x^\alpha)(0) = 0$.*
 ii) *For every $u \in U_p$, $(u \cdot f)(0) = 0$ (cf. (8.3b)).*
iii) *For every $S \in D^{an}(\mathbb{R}^n)_p$, $(S \cdot f)(0) = 0$ (cf. (8.3c)).*
 iv) *The Taylor series $T(f)$ (cf. (8.1b)) belongs to $\mathbb{C}[[x_1, \ldots x_n]]_p$.*
  v) *The Taylor series $\tau(f)$ (cf. (8.1e)) belongs to $\operatorname{Hom}_{\mathbb{C}}(U, \mathbb{C})_p$.*

This is elementary. When the conditions are satisfied, we say that f *vanishes to order p at zero*, and we write

$$f \in C^\infty(\mathbb{R}^n)_p. \qquad (8.4)$$

Because of the proposition, we may refer to (say) $\operatorname{Hom}_{\mathbb{C}}(U, \mathbb{C})_p$ as the *space of Taylor series vanishing to order p*. The first property distinguishing differential operators among all endomorphisms of Taylor series is this.

Proposition 8.2 *Suppose $f \in C^\infty(\mathbb{R}^n)_p$ (cf. (8.4)) and $S \in D_q^{an}(\mathbb{R}^n)$ (cf. (8.3c)). Then $S \cdot f \in C^\infty(\mathbb{R}^n)_{p-q}$. Here, we write $C^\infty(\mathbb{R}^n)_r = C^\infty(\mathbb{R}^n)$ for $r < 0$.*
  Similarly, suppose $\mu \in \operatorname{Hom}_{\mathbb{C}}(U, \mathbb{C})_p$ is a Taylor series vanishing to order p (cf. (8.3b)), and $S \in D_q^{form}(\mathbb{R}^n)$. Then $S \cdot f \in \operatorname{Hom}_{\mathbb{C}}(U, \mathbb{C})_{p-q}$ is a Taylor series vanishing to order $p - q$. Here, we write $\operatorname{Hom}_{\mathbb{C}}(U, \mathbb{C})_r = \operatorname{Hom}_{\mathbb{C}}(U, \mathbb{C})$ for $r < 0$.

This is an immediate consequence of the definitions and of the fact that the filtrations in (8.3) respect the algebra structures:

$$\begin{aligned} U_q U_r &\subset U_{q+r}, \\ D_q^{an}(\mathbb{R}^n) D_r^{an}(\mathbb{R}^n) &\subset D_{q+r}^{an}(\mathbb{R}^n), \\ D_q^{form}(\mathbb{R}^n) D_r^{form}(\mathbb{R}^n) &\subset D_{q+r}^{form}(\mathbb{R}^n). \end{aligned} \qquad (8.5)$$

With this property in mind, we say that a linear transformation A of $\text{Hom}_{\mathbb{C}}(U, \mathbb{C})$ is *weakly of order q* if

$$A \cdot \text{Hom}_{\mathbb{C}}(U, \mathbb{C})_p \subset \text{Hom}_{\mathbb{C}}(U, \mathbb{C})_{p-q} \qquad (8.6)$$

for all $p \geq 0$. (The condition is nonempty only for $p \geq q$.)

The most obvious linear transformations weakly of order q are the differential operators of order q with formal power series coefficients. There are more, however. When $n = 1$, the differential operator $A = x \frac{\partial}{\partial x}$ acts on formal power series by

$$A\Big(\sum_{j \geq 0} c_j x^j \Big) = \sum_{j \geq 0} j c_j x^j.$$

This action obviously preserves the subspaces $\mathbb{C}[[x]]_p$ of (8.3f), and so A is weakly of order 0. As a differential operator, however, A is first order. A little reflection shows that the problem is the vanishing at zero of the coefficient of the highest derivative in A. Here is a precise statement.

Lemma 8.3 *Suppose*

$$S = \sum_{|\alpha| \leq p} \sigma_\alpha(x) \frac{\partial^\alpha}{\partial x^\alpha} \qquad (\sigma_\alpha \in \mathbb{C}[[x_1, \ldots, x_n]])$$

is a differential operator with formal power series coefficients of order at most p. Assume that there is a β with $|\beta| = p$ and $\sigma_\beta(0) \neq 0$. (Here, $\sigma_\beta(0)$ means the constant term of the formal power series σ_β.) Then S is not weakly of order $p - 1$.

Proof: The formal power series $f = x^\beta$ vanishes to order $p - 1$ at 0; but $(S \cdot f)(0) = \beta! \, \sigma_\beta(0)$ is not zero. So $S \cdot f$ does not vanish to order 0 at 0, and so S cannot be weakly of order $p - 1$. $\square$

The lemma says that the notion of weak order for endomorphisms of Taylor series allows us to detect the order of differential operators the leading coefficients of which do not vanish at zero. To continue, we need a way to construct such differential operators from arbitrary ones (with formal power series coefficients). That is, we need a way to reduce the order of vanishing of the leading coefficients. The way to do *that* is to differentiate those coefficients. The next lemma provides a way to differentiate leading coefficients while still thinking of the operators as acting on Taylor series. Recall that a *vector field on $\mathbb{R}^n$ with formal power series coefficients* is a first order differential operator ξ that annihilates the constant function:

$$\xi = \sum_{i=1}^{n} f_i(x) \frac{\partial}{\partial x_i} \qquad (f_i \in \mathbb{C}[[x_1, \ldots, x_n]]). \tag{8.7a}$$

The *symbol of ξ at 0* is the vector

$$\sigma(\xi) = (f_i(0)) \in \mathbb{C}^n. \tag{8.7b}$$

We fix now a set $\{\xi_1, \ldots, \xi_N\}$ of vector fields, and assume that

$$\text{the symbols } \sigma(\xi_1), \ldots, \sigma(\xi_N) \text{ span } \mathbb{C}^n. \tag{8.7c}$$

(For the purposes of this section we could just take the n vector fields $\partial/\partial x_i$, but for the applications in Section 9 the more general assumption will be useful.) As a consequence of (8.7c), we can draw the following conclusion. Suppose f is a nonzero formal power series. Then, there is an integer $r \geq 0$ so that f vanishes to order $r - 1$ at 0, but f does not vanish to order r. For this value of r, we can find a sequence $(i_1, \ldots, i_r) \in \{1, \ldots, N\}^r$ so that

$$(\xi_{i_1} \cdots \xi_{i_r} f)(0) \neq 0. \tag{8.7d}$$

Here is the lemma that allows us to apply these ideas to differential operators.

Lemma 8.4 *Suppose*

$$S = \sum_{|\alpha| \leq p} \sigma_\alpha(x) \frac{\partial^\alpha}{\partial x^\alpha} \qquad (\sigma_\alpha \in \mathbb{C}[[x_1, \ldots, x_n]])$$

is a differential operator with formal power series coefficients of order at most p; and suppose

$$\xi = \sum_{i=1}^{n} f_i(x) \frac{\partial}{\partial x_i} \qquad (f_i \in \mathbb{C}[[x_1, \ldots, x_n]])$$

is a vector field. Write

$$T = [\xi, S] = \xi \circ S - S \circ \xi$$

for the commutator of S and ξ. Then T is a differential operator of order at most p with formal power series coefficients:

$$T = \sum_{|\alpha| \le p} \tau_\alpha(x) \frac{\partial^\alpha}{\partial x^\alpha} \qquad (\tau_\alpha \in \mathbb{C}[[x_1, \dots, x_n]]).$$

Assume that the leading coefficients of S all vanish to order k at x = 0: that is, that

$$\sigma_\alpha \in \mathbb{C}[[x_1, \dots, x_n]]_k \qquad \text{whenever } |\alpha| = p.$$

If |β| = p, then

$$\tau_\beta - \xi \cdot \sigma_\beta \in \mathbb{C}[[x_1, \dots, x_n]]_k.$$

That is, the leading coefficients of T are obtained from those of S by applying the differential operator ξ, up to terms vanishing to order k at x = 0.

This can be proved by a straightforward computation, which we omit.

Corollary 8.5   *Suppose S is a differential operator of order at most p with formal power series coefficients. Assume that the leading coefficients of S vanish to order r − 1 at x = 0, but that some leading coefficient does not vanish to order r. Then we can find a sequence $(i_1, \dots, i_r) \in \{1, \dots, N\}^r$ so that some leading coefficient of the iterated commutator*

$$T = [\xi_{i_1}, \cdots [\xi_{i_r}, S] \cdots]$$

has a leading coefficient that does not vanish at x = 0.

This is immediate from the lemma and (8.7d).

Corollary 8.5 suggests how to refine the definition of weak order q given at (8.6). In the setting of (8.7), we say that an endomorphism A of $\mathrm{Hom}_{\mathbb{C}}(U, \mathbb{C})$ is *of order q* if for every sequence $(i_1, \dots, i_r) \in \{1, \dots, N\}^r$, the iterated commutator

$$[\xi_{i_1}, \cdots [\xi_{i_r}, A] \cdots] \tag{8.8}$$

is weakly of order q. The definition appears to depend on the choice of the vector fields ξ_i, but Theorem 8.7 will show (subject to the assumption (8.7c)) that it does not.

Lemma 8.6   *Suppose S is a nonzero differential operator with formal power series coefficients. Then the order of S as a differential operator is equal to its order as an endomorphism of the space* $\mathrm{Hom}_{\mathbb{C}}(U, \mathbb{C})$ *of formal power series (cf. (8.8)).*

This follows from Corollary 8.5, Lemma 8.3, and the definitions.

Here is the main theorem of this section.

Theorem 8.7   *Suppose T is an endomorphism of Taylor series of order less than or equal to p (cf. (8.8)). Then T is a differential operator with formal power series coefficients of order less than or equal to p.*

Sketch of proof: By an elementary calculation, we can find a differential operator D of order less than or equal to p, with formal power series coefficients, having the property that

$$D(x^{\alpha}) = T(x^{\alpha}) \qquad (|\alpha| \leq p).$$

(This amounts to solving a finite system of linear equations in the ring of formal power series. The coefficient matrix is square and upper triangular with nonzero integers on the diagonal; so the system is solvable.)   After replacing T by $T - D$, we may therefore assume that

$$T(x^{\alpha}) = 0 \qquad (|\alpha| \leq p). \tag{8.9a}$$

On the other hand, if $|\beta| \geq p + 1$, then x^{β} vanishes to order at least p at 0, so $T(x^{\beta})$ vanishes to order at least 0. That is,

$$T(x^{\beta})(0) = 0 \qquad (|\beta| \geq p + 1). \tag{8.9b}$$

We are now trying to show that $T = 0$. The first problem is to deduce from the preceding formulas a statement about the effect of T on arbitrary formal power series. For that we use the natural topology on formal power series, in which the various subspaces $\mathbb{C}[[x_1, \ldots, x_n]]_p$ of series vanishing to order p form a neighborhood base at 0. This means that a sequence f_m of formal power series converges to f if $f - f_m$ vanishes to order at least r_m, and r_m goes to infinity with m. In this topology, a formal power series

$\sum c_\alpha x^\alpha$ is the limit of its partial sums $\sum_{|\alpha| \le m} c_\alpha x^\alpha$. An operator A that is weakly of order p is continuous, and therefore

$$A \left(\sum c_\alpha x^\alpha \right) = \lim_{m \to \infty} \sum_{|\alpha| \le m} c_\alpha A(x^\alpha).$$

This means that any fixed coefficient of the formal power series $A(\sum c_\alpha x^\alpha)$ is equal to the corresponding coefficient of $\sum_{|\alpha| \le m} c_\alpha A(x^\alpha)$ for m sufficiently large.

Bearing in mind these remarks, we find that (8.9a) and (8.9b) imply

$$T(f)(0) = 0 \qquad (f \in \mathbb{C}[[x_1, \ldots, x_n]]). \tag{8.9c}$$

We are trying to show that $T = 0$. Because of (8.7d), it suffices to show that if $(i_1, \ldots, i_r) \in \{1, \ldots, N\}^r$, then

$$(\xi_{i_1} \cdots \xi_{i_r} Tf)(0) = 0. \tag{8.9d}$$

We will prove this statement by induction on r, simultaneously with

$$([\xi_{i_1}, [\cdots [\xi_{i_r}, T]] \cdots]f)(0) = 0. \tag{8.9e}$$

The case $r = 0$ is precisely (8.9c); so suppose $r \ge 1$, and that (8.9d) and (8.9e) are known for $r - 1$. The iterated commutator in (8.9e) may be expanded as a sum of terms of the form $\pm(\xi_I T \xi_J f)(0)$. Here, I and J are disjoint ordered sets with union $\{i_1, \ldots, i_r\}$, and ξ_I is the composition of the vector fields ξ_i (with $i \in I$) in the order determined by I. Just one of these terms has $|I| = r$: It is exactly the term appearing in (8.9d). All of the other terms have $|I| < r$, and therefore they vanish by inductive hypothesis. It follows that (for a fixed f) (8.9d) and (8.9e) are equivalent.

To prove one of these statements, it is enough (by the foregoing remarks about topology) to take $f = x^\alpha$. If $|\alpha| \le p$, then (8.9d) follows from (8.9a). So suppose $|\alpha| > p$, then x^α vanishes to order at least p at 0. By hypothesis, the iterated commutator in (8.9e) is weakly of order at most p; so the iterated commutator applied to x^α vanishes to order at least 0 at 0, as required by (8.9e). This completes the induction and therefore the proof. □

9   Twisted differential operator algebras

Problem 3.10 suggests that we should construct a representation by first constructing a Dixmier algebra (A, ϕ), and that the action of $U(\mathfrak{g})$ on the representation should be given by ϕ and an action of A. The construction given in Section 7, which may be summarized as $\operatorname{Ind}_{P_X}^G (\tau_{P_X})$, does not proceed in this way; but it can be rearranged to fit better. For the moment we retain the setting of Section 7, although that will soon be modified. The Hilbert space of the induced representation is $L^2(G/P_X, \mathcal{H}_\tau \otimes D_{1/2})$. The group G acts by left translation of sections of an equivariant bundle $\mathcal{V}_\tau$ (whose fiber at eP_X is $\mathcal{H}_\tau \otimes D_{1/2}$). Consequently, elements of the complexified Lie algebra $\mathfrak{g}$ act by first order real analytic differential operators on $\mathcal{V}_\tau$; and elements of $U(\mathfrak{g})$ act by real analytic differential operators on $\mathcal{V}_\tau$. (Notice that general elements of the Hilbert space of the representation are only L^2 sections of $\mathcal{V}_\tau$, rather than smooth ones; so the differential operators will not really act on them. But it turns out (since G/P_X is compact) that the smooth vectors of the representation (Definition 2.4) are precisely the smooth sections of $\mathcal{V}_\tau$, and differential operators do act on these.)

This discussion suggests a candidate for the Dixmier algebra A: It might be the algebra D_τ^{an} of all real analytic differential operators on $\mathcal{V}_\tau$. We have seen that there is a natural algebra homomorphism $\phi\colon U(\mathfrak{g}) \to D_\tau^{\mathrm{an}}$. The action of G on G/P_X provides an action Ad of G on D_τ^{an} by algebra automorphisms, and this is easily seen to be compatible with ϕ and the adjoint action on $U(\mathfrak{g})$.

The first problem is that the action of G on D_τ^{an} is not algebraic in any sense. A basic property of algebraic representations is that they are locally finite: Any vector is contained in a finite dimensional G-invariant subspace. The differential operators of order zero include the multiplication operators m_f by analytic functions f on G/P_X. The adjoint action on m_f is by translation of the function f. If f is not locally constant, then its translates can never span a finite dimensional representation of G. (Proving that statement is a good exercise in elementary representation theory.) One way out of this difficulty is to define D_τ^{alg} as the largest subspace of D_τ^{an} on which the adjoint action is algebraic. Since the adjoint action on $U(\mathfrak{g})$ is algebraic, we will have $\phi\colon U(\mathfrak{g}) \to D_\tau^{\mathrm{alg}}$. It is possible to make sense of this idea, but it will be convenient for us to adopt a slightly less direct approach.

We are now prepared to formulate working hypotheses for the section. Suppose that

$$H_{\mathbb{C}} \subset G_{\mathbb{C}} \text{ are complex connected algebraic groups.} \qquad (9.1a)$$

We write $\mathfrak{h} \subset \mathfrak{g}$ for the corresponding Lie algebras. Fix a Lie algebra homomorphism

$$\lambda \colon \mathfrak{h} \to \mathbb{C}. \qquad (9.1b)$$

This amounts to a one-dimensional representation

$$\lambda \colon \mathfrak{h} \to \operatorname{End}(\mathbb{C}_\lambda). \qquad (9.1b')$$

One particular homomorphism plays a special rôle:

$$\delta \colon \mathfrak{h} \to \mathbb{C}, \qquad \delta(X) = \frac{1}{2}\operatorname{tr}(\operatorname{ad}(X) \text{ on } \mathfrak{g}/\mathfrak{h}). \qquad (9.1c)$$

Eventually, we will assume

$$G_{\mathbb{C}} \text{ is reductive and } H_{\mathbb{C}} \text{ is a parabolic subgroup.} \qquad (9.1d)$$

These are all the assumptions and data required for the main construction. To motivate the construction, however, it will sometimes be helpful to assume more. For those purposes, we may sometimes assume

$$G \text{ is a real form of } G_{\mathbb{C}}, \text{ and } H = H_{\mathbb{C}} \cap G \text{ is a real form of } H_{\mathbb{C}}; \qquad (9.2a)$$

and

$$i\lambda \text{ is the differential of a one-dimensional unitary character } (\Lambda, \mathbb{C}_\Lambda) \text{ of } H. \qquad (9.2b)$$

We will then write $\mathfrak{h}_0 \subset \mathfrak{g}_0$ for the real Lie algebras. Under these assumptions, the real analytic manifold G/H is a real form of the complex manifold (actually, a complex algebraic variety) $G_{\mathbb{C}}/H_{\mathbb{C}}$. Then, H acts by the adjoint action on the real vector space $\mathfrak{g}_0/\mathfrak{h}_0$ (the real tangent space at the base point to G/H); so H acts on the one-dimensional real vector space $D^{1/2}(\mathfrak{g}_0/\mathfrak{h}_0)$. We get

$$\delta \text{ is the differential of the character } \Delta \text{ of } H \text{ on } D^{1/2}(\mathfrak{g}_0/\mathfrak{h}_0). \qquad (9.2c)$$

Using this character, we can define

$$\mathcal{L}_\Lambda = \text{ line bundle on } G/H \text{ induced by } \Lambda \otimes \Delta, \qquad (9.2d)$$
$$C_{0,\Lambda}^\infty = \text{ space of compactly supported smooth sections of} \mathcal{L}_\Lambda. \quad (9.2e)$$

In the setting of (4.9), the Hilbert space of the unitary representation $\text{Ind}_H^G(\Lambda)$ is constructed as a completion of $C_{0,\Lambda}^\infty$.

Here is the main result. It is a folk theorem, much harder to attribute correctly than to prove. The main ideas go back at least to Kirillov's work in the 1960s on nilpotent groups, and probably much earlier. The only difficult part of this formulation (that D_λ is a Dixmier algebra in the setting of (9.1d)) can be deduced from [CBD]. The paper [BB] demonstrated the particular importance of twisted differential operators on flag varieties, and [BoB1] and [BoB2] provide a thorough treatment.

Theorem 9.1 *Suppose we are in the setting of (9.1a,b). Then there is a completely prime algebra $D(G_{\mathbb{C}}/H_{\mathbb{C}})_\lambda = D_\lambda$, called a twisted differential operator algebra. This algebra is endowed with an algebraic action Ad of $G_{\mathbb{C}}$ by algebra automorphisms, and with an algebra homomorphism*

$$\phi_\lambda : U(\mathfrak{g}) \to D_\lambda.$$

The adjoint actions of $G_{\mathbb{C}}$ on $U(\mathfrak{g})$ and D_λ are compatible with ϕ_λ. The differential ad *of the adjoint action is the difference of the left and right actions of $\mathfrak{g}$ defined by ϕ_λ:*

$$\text{ad}(X)(T) = \phi_\lambda(X)T - T\phi_\lambda(X) \qquad (X \in \mathfrak{g}, \ T \in D_\lambda).$$

If in addition (9.1d) is satisfied, then $(D_\lambda, \phi_\lambda)$ is a Dixmier algebra for $G_{\mathbb{C}}$ (Definition 3.5).

Suppose that the auxiliary conditions (9.2a,b) are satisfied. Then D_λ is isomorphic to a subalgebra of the analytic differential operators on sections of $\mathcal{L}_\Lambda$ (cf. (9.2d,e)). The adjoint action of G arises by change of variables from the action of G on G/H by left translation; and the homomorphism ϕ_λ arises from the natural action of $U(\mathfrak{g})$ by differential operators on $C_{0,\Lambda}^\infty$.

Proof: The statements in the last paragraph are intended to guide the construction of D_λ in general. The problem we face is essentially to find a description of the differential operators on $\mathcal{L}_\Lambda$ that refers only to G, H, and λ. To that end, recall that the differential operators are certain endomorphisms of $C_{0,\Lambda}^\infty$. The operators we want are actually going to have analytic coefficients. This suggests the possibility of studying them by means of Taylor series expansions. As a first step, we need to understand Taylor series for sections of $\mathcal{L}_\Lambda$. Now, the Taylor series of a function f at a point p can be thought of as a list of all the values at p of derivatives of f. If f is not a function but a section of a bundle, then we should apply differential operators on sections of the bundle, and the values will lie in the fiber of the bundle at p. Here is a description of Taylor series for sections of $\mathcal{L}_\Lambda$.

Lemma 9.2   *In the setting of (9.2), there is a surjective linear map (Taylor series at eH)*

$$\tau \colon C_{0,\Lambda}^\infty \to \operatorname{Hom}_{U(\mathfrak{h})}(U(\mathfrak{g}), \mathbb{C}_{i\lambda+\delta}).$$

(Here, the Hom *is defined using the left action of $U(\mathfrak{h})$ on $U(\mathfrak{g})$.) The kernel of τ consists of all sections vanishing to infinite order at eH.*

Sketch of proof: The fiber of $\mathcal{L}_*$ at eH is the one-dimensional representation $\mathbb{C}_\Lambda \otimes D^{1/2}(\mathfrak{g}_0/\mathfrak{h}_0)$ of H. Because of (9.2b,c), the Lie algebra $\mathfrak{h}$ acts on this fiber by $i\lambda + \delta$; so (after fixing a choice of a half-density on $\mathfrak{g}_0/\mathfrak{h}_0$) we may identify the fiber with $\mathbb{C}_{i\lambda+\delta}$. The space $C_{0,\Lambda}^\infty$ may be identified with a certain space of smooth functions on G with values in $\mathbb{C}_{i\lambda+\delta}$ (transforming appropriately under H on the right, and satisfying a support condition). The group G acts on this space by left translation, and the Lie algebra $\mathfrak{g}_0$ acts by right invariant vector fields. Explicitly,

$$(X{\cdot}\sigma)(g) = \frac{d}{dt}(\sigma(\exp(-tX)g))|_{t=0} \qquad (\sigma \in C_{0,\Lambda}^\infty,\ X \in \mathfrak{g}_0,\ g \in G). \quad (9.3a)$$

This is a Lie algebra representation, and so it extends to an action of the universal enveloping algebra on $C_{0,\Lambda}^\infty$. (This is the action mentioned at the end of Theorem 9.1.) The Taylor series map τ is defined by

$$\tau(\sigma)(u) = (u \cdot \sigma)(1) \qquad (\sigma \in C_{0,\Lambda}^\infty,\ u \in U(\mathfrak{g})). \quad (9.3b)$$

If $Z \in \mathfrak{h}_0$, then

$$
\begin{aligned}
\tau(\sigma)(Zu) &= (Z \cdot u \cdot \sigma)(1) \\
&= \frac{d}{dt}(u \cdot \sigma)(\exp(-tZ))|_{t=0} \\
&= \frac{d}{dt}[(\Lambda \otimes \Delta)(\exp tZ)] \cdot (u \cdot \sigma)(1)|_{t=0} \\
&= [(i\lambda + \delta)(Z)] \cdot (u \cdot \sigma(1)) \\
&= [(i\lambda + \delta)(Z)] \cdot \tau(\sigma)(u).
\end{aligned}
$$

(The third equality follows from the transformation property of $u \cdot \sigma$ under H on the right.) This shows that τ maps to the correct Hom space. Surjectivity of τ is equivalent (after introducing appropriate local coordinates on G/H and using the Poincare–Birkhoff–Witt theorem) to the fact that every formal power series on $\mathbb{R}^n$ is the Taylor series at 0 of a smooth function. The assertion about the kernel of τ is clear from (9.3b). $\square$

We want to investigate the space of Taylor series more closely. Notice first that

$$ M_\lambda = \mathrm{Hom}_{U(\mathfrak{h})}(U(\mathfrak{g}), \mathbb{C}_{i\lambda+\delta}) \qquad (9.4a) $$

may be defined in the setting (9.1a–d); the additional structure of (9.2) is not used. The complex vector space M_λ carries a $U(\mathfrak{g})$-module structure:

$$ (v \cdot \mu)(u) = \mu(uv) \qquad (u, v \in U(\mathfrak{g}),\ \mu \in M_\lambda). \qquad (9.4b) $$

Sometimes it will be convenient to write this action as an algebra homomorphism:

$$ \phi_\lambda \colon U(\mathfrak{g}) \to \mathrm{End}(M_\lambda), \qquad (\phi_\lambda(v)\mu)(u) = \mu(uv). \qquad (9.4b') $$

Because we use the right action of $U(\mathfrak{g})$ on itself, $v \cdot \mu$ inherits from μ the transformation property under $U(\mathfrak{h})$ on the left, and so it belongs to M_λ. Next, recall the standard filtration on $U(\mathfrak{g})$: $U_n(\mathfrak{g})$ is the span of all products of less than or equal to n elements of $\mathfrak{g}$. In terms of the realization of $U(\mathfrak{g})$ as right invariant differential operators on a real form G, the subspace $U_n(\mathfrak{g})$ corresponds to the differential operators of order at most n. We define the *order of vanishing filtration of M_λ* by

$$ M_{\lambda,n} = \{\mu \in M_\lambda \mid \mu(U_n(\mathfrak{g})) = 0\}. \qquad (9.4c) $$

This is a decreasing filtration of M_λ:

$$M_\lambda \supset M_{\lambda,0} \supset M_{\lambda,1} \supset \cdots, \qquad \bigcap_{n=0}^{\infty} M_{\lambda,n} = 0. \qquad (9.4d)$$

(It is often useful to define $U_{-1}(\mathfrak{g}) = 0$, and $M_{\lambda,-1} = M_\lambda$.) The action of $U(\mathfrak{g})$ is compatible with the filtrations in the following sense:

$$U_p(\mathfrak{g}) \cdot M_{\lambda,q} \subset M_{\lambda,q-p}. \qquad (9.4e)$$

Lemma 9.3 *In the setting of (9.2), the Taylor series map τ of Lemma 9.2 intertwines the action of $U(\mathfrak{g})$ on $C^\infty_{0,\Lambda}$ (cf. (9.3a)) and on M_λ (cf. (9.4b)). The subspace $M_{\lambda,n}$ is precisely the image under τ of sections vanishing to order at least n at eH.*

Here, we say that a section σ vanishes to order zero at a point p if $\sigma(p) = 0$; and we say that σ vanishes to order n if $(T\sigma)(p) = 0$ for every differential operator T of order at most n. We omit the simple proof.

Now that we have a space of Taylor series, Theorem 8.7 suggests how to define differential operators (with formal power series coefficients). The distinguished vector fields used in (8.8) are provided here by the action of $\mathfrak{g}$.

Definition 9.4 In the setting of (9.1) and (9.4), an endomorphism T of M_λ is said to be *weakly of order q* if $T(M_{\lambda,p}) \subset M_{\lambda,p-q}$ for all $p \geq 0$ (cf. (8.6)). It is *of order q* if for every sequence $(X_1, \ldots, X_r)$ of elements of $\mathfrak{g}$, the iterated commutator $[\phi_\lambda(X_1), [\cdots [\phi_\lambda(X_r), T]] \cdots]$ is weakly of order q. We write $D^{\mathrm{form}}_{\lambda,q}$ for the collection of endomorphisms of order q, and

$$D^{\mathrm{form}}_\lambda = \bigcup_q D^{\mathrm{form}}_{\lambda,q}.$$

It is not difficult to see that $D^{\mathrm{form}}_\lambda$ is a filtered algebra, isomorphic to the algebra of differential operators in $\dim \mathfrak{g}/\mathfrak{h}$ variables with formal power series coefficients. (Consequently, $D^{\mathrm{form}}_\lambda$ is completely prime.) As a consequence of (9.4e), the map of (9.4b') restricts to a filtered algebra homomorphism

$$\phi_\lambda : U(\mathfrak{g}) \to D^{\mathrm{form}}_\lambda. \qquad (9.5a)$$

We make $\mathfrak{g}$ act on D_λ^{form} by

$$\text{ad}(X)(T) = [\phi_\lambda(X), T]. \qquad (9.5b)$$

Because ϕ_λ is a Lie algebra homomorphism, ad is a Lie algebra representation of $\mathfrak{g}$ (by derivations) on D_λ^{form}. It therefore extends to an associative algebra homomorphism

$$\text{ad}: U(\mathfrak{g}) \to \text{End}(D_\lambda^{\text{form}}). \qquad (9.5c)$$

The algebra D_λ^{form} is close to the requirements of Theorem 9.1. What is missing is the action Ad of $G_{\mathbb{C}}$ by algebra automorphisms. This should be an exponentiated form of the action ad of $\mathfrak{g}$. We cannot perform this exponentiation on all of D_λ^{form}. Using an idea of Zuckerman, we will essentially define D_λ to be the largest subalgebra of D_λ^{form} on which the exponentiated action makes sense. Here is a simple version of Zuckerman's idea.

Definition 9.5 (see [V1, Definition 6.2.4]) Suppose $G_{\mathbb{C}}$ is a connected complex algebraic group with Lie algebra $\mathfrak{g}$, and V is a representation of $\mathfrak{g}$. We will define an algebraic representation $\Gamma^{G_{\mathbb{C}}}(V) = \Gamma V$ of $G_{\mathbb{C}}$. (Recall that this means that every element of ΓV is to lie in a finite dimensional $G_{\mathbb{C}}$-invariant subspace, on which the action of $G_{\mathbb{C}}$ is algebraic.) As a vector space, ΓV will be a subspace of V; and the differential of the representation of $G_{\mathbb{C}}$ is the original action of $\mathfrak{g}$. To do this, let $\tilde{G}_{\mathbb{C}}$ be the universal covering group of $G_{\mathbb{C}}$. There is a short exact sequence

$$1 \to Z \to \tilde{G}_{\mathbb{C}} \to G_{\mathbb{C}} \to 1.$$

Here Z is a discrete central subgroup of $\tilde{G}_{\mathbb{C}}$. Define

$$\tilde{\Gamma} V = \{v \in V \mid \dim U(\mathfrak{g})v < \infty\}.$$

This is a $\mathfrak{g}$-stable subspace of V, on which the action of $\mathfrak{g}$ is locally finite. By the dictionary between finite dimensional Lie group and Lie algebra representations, it follows that $\tilde{\Gamma} V$ carries a locally finite representation of $\tilde{G}_{\mathbb{C}}$, with differential given by the action of $\mathfrak{g}$. Set

$$\Gamma_0 V = \{v \in \tilde{\Gamma} V \mid z \cdot v = v, \text{ all } z \in Z\}.$$

Because Z is normal in $G_{\mathbb{C}}$, the subspace $\Gamma_0 V$ of $\tilde{\Gamma} V$ is $\tilde{G}_{\mathbb{C}}$-invariant; and the representation of $\tilde{G}_{\mathbb{C}}$ on $\Gamma_0 V$ obviously factors to $G_{\mathbb{C}}$. Finally, define

$$\Gamma V = \{v \in \Gamma_0 V \mid \text{the function } g \mapsto g \cdot v \text{ from } G_{\mathbb{C}} \text{ to } V \text{ is algebraic}\}.$$

This makes sense because the function takes values in a finite dimensional subspace of V. It is more or less obvious that ΓV is an invariant subspace of $\Gamma_0 V$, and that the representation of $G_{\mathbb{C}}$ on ΓV is algebraic.

We now define

$$D_\lambda = \Gamma^{G_{\mathbb{C}}}(D_\lambda^{\text{form}}), \tag{9.6a}$$

and write Ad for the representation of $G_{\mathbb{C}}$ on D_λ. Because $\mathfrak{g}$ acts by derivations, it is straightforward to check that $\tilde{\Gamma} D_\lambda$ is a subalgebra of D_λ^{form} on which $\tilde{G}_{\mathbb{C}}$ acts by algebra automorphisms. It follows easily that D_λ is a subalgebra on which $G_{\mathbb{C}}$ acts algebraically by algebra automorphisms. Because the adjoint action of $G_{\mathbb{C}}$ on $U(\mathfrak{g})$ is algebraic, the image of ϕ_λ (cf. (9.5a)) is contained in D_λ:

$$\phi_\lambda : U(\mathfrak{g}) \to D_\lambda. \tag{9.6b}$$

The algebra D_λ is completely prime because it is a subalgebra of the completely prime algebra D_λ^{form}.

This completes the verification of the statements in the first paragraph of Theorem 9.1. That D_λ is a Dixmier algebra under the hypothesis (9.1d) — that is, that the finiteness requirements of Definition 3.5 are satisfied — is well known, going back at least to [CBD]. (There is an explicit verification in [V7, Corollary 4.17]. One can also find there a discussion of the symbol calculus for D_λ.) The assertions about the embedding D_λ in an algebra of analytic differential operators we leave to the reader; the discussion in Section 8 should help to make them plausible. $\square$

Example 9.6   Suppose $G_{\mathbb{C}} = \mathbb{C}^n$ and $H_{\mathbb{C}}$ is trivial, and $\lambda = 0$. Then

$$M_\lambda \simeq \mathbb{C}[[x_1, \ldots, x_n]]$$

(see (8.1g) and (9.4a)). According to Theorem 8.7 and Definition 9.4,

$$D_\lambda^{\text{form}} \simeq D^{\text{form}}(\mathbb{R}^n),$$

the algebra of differential operators on $\mathbb{R}^n$ with formal power series coefficients. We also have

$$U(\mathfrak{g}) \simeq \mathbb{C}[\partial/\partial x_1, \ldots, \partial/\partial x_n],$$

and the map ϕ_λ from $U(\mathfrak{g})$ to $D_\lambda^{\mathrm{form}}$ is the natural embedding. It is therefore easy to compute the adjoint action defined in (9.5): If $T \in U(\mathfrak{g})$ is a constant coefficient differential operator, then

$$\mathrm{ad}(T) \left(\sum_\alpha f_\alpha \frac{\partial^\alpha}{\partial x^\alpha} \right) = \sum_\alpha (T f_\alpha) \frac{\partial^\alpha}{\partial x^\alpha}.$$

In the setting of Definition 9.5, it follows that $\tilde{\Gamma}(D_\lambda^{\mathrm{form}})$ consists of those differential operators the coefficients f_α of which have the following property: The space of all derivatives of f_α is finite dimensional. (Here, "derivatives" refers to applying constant coefficient differential operators.) Using linear algebra and calculus, it is not hard to show that these functions are all finite linear combinations of polynomials times exponentials:

$$f_\alpha(x) = \sum_{\xi \in \mathbb{C}^n} p_{\alpha,\xi}(x) \exp(\sum \xi_j x_j),$$

with each $p_{\alpha,\xi}$ a polynomial in x. The group $G_{\mathbb{C}}$ is simply connected, and so $\tilde{\Gamma}(D_\lambda^{\mathrm{form}}) = \Gamma_0(D_\lambda^{\mathrm{form}})$. The adjoint action of $G_{\mathbb{C}}$ is by (complex) translation of the coefficients:

$$\mathrm{Ad}(z) \left(\sum_\alpha f_\alpha \frac{\partial^\alpha}{\partial x^\alpha} \right) = \sum_\alpha (\rho(z) f_\alpha) \frac{\partial^\alpha}{\partial x^\alpha},$$

with

$$(\rho(z) f_\alpha)(x) = f_\alpha(x + z) = \sum_{\xi \in \mathbb{C}^n} p_{\alpha,\xi}(x + z) \exp(\sum \xi_j (x_j + z_j)).$$

The algebraic functions on $\mathbb{C}^n$ are the polynomials. From this, it follows that $\Gamma D_\lambda^{\mathrm{form}}$ consists of differential operators with polynomial coefficients:

$$D_\lambda \simeq \mathbb{C}[x_1, \ldots, x_n, \partial/\partial x_1, \ldots, \partial/\partial x_n].$$

To complete this section, we describe the coadjoint orbits to which the Dixmier algebras D_λ should correspond in Conjecture 3.6. Assume, then, that we are in the setting (9.1a–d). We follow Section 3 of [V7], which the reader may consult for more details and generalizations. Define

$$\Sigma_{\lambda, G_\mathbb{C}/H_\mathbb{C}} = \{\xi \in \mathfrak{g}^* \mid \xi|_\mathfrak{h} = \lambda\}. \tag{9.7a}$$

This is an $H_\mathbb{C}$-stable affine subspace of $\mathfrak{g}^*$, of dimension equal to the codimension of $\mathfrak{h}$ in $\mathfrak{g}$. We may therefore use it to construct the fiber product

$$\Sigma_\lambda = G_\mathbb{C} \times_{H_\mathbb{C}} \Sigma_{\lambda, G_\mathbb{C}/H_\mathbb{C}}, \tag{9.7b}$$

an affine bundle over the projective variety $G_\mathbb{C}/H_\mathbb{C}$. There is a natural "moment map"

$$\mu_\lambda : \Sigma_\lambda \to \mathfrak{g}^*, \qquad \mu_\lambda(\text{equivalence class of } (g, \xi)) = \mathrm{Ad}^*(g)(\xi). \tag{9.7c}$$

Obviously, the image of μ_λ is a union of coadjoint orbits.

Proposition 9.7   *In the setting of (9.1a–d) and (9.7), the moment map μ_λ is proper and generically finite; its image is the closure of a single coadjoint orbit $\mathcal{O}_\lambda$. More precisely, the parabolic subgroup $H_\mathbb{C}$ has an open orbit $\mathcal{O}_{\lambda, G_\mathbb{C}/H_\mathbb{C}}$ on $\Sigma_{\lambda, G_\mathbb{C}/H_\mathbb{C}}$. This defines an open subvariety*

$$\tilde{\mathcal{O}}_\lambda = G_\mathbb{C} \times_{H_\mathbb{C}} \mathcal{O}_{\lambda, G_\mathbb{C}/H_\mathbb{C}}$$

of Σ_λ. The restriction of the moment map μ_λ to $\tilde{\mathcal{O}}_\lambda$ is a finite covering map onto $\mathcal{O}_\lambda$.

The most difficult case of the proposition is $\lambda = 0$. In that case it is essentially due to Richardson: Σ_λ is the cotangent bundle of the partial flag variety $G_\mathbb{C}/H_\mathbb{C}$, and $\mathcal{O}$ is the Richardson nilpotent orbit attached to $H_\mathbb{C}$. For the general case, we simply refer to [V7].

It is now more or less clear that our constructions for hyperbolic orbits in Section 7 fit into the framework of Problem 3.10. Here is a specific statement.

Corollary 9.8 *Suppose G is a real reductive group of inner type $G_{\mathbb{C}}$ (Definition 3.8), and that $\xi \in \mathfrak{g}_0^*$ is a hyperbolic element (Definition 5.8). Define a parabolic subgroup $P = LN$ as in Proposition 7.1, and let $P_{\mathbb{C}} \subset G_{\mathbb{C}}$ be its complexification.*

i) The linear functional ξ restricts to a character $\lambda(\xi)$ of the Lie algebra $\mathfrak{p}$. Write $D_\xi = D_{\lambda(\xi)}$ for the twisted differential operator algebra on $G_{\mathbb{C}}/Q_{\mathbb{C}}$ attached to $\lambda(\xi)$ in (9.6).

ii) Suppose τ is an integral orbit datum at ξ (Definition 4.6 and Proposition 7.2), and $(\pi(\tau), \mathcal{H})$ is the corresponding unitary representation of G (cf. (4.9)). Then the action of $U(\mathfrak{g})$ on the smooth vectors $\mathcal{H}^\infty$ extends naturally to an action of D_ξ.

iii) In the notation of (9.7) and Proposition 9.7, $\Sigma_{\lambda(\xi)} = \tilde{\mathcal{O}}_\lambda(\xi) \simeq \mathcal{O}_\lambda = G_{\mathbb{C}} \cdot \xi$.

The assertion in (ii) follows from the last part of Theorem 9.1. That in (iii) can be deduced from Proposition 7.1(iv) applied to $G_{\mathbb{C}}$. We leave the details to the reader, along with such tasks as the construction of a Hermitian transpose on D_ξ.

In the setting of (9.1), the correspondence of Conjecture 3.6 should carry the orbit cover $\tilde{\mathcal{O}}_\lambda$ (Proposition 9.7) to the Dixmier algebra D_λ (cf. (9.6)). For $G_{\mathbb{C}} = \mathrm{GL}(n, \mathbb{C})$, every equivariant orbit cover is of the form $\tilde{\mathcal{O}}_\lambda$ for some parabolic $H_{\mathbb{C}}$ and character λ. (Actually, the covers are necessarily trivial in this case.) The same orbit may arise in several different ways, but Borho has shown in [Bo] that then the various D_λ are all isomorphic. (To be precise, one needs in addition to [Bo] the result from [BoB1] that the maps ϕ_λ are always surjective for $\mathrm{GL}(n, \mathbb{C})$.) So there is a well defined Dixmier correspondence for $\mathrm{GL}(n, \mathbb{C})$. The injectivity of the correspondence is established in [BoJ], proving Conjecture 3.6 for $\mathrm{GL}(n, \mathbb{C})$.

For most other cases, there are equivariant orbit covers not of the form $\tilde{\mathcal{O}}_\lambda$. (Such covers exist exactly when $G_{\mathbb{C}}$ has a simple factor not of type A, or $G_{\mathbb{C}}$ has disconnected center.) In those cases, the twisted differential operator algebras D_λ do not suffice to prove Conjecture 3.6. A discussion of what else is needed may be found in [V7].

10   Elliptic orbits, complex polarizations, and admissibility

In Section 7 we gave a construction of unitary representations of a reductive group G attached to hyperbolic coadjoint orbits. This construction is very nice as far as it goes, but it does not apply to other coadjoint orbits. In this section we will introduce an analogous construction for elliptic orbits. We begin with some structure theory along the lines of Proposition 7.1.

Proposition 10.1   *In the setting of (7.1), suppose that* $\mathrm{ad}(X)$ *is diagonalizable on the complexified Lie algebra* $\mathfrak{g}$, *with purely imaginary eigenvalues. (This happens in particular if X is elliptic (Lemma 5.10).) Write $\mathfrak{g}^t$ for the t-eigenspace of* $\mathrm{ad}(iX)$, *so that*

$$\mathfrak{g} = \sum_{t \in \mathbb{R}} \mathfrak{g}^t, \qquad \mathfrak{g}_X = \mathfrak{g}^0. \tag{10.1a}$$

Define

$$\mathfrak{p}_X = \sum_{t \geq 0} \mathfrak{g}^t, \qquad \mathfrak{n}_X = \sum_{t > 0} \mathfrak{g}^t. \tag{10.1b}$$

i) *The decomposition (10.1a) makes* $\mathfrak{g}$ *an* $\mathbb{R}$-*graded Lie algebra:*

$$[\mathfrak{g}^s, \mathfrak{g}^t] \subset \mathfrak{g}^{s+t}.$$

ii) *The subspace* $\mathfrak{g}^t$ *is orthogonal to* $\mathfrak{g}^s$ *with respect to* ω_X *unless* $s = -t$.

iii) *Complex conjugation on* $\mathfrak{g}$ *with respect to the real form* $\mathfrak{g}_0$ *carries* $\mathfrak{g}^t$ *onto* $\mathfrak{g}^{-t}$. *In particular,*

$$\overline{\mathfrak{p}}_X = \sum_{t \leq 0} \mathfrak{g}^t, \qquad \overline{\mathfrak{n}}_X = \sum_{t < 0} \mathfrak{g}^t,$$

so that

$$\mathfrak{p}_X + \overline{\mathfrak{p}}_X = \mathfrak{g}, \qquad \mathfrak{p}_X \cap \overline{\mathfrak{p}}_X = \mathfrak{g}^0.$$

iv) *The adjoint action of* G_X *preserves each subspace* $\mathfrak{g}^t$, *and so preserves the subalgebra* $\mathfrak{p}_X$.

v) *Suppose* (γ, F) *is a finite dimensional representation of* G_X, *and that* $d\gamma(X)$ *(the differentiated representation applied to the Lie algebra element X) is a scalar operator. Then $d\gamma$ extends uniquely to a representation* ϕ *of* $\mathfrak{p}_X$. *This extension satisfies* $\phi|_{\mathfrak{n}_X} = 0$, *and*

$$\phi(\mathrm{Ad}(g)Z) = \gamma(g)\phi(Z)\gamma(g^{-1}) \qquad (g \in G_X, \ Z \in \mathfrak{p}_X). \tag{10.1c}$$

326 *D. Vogan, Jr.*

Proof: Parts (i) and (ii) are proved just as in Proposition 7.1. (In fact, iX is a hyperbolic element of $\mathfrak{g}$, so (i) and (ii) may be regarded as special cases of Proposition 7.1.) For (iii), use the fact that $\overline{iX} = -iX$ and the fact that complex conjugation is a Lie algebra automorphism. Part (iv) is obvious. For (v), the property of vanishing on $\mathfrak{n}_X$ is easily seen to define an extension ϕ satisfying (10.1c). For the uniqueness, suppose that ϕ is any representation of $\mathfrak{p}_X$. Then, if $Z \in \mathfrak{g}^t$, $\phi(Z)$ must carry the λ eigenspace of $\phi(iX)$ into the $\lambda + t$ eigenspace. In our case $\phi(iX)$ is assumed to act by scalars, and so $\mathfrak{n}_X$ must act by zero. $\square$

The idea is that the complex Lie algebra $\mathfrak{p}_X$ is something like a polarization at ξ_X (Definition 4.7). In fact, $\mathfrak{p}_X/\mathfrak{g}_X$ is a complex Lagrangian subspace of the complexified tangent space $T_{\xi_X}(G \cdot \xi_X)_{\mathbb{C}}$. There is no subgroup of G with Lie algebra $\mathfrak{p}_X$; but we observed in (4.7c) that it was possible to write down an interesting representation space for G using only G_X and the Lie algebra of the polarization. In the setting of Proposition 10.1(v), the space in question is

$$
\begin{aligned}
\Gamma(G/G_X, \mathfrak{p}_X; F) \\
= \{f \in C^\infty(G, F) \mid f(gh) = \gamma(h)^{-1}f(g), \ \rho(Z)f = -\phi(Z)f \\
(g \in G, \ h \in G_X, \ Z \in \mathfrak{p}_X)\}.
\end{aligned}
\tag{10.2}
$$

That is, we are beginning with sections of the smooth vector bundle on G/G_X defined by (γ, F) and imposing a family of first order differential equations corresponding to elements of $\mathfrak{n}_X$. Proposition 10.1(iii) suggests that these differential equations resemble the Cauchy–Riemann equations. Here is a precise statement.

Proposition 10.2 *In the setting of (7.1), suppose that* $\mathrm{ad}(X)$ *is diagonalizable with purely imaginary eigenvalues. Then there is a distinguished G-invariant complex structure on the coadjoint orbit $G \cdot \xi_X \simeq G/G_X$. It may be characterized by the requirement that $\mathfrak{p}_X/\mathfrak{g}_X$ (cf. (10.1b)) is the antiholomorphic tangent space at the identity coset (corresponding to the point ξ_X).*

In the setting of Proposition 10.1(v), the Lie algebra representation ϕ defines a G-invariant complex structure on the vector bundle $\mathcal{F} = G \times_{G_X} F$

over $G \cdot \xi_X$. The space (10.2) may be identified with the space of holomorphic sections of $\mathcal{F}$.

A somewhat more detailed discussion of this result may be found in [V5, Propositions 1.19 and 1.21].

Proposition 10.2 suggests that any generalization of Corollary 9.8 to elliptic coadjoint orbits will need to use complex analysis. To begin, we first consider exactly which (holomorphic) vector bundles on G/G_X are relevant. The most obvious possibility is to use the bundle defined by an integral orbit datum $(\gamma, \mathcal{H}_\gamma)$ at ξ_X (Definition 4.6). That is, we might require

$$d\gamma(Y) = i\xi_X(Y) = i\langle X, Y \rangle \qquad (Y \in \mathfrak{g}_X).$$

But this is *not* precisely analogous to what we did in the hyperbolic case (cf. (4.9)). There the integral orbit datum was first twisted by the character of the isotropy group on half-densities. Explicitly (for X hyperbolic and $g \in G_X$), $\Delta_X(g)$ is the square root of the absolute value of the determinant of the adjoint action of g on $(\mathfrak{g}_0/\mathfrak{p}_{X,0})^*$. The differential of Δ_X is a character δ_X of $\mathfrak{g}_{X,0}$: $\delta_X(Y)$ is half the trace of the adjoint action of Y on $(\mathfrak{g}_0/\mathfrak{p}_{X,0})^*$.

For X elliptic, the quotient $\mathfrak{g}/\mathfrak{p}_X$ is only a complex vector space, lacking a G_X-invariant real form. There is accordingly no natural analog of half-densities, or of the character Δ_X. But δ_X still makes sense: We can compute traces in the complexification of a vector space, and so it is consistent with the hyperbolic case to define a character of $\mathfrak{g}_X$ by

$$\delta_X(Y) = \frac{1}{2}\mathrm{tr}(\mathrm{ad}(Y) \text{ on } (\mathfrak{g}/\mathfrak{p}_X)^*) = \frac{1}{2}\mathrm{tr}(\mathrm{ad}(Y) \text{ on } \mathfrak{n}_X) \qquad (10.3a)$$

This need not be the differential of a character of G_X. Nevertheless, a better analog of (4.9) is to consider bundles on G/G_X corresponding to irreducible unitary representations (γ, F_γ) of G_X with the property that

$$d\gamma(Y) = i\xi_X(Y) + \delta_X(Y). \qquad (10.3b)$$

Such a representation γ is called an *admissible orbit datum at ξ*. The orbit $G \cdot \xi$ is called *admissible* if there is an admissible orbit datum at ξ. (Compare the definition of integral orbit datum in Definition 4.6. If there is a one-dimensional character Δ_X of G_X with differential δ_X, then tensoring with Δ_X defines a bijection between integral orbit data and admissible orbit data. But, in general, the two notions are simply different.)

The definition of admissible given here makes sense only for elliptic orbits of reductive groups. There is a more sophisticated notion, due to Duflo, that makes sense for arbitrary coadjoint orbits of arbitrary Lie groups. We refer to [V5, Definition 10.16] for Duflo's definition; one can also find there a discussion of how it reduces to (10.3) in the elliptic case.

According to Proposition 10.2, an admissible orbit datum (γ, F_γ) defines a holomorphic vector bundle $\mathcal{F}_\gamma$ over the coadjoint orbit $G \cdot \xi_X$. The unitary structure on the representation γ provides a G-invariant Hermitian structure on the vector bundle, and the symplectic structure on the orbit provides a canonical G-invariant measure. Accordingly, there is a natural analog of the unitary representation defined in (4.9) for a real polarization: The Hilbert space is the space $L^2(G \cdot \xi_X, \mathfrak{p}_X; \mathcal{F}_\gamma)$ of square-integrable holomorphic sections of the vector bundle (compare (10.2) and (4.7c)).

Unfortunately, this space is almost always zero: In fact, $\mathcal{F}_\gamma$ usually has no nontrivial holomorphic sections at all. This is most easily seen when G is compact. Taking, for example, $G = \mathrm{SU}(2)$ and ξ_X any nonzero admissible element of $\mathfrak{g}_0^*$, we find that $G \cdot \xi_X$ is isomorphic to $\mathbb{C}P^1$, and that the line bundle $\mathcal{F}_\gamma$ is one usually denoted $\mathcal{O}(-n-1)$, with n a positive integer. (Essentially, $-n$ arises from the term $i\xi_X$ in (10.3b), and -1 from δ_X.) The line bundle $\mathcal{O}(-n-1)$ has only the zero section. What is interesting is its first cohomology:

$$H^1(G \cdot \xi_X, \mathcal{F}_\gamma) \simeq \mathbb{C}^n,$$

the unique irreducible representation of G of dimension n. At first glance, this suggests that we have chosen the wrong complex structure on $G \cdot \xi_X$. Indeed, everything in this section works with only trivial changes if we reverse the rôles of $\mathfrak{p}_X$ and $\bar{\mathfrak{p}}_X$. With the new complex structure the line bundle $\mathcal{F}_\gamma$ becomes $\mathcal{O}(n-1)$, still with n a positive integer. This bundle *does* have holomorphic sections:

$$H^0(G \cdot \xi_X, \mathcal{F}_\gamma) \simeq \mathbb{C}^n,$$

the n-dimensional irreducible representation.

There is a price to be paid for such a change, however. If instead we consider the group $G = SL(2, \mathbb{R})$ and take ξ_X to be a nonzero admissible elliptic element, then $G \cdot \xi_X$ is the upper half plane with the usual action of

G by linear fractional transformations. The line bundle $\mathcal{F}_\gamma$ is holomorphically trivial, and therefore it has lots of holomorphic sections (and vanishing higher cohomology). When the complex structure is chosen as in Proposition 10.2, there are even square-integrable holomorphic sections; so we get an interesting unitary representation of G (a holomorphic discrete series representation). But if the complex structure is defined instead using the opposite parabolic subalgebra, the only square-integrable section is zero.

The conclusion that we draw from these two examples is that one should use the complex structure specified in Proposition 10.2, and look for a unitary representation in some higher cohomology of $G \cdot \xi_X$ with coefficients in $\mathcal{F}_\gamma$. The question of exactly which cohomology to consider is illuminated by the following vanishing theorem of Schmid and Wolf. Recall from Theorem 6.5 and (7.1) that any elliptic orbit in $\mathfrak{g}_0^*$ has a representative ξ_X with $X \in \mathfrak{k}_0$.

Theorem 10.3 ([SW])   *Suppose G is a real reductive group with maximal compact subgroup K, and $X \in \mathfrak{k}_0$. Put on $D = G \cdot \xi_X$ the complex structure defined by Proposition 10.2. Then $Z = K \cdot \xi_X$ is a compact complex subvariety of D; write s for its complex dimension. The variety D is $(s+1)$-complete in the sense of Andreotti and Grauert. In particular, this means that Z is a compact complex subvariety of maximal dimension in D; and that if $\mathcal{F}$ is any coherent analytic sheaf on D, then $H^q(D, \mathcal{F}) = 0$ for $q > s$.*

In the first example just discussed (with $G = \mathrm{SU}(2)$), we have $Z = \mathbb{C}P^1$ and $s = 1$. In the second example (with $G = \mathrm{SL}(2, \mathbb{R})$), Z is a point and $s = 0$.

We can now say what representation ought to be attached to an admissible elliptic coadjoint orbit $G \cdot \xi$. By Theorem 6.5, there is no loss of generality in assuming that $\theta \xi = \xi$. We begin with an admissible orbit datum γ as in (10.3), and form the corresponding holomorphic vector bundle $\mathcal{F}_\gamma$ over the complex manifold $G \cdot \xi$ (Proposition 10.2). Theorem 10.3 says that $K \cdot \xi$ is a compact complex submanifold of dimension s. Form the cohomology group $H^s(G \cdot \xi, \mathcal{F}_\gamma)$. There are at least two important ways to think of this space. (The isomorphism between them is Dolbeault's theorem.) One is as a Dolbeault cohomology group, the quotient of closed $(0, s)$ forms on $G \cdot \xi$ with values in $\mathcal{F}_\gamma$ by exact forms. This shows first of all that the cohomology group carries a natural representation of G (by translation of $(0, s)$ forms). At the same time, it emphasizes the central difficulty in putting a nice topology on the space: It is not obvious that the space of exact forms is closed. (This fact was proved in general by H. Wong in his 1992 Harvard thesis [Wo].) It follows that $H^s(G \cdot \xi, \mathcal{F}_\gamma)$ has a natural complete locally convex Hausdorff topology.

A second way to think of $H^s(G \cdot \xi, \mathcal{F}_\gamma)$ is as a Čech cohomology group with coefficients in the sheaf of germs of holomorphic sections of $\mathcal{F}_\gamma$. From this point of view an element is represented by a family of holomorphic sections of $\mathcal{F}_\gamma$, each defined over some small open set (say an intersection of $s+1$ elements from a covering of $G \cdot \xi$ by Stein open sets). The advantage of this point of view is that every holomorphic differential operator on sections of $\mathcal{F}_\gamma$ clearly acts on the cohomology.

Suppose, for example, that G is of inner type $G_\mathbb{C}$ (Definition 3.8). Then, $\mathfrak{p}_\xi$ (Proposition 10.1) is the Lie algebra of a parabolic subgroup $P_{\xi,\mathbb{C}}$ of $G_\mathbb{C}$, and it is not difficult to show that

$$G_\xi = \{g \in G \mid \mathrm{Ad}(g) \in P_{\xi,\mathbb{C}}\}. \tag{10.4a}$$

Consequently, $G \cdot \xi \simeq G/G_\xi$ is an open submanifold of the flag variety $G_\mathbb{C}/P_{\xi,\mathbb{C}}$. Comparing (10.3) with the definitions of Section 9 (particularly (9.4a)), we find that the twisted differential operator algebra D_ξ (attached to the character ξ of $\mathfrak{p}_\xi$ by (9.6)) may be regarded as an algebra of holomorphic differential operators on $\mathcal{F}_\gamma$. Therefore,

$$D_\xi \text{ acts naturally on } H^s(G \cdot \xi, \mathcal{F}_\gamma). \tag{10.4b}$$

The Hermitian structure on F_γ gives rise to a natural Hermitian transpose on differential operators on $\mathcal{F}_\gamma$. This transpose preserves D_ξ, defining a Hermitian transpose of Dixmier algebras in the sense of Definition 3.8.

We are now getting close to having the structure required by Problem 3.10. The Dixmier algebra D_ξ and the group K (or even the larger group G) both act on $H^s(G \cdot \xi, \mathcal{F}_\gamma)$. This space fails to be a (D_ξ, K)-module only because the action of K is not locally finite. We have already met such a problem in connection with the construction of D_ξ, and we adopt the same solution here: We pass to the subspace of K-finite vectors. Define

$$\begin{aligned} V_{\xi,\gamma} &= H^s(G \cdot \xi, \mathcal{F}_\gamma)^K \\ &= \{v \in H^s(G \cdot \xi, \mathcal{F}_\gamma) \mid \dim(\mathrm{span}(K \cdot v)) < \infty\}. \end{aligned} \tag{10.4c}$$

It is not difficult to show that $V_{\xi,\gamma}$ is preserved by D_ξ and K, and that it is a (D_ξ, K)-module in the sense of Definition 3.8.

To complete the requirements of Problem 3.10, we need a unitary structure on $V_{\xi,\gamma}$. At this point, the analytic ideas that have brought us so far seem to fail: No general construction of such a unitary structure is known. A great deal is known about special cases; two entry points to the literature are [Z] and [RSW]. Briefly, one seeks a unitary inner product defined by integrating certain distinguished Dolbeault cohomology classes. In its simplest form, this program cannot succeed for arbitrary admissible elliptic orbits. To see why, consider elliptic orbits for the group $Sp(2n, \mathbb{R})$ with $G_\xi \simeq U(n)$. Such orbits are parametrized by the nonzero real numbers; the orbit $G \cdot \xi_t$ with parameter t is admissible if and only if $2t+n+1$ is an even integer. (The corresponding character γ of $U(n)$ is $\det^{(2t+n+1)/2}$.) The number s is equal to zero, and so $H^s(G \cdot \xi_t, \mathcal{F}_\gamma)$ is the space of holomorphic sections of the line bundle on $Sp(2n, \mathbb{R})/U(n)$ corresponding to the character γ. The unitary structure proposed in [RSW] is just integration of holomorphic sections over the orbit. The convergence of these integrals was studied in [HC2]. For $t > (n-1)/2$, all the K-finite sections are square-integrable; but for $0 < t \leq (n-1)/2$, there are no nonzero square-integrable holomorphic sections. As soon as n is at least two, therefore, we find cases where the inner product does not arise by integration. (It seems likely that the integrals will converge in general for "most" admissible elliptic orbits, as they do in this example. No such convergence has yet been proved, however.)

Despite these difficulties, there is a strong positive result.

Theorem 10.4 ([V2])   *Suppose G is a real reductive group of inner type $G_\mathbb{C}$, and that $\xi \in \mathfrak{g}_0^*$ satisfies $\theta\xi = \xi$ (so that $G \cdot \xi$ is a typical elliptic coadjoint orbit). Assume that $G \cdot \xi$ is admissible, with (γ, F_γ) an admissible orbit datum (cf. (10.3)). Define a (D_ξ, K)-module $V_{\xi,\gamma}$ as in (10.4). Then $V_{\xi,\gamma}$ carries a natural unitary structure (Definition 3.8). In particular, it is the Harish-Chandra module of a unitary representation $\pi(\xi, \gamma)$ of G.*

This result provides unitary representations attached to elliptic coadjoint orbits. The representation $V_{\xi,\gamma}$ turns out to be irreducible as a (D_ξ, K)-module but not necessarily as a $(\mathfrak{g}, K)$-module; so the unitary representation $\pi(\xi, \gamma)$ of G may be reducible.

We will now outline part of the proof of this theorem. The main point is to give (following Zuckerman) an algebraic construction analogous to the geometric one in (10.4). Roughly speaking, we replace holomorphic functions by formal power series. In the algebraic setting, an invariant Hermitian form can be constructed without much difficulty, and (with a little more difficulty) the positivity of the form can be proved. This is all that is required to attach a unitary representation to the orbit $G \cdot \xi$ (by Theorem 3.9). To complete the proof of Theorem 10.4 as stated, one must also identify the algebraic construction with the geometric one. Roughly speaking, this amounts to proving the convergence of some formal power series solutions of differential equations. The necessary ideas go back to Schmid's 1967 thesis [S1]; the result was proved completely in [Wo].

We begin with the finite dimensional module F_γ for G_ξ. By Corollary 5.13, G_ξ is a real reductive group, with Cartan involution the restriction of θ and maximal compact subgroup $K_\xi = G_\xi \cap K$. In particular, K_ξ acts on F_γ. At the same time, Proposition 10.1(v) provides a representation of the complex Lie algebra $\mathfrak{p}_\xi$ on F_γ. The same result ensures that the representations of $\mathfrak{p}_\xi$ and K_ξ enjoy a compatiblity analogous to that required for $(\mathfrak{g}, K)$-modules in Definition 2.16; we call F_γ a $(\mathfrak{p}_\xi, K_\xi)$-module accordingly. Recall now that the differential of γ is $i\xi + \delta$ (cf. (10.3)). In analogy with the definition of M_λ in (9.4), we therefore set

$$W_\gamma = \mathrm{Hom}_{U(\mathfrak{p}_\xi)}(U(\mathfrak{g}), F_\gamma). \tag{10.5a}$$

The Lie algebra $\mathfrak{g}$ acts on W_γ by right multiplication on $U(\mathfrak{g})$. The group K_ξ acts by combining the adjoint action on $U(\mathfrak{g})$ with the action on F_γ:

$$(h \cdot w)(u) = h \cdot (w(\mathrm{Ad}(h^{-1})u)) \qquad (h \in K_\xi,\ w \in W_\gamma,\ u \in U(\mathfrak{g})). \tag{10.5b}$$

This action need not be locally finite, and so we define

$$W_\gamma^{K_\xi} = \{w \in W_\gamma \mid \dim(\mathrm{span}(K_\xi \cdot w)) < \infty\}. \tag{10.5c}$$

Up to this point, there is no need to restrict to the compact subgroup K_ξ; we could just as well have kept track of an action of G_ξ. (The K_ξ-finite vectors turn out automatically to be G_ξ finite.) But it is the K_ξ action we will soon need.

Lemma 10.5   *The space W_γ may be identified with the space of formal power series for sections of the vector bundle $\mathcal{F}_\gamma$ at the base point $\xi \in G \cdot \xi$. The actions of $\mathfrak{g}$ and K_ξ satisfy compatibility conditions analogous to those in Definition 2.16, making $W_\gamma^{K_\xi}$ a $(\mathfrak{g}, K_\xi)$-module. If G is of inner type $G_{\mathbb{C}}$, then the Dixmier algebra D_ξ acts naturally on W_γ, extending the action of $U(\mathfrak{g})$ and preserving the subspace $W_\gamma^{K_\xi}$; in this way, $W_\gamma^{K_\xi}$ becomes a (D_ξ, K_ξ)-module (Definition 3.8).*

This elementary result is a holomorphic version of Lemma 9.2. In the setting of Problem 3.10, one can take $W_\gamma^{K_\xi}$ as one of the D_ξ-modules $W_i(G \cdot \xi)$.

The next step is to construct from $W_\gamma^{K_\xi}$ a (D_ξ, K)-module. We will use a functor $\Gamma = \Gamma_{\mathfrak{k}, K_\xi}^{\mathfrak{k}, K}$ introduced by Zuckerman for passing to the subspace $\Gamma W_\gamma^{K_\xi}$ of K-finite vectors in $W_\gamma^{K_\xi}$. One way to understand this approach is in terms of Theorem 10.3. The "$(s+1)$-completeness" property says roughly that $G \cdot \xi$ looks like a Stein manifold away from the compact subvariety $K \cdot \xi$. Holomorphic bundles on Stein manifolds have many global sections. This means that (morally) the obstruction to globalizing a formal power series section of $\mathcal{F}_\gamma$ is mostly in the direction of $K \cdot \xi$. Now, it is not difficult to see that a K-finite formal power series section of $\mathcal{F}_\gamma$ must represent a holomorphic section over $K \cdot \xi$. In light of the $(s + 1)$-completeness, this suggests that elements of $\Gamma W_\gamma^{K_\xi}$ should represent global holomorphic sections of $\mathcal{F}_\gamma$. That is, we might expect

$$\Gamma W_\gamma^{K_\xi} \simeq H^0(G \cdot \xi, \mathcal{F}_\gamma)^K.$$

This turns out to be true, and not too difficult to prove. The problem, as we already observed after (10.3), is that both sides are usually zero; we need analogs not of holomorphic sections but of higher cohomology.

Zuckerman's great observation was that the functor Γ is only left exact, and that it has right derived functors Γ^i. Although it is harder to justify precisely, one might still hope formally that

$$\Gamma^i W_\gamma^{K_\xi} \simeq H^i(G \cdot \xi, \mathcal{F}_\gamma)^K.$$

This statement was proved in [Wo].

Here is the definition of Γ. (Compare Definition 9.5, which is similar but simpler.)

Definition 10.6 ([V1, Definition 6.2.4]) Suppose K is a compact Lie group with complexified Lie algebra $\mathfrak{k}$, and H is a closed subgroup of K. Suppose W is a $(\mathfrak{k}, H)$-module (defined as in Definition 2.16). We want to define a $(\mathfrak{k}, K)$-module $\Gamma_{(\mathfrak{k},H)}^{(\mathfrak{k},K)} W = \Gamma W$. To begin, let $\tilde{K}_0$ be the universal cover of the identity component of K, so that we have

$$1 \to Z \to \tilde{K}_0 \to K_0 \to 1.$$

Here Z is a discrete central subgroup of $\tilde{K}_0$. Define

$$\tilde{\Gamma} W = \{w \in W \mid \dim U(\mathfrak{k})w < \infty\}.$$

As in Definition 9.5, $\tilde{\Gamma} W$ carries a locally finite representation of $\tilde{K}_0$; and we can define

$$\Gamma_0 W = \{w \in \tilde{\Gamma} W \mid z \cdot w = w, \text{ all } z \in Z\}.$$

This is a subspace of W carrying a locally finite representation π_0 of K_0; it is also preserved by the representation τ of H on W. We may therefore define

$$\Gamma_1 W = \{w \in \Gamma_0 W \mid \pi_0(h)w = \tau(h)w, \text{ all } h \in H \cap K_0\}.$$

This subspace is invariant under the representations π_0 and τ. Now, define K_1 to be the subgroup of K generated by K_0 and H. There is a unique representation π_1 of K_1 on $\Gamma_1 W$ that extends both π_0 and τ. Finally, set

$$\Gamma W = \operatorname{Ind}_{K_1}^K \Gamma_1 W,$$

a locally finite representation of K.

The reader may try to understand geometrically each of the steps in the construction of ΓW when (for example) W is the space of H-finite formal power series sections of a bundle $\mathcal{E}$ on K/H. In this case ΓW may be identified with the space of K-finite global sections of $\mathcal{E}$. One interesting step is the last one, of induction from K_1 to K. The point there is that the index m of K_1 in K is just the number of connected components of the homogeneous space K/H. Ordinarily, one could not hope to understand sections on different connected components using Taylor series; but the group action allows us to do just that. The space $\Gamma_1 W$ may be identified with K_1-finite sections of $\mathcal{E}$ over the identity component $K_1/H \simeq K_0/(K_0 \cap H)$. Induction more or less replaces this space by a sum of m copies of it; each copy corresponds to sections supported on one of the components of K/H.

Proposition 10.7 (Zuckerman; see [V1, Chapter 6]) . *The functor Γ of Definition 10.6 is a left exact functor from the category of $(\mathfrak{k}, H)$-modules to the category of $(\mathfrak{k}, K)$-modules. It has right derived functors Γ^i, which are nonzero exactly for $0 \le i \le \dim_{\mathbb{R}}(K/H)$.*

Suppose now that K is the maximal compact subgroup of a reductive group G. If W is a $(\mathfrak{g}, H)$-module (defined in analogy with Definition 2.16), then $\Gamma^i W$ becomes naturally a $(\mathfrak{g}, K)$-module. Similarly, if A is a Dixmier algebra and W is an (A, H)-module, then $\Gamma^i W$ is naturally an (A, K)-module.

The last assertion (about Dixmier algebras) is not part of Zuckerman's original ideas; it is more or less a folk theorem from the early 1980s. It may be proved using the method of [W, Section 6.3].

Theorem 10.8 (see [V2])   *In the setting of (10.5), write $\Gamma = \Gamma_{\mathfrak{k},K_\xi}^{\mathfrak{k},K}$ for the functor of Definition 10.6. Furthermore, write $s = \dim_{\mathbb{C}}(K \cdot \xi) = \frac{1}{2}\dim_{\mathbb{R}}(K/K_\xi)$.*

i) $\Gamma^i W_\gamma^{K_\xi} = 0$ for $i \ne s$.
ii) $\Gamma^s W_\gamma^{K_\xi}$ carries a natural nondegenerate $(\mathfrak{g}, K)$-invariant Hermitian form.
iii) The form in (ii) is positive definite.
iv) Suppose G is of inner type $G_{\mathbb{C}}$. Then there is a natural Hermitian transpose on D_ξ making $\Gamma^s W_\gamma^{K_\xi}$ into a unitary (D_ξ, K)-module.
v) $\Gamma^s W_\gamma^{K_\xi}$ is irreducible as a (D_ξ, K)-module.

In lieu of a proof, here are some historical remarks. The vanishing theorem in (i) is due to Zuckerman for most γ; a proof may be found in [V1]. The proof for all γ as in (10.3) appears first in [V2]. The form in (ii) was constructed by Zuckerman, but his proof of its invariance under $\mathfrak{g}$ was incomplete. Repairs were provided in [EW].

Part (iii) is (a special case of) the main result of [V2]. The proof there is a reduction to the special case when G_ξ is a compact Cartan subgroup of G. In that case, Schmid in [S2] had essentially identified $\Gamma^s W_\gamma^{K_\xi}$ with one of the discrete series representations constructed by Harish-Chandra (using deep analytic techniques). Harish-Chandra's discrete series representations are irreducible and unitary; so any nonzero invariant Hermitian form on them must be definite. A purely algebraic proof of (iii) was later found by Wallach (see [W]).

Part (iv) is more or less a folk theorem; it can be proved in the same way as (ii). The irreducibility of $\Gamma^s W_\gamma^{K_\xi}$ even as a $(\mathfrak{g}, K)$-module was proved by Zuckerman for most γ (see [V1]). The general result in (v) is perhaps an unpublished result of Bernstein; it follows from the generic case using the translation technique in [V7, Corollary 7.14].

Theorem 10.4 follows from Theorem 10.8 and the result of Wong stated before Definition 10.6.

Acknowledgements

During the summer of 1994, I lectured on this material at the Nankai Institute of Mathematics; at the Instituto de Matemáticas at UNAM; at CIMAT in Guanajuato; and finally at the European School of Group Theory. The comments of all of those audiences have been a great help to me in preparing these notes. Henrik Schlichtkrull offered many improvements and corrections to the manuscript. For the flaws that remain, I must of course assume the responsibility.

References

[AM] R. Abraham and J. Marsden, *Foundations of Mechanics*, Benjamin/Cummings, Reading, Massachusetts, 1978.

[Ar] V. Arnold, *Mathematical Methods of Classical Mechanics*, Graduate Texts in Mathematics **60**, Springer-Verlag, New York Heidelberg Berlin, 1978.

[BB] A. Beilinson and J. Bernstein, *Localisation de $\mathfrak{g}$-modules*, C. R. Acad. Sci. Paris **292** (1981), 15–18.

[BW] A. Borel and N. Wallach, *Continuous Cohomology, Discrete Subgroups, and Representations of Reductive Groups*, Annals of Mathematics Studies **94**, Princeton University Press, Princeton, New Jersey, 1980.

[Bo]   W. Borho, *Definition einer Dixmier–Abbildung für* $\mathfrak{sl}(n, \mathbb{C})$, Invent. Math. **40** (1977), 143–169.

[BoB1] W. Borho and J.-L. Brylinski, *Differential operators on homogeneous spaces I*, Invent. Math. **69** (1982), 437–476.

[BoB2] _____ , *Differential operators on homogeneous spaces III*, Invent. Math. **80** (1985), 1–68.

[BoJ] W. Borho and J. C. Jantzen, *Über primitive Ideale in der Einhüllenden einer halbeinfachen Lie-Algebra,* Invent. Math. **39** (1977), 1–53.

[CM]   D. Collingwood and W. McGovern, *Nilpotent Orbits in Semisimple Lie Algebras*, Van Nostrand Reinhold, New York, 1993.

[CBD] N. Conze-Berline and M. Duflo, *Sur les représentations induites des groupes semi-simples complexes*, Comp. Math. **34** (1977), 307–336.

[D]   J. Dixmier, *Ideaux primitifs dans les algèbres enveloppantes*, J. Algebra **48** (1978), 96–112.

[EW]   T. Enright and N. Wallach, *Notes on homological algebra and representations of Lie algebras*, Duke Math. J. **47** (1980), 1–15.

[G]   A. Guichardet, *Théorie de Mackey et méthode des orbites selon M. Duflo*, Expo. Math. **3** (1985), 303–346.

[HC1] Harish-Chandra, *Representations of a semisimple Lie group on a Banach space, I,* Trans. Amer. Math. Soc. **75** (1953), 185–243.

[HC2] _____ , *Representations of semisimple Lie groups, V,* Amer. J. Math. **78** (1956), 1–41.

[Hu]   J. E. Humphreys, *Introduction to Lie Algebras and Representation Theory*, Springer-Verlag, Berlin Heidelberg New York, 1972.

[J]   A. Joseph, *Kostant's problem and Goldie rank*, pp. 249–266 in *Noncommutative Harmonic Analysis and Lie Groups*, (J. Carmona and M. Vergne, eds.), Lecture Notes in Mathematics **880**, Springer-Verlag, Berlin Heidelberg New York, 1981.

[KV]   A. Knapp and D. Vogan, *Cohomological Induction and Unitary Representations*, Princeton University Press, Princeton, New Jersey, 1995.

[Ko]   B. Kostant, *Quantization and unitary representations*, in *Lectures in Modern Analysis and Applications*, (C. Taam, ed.), Lecture Notes in Mathematics **170**, Springer-Verlag, Berlin Heidelberg New York, 1970.

[L]   J. Lepowsky, *Algebraic results on representations of semisimple Lie groups*, Trans. Amer. Math. Soc. **176** (1973), 1–44.

[RSW]   J. Rawnsley, W. Schmid, and J. Wolf, *Singular unitary representations and indefinite harmonic theory*, J. Funct. Anal. **51** (1983), 1–114.

[S1]   W. Schmid, *Homogeneous complex manifolds and representations of semisimple Lie groups*, pp. 223–286 in *Representation Theory and Harmonic Analysis on Semisimple Lie Groups*, (P. Sally and D. Vogan, eds.), Mathematical Surveys and Monographs **31**, Amer. Math. Soc., Providence, Rhode Island, 1989.

[S2]   ———, *Some properties of square-integrable representations of semisimple Lie groups*, Ann. of Math. **102** (1975), 535–564.

[SW]   W. Schmid and J. Wolf, *A vanishing theorem for open orbits on complex flag manifolds*, Proc. Amer. Math. Soc. **92** (1984), 461–464.

[V1]   D. Vogan, *Representations of Real Reductive Lie Groups*, Birkhäuser, Boston Basel Stuttgart, 1981.

[V2]   ———, *Unitarizability of certain series of representations*, Ann. of Math. **120** (1984), 141–187.

[V3]   ———, *The orbit method and primitive ideals for semisimple Lie algebras*, in *Lie Algebras and Related Topics*, CMS Conference Proceedings **5** (D. Britten, F. Lemire, and R. Moody, eds.), Amer. Math. Soc., Providence, Rhode Island, 1986.

[V4]   ———, *Representations of reductive Lie groups*, pp. 245–266 in *Proceedings of the International Congress of Mathematicians 1986*, Vol. I, Amer. Math. Soc., Providence, Rhode Island, 1987.

[V5]   ———, *Unitary Representations of Reductive Lie Groups*, Annals of Mathematics Studies **118**, Princeton University Press, Princeton, New Jersey, 1987.

[V6]   ———, *Noncommutative algebras and unitary representations*, pp. 35–60 in *The Mathematical Heritage of Hermann Weyl*, Proceedings of Symposia in Pure Mathematics **48**, Amer. Math. Soc., Providence, Rhode Island, 1988.

[V7] _____, *Dixmier algebras, sheets, and representation theory*, pp. 333-395 in *Operator Algebras, Unitary Representations, Enveloping Algebras, and Invariant Theory*, (A. Connes, M. Duflo, A. Joseph, and R. Rentschler, eds.), Birkhäuser, Boston, 1990.

[V8] _____, *The unitary dual of G_2*, Invent. Math. **116** (1994), 677–791.

[W]   N. Wallach, *Real Reductive Groups I*, Academic Press, San Diego, 1988.

[Wo] H. Wong, *Dolbeault cohomologies and Zuckerman modules associated with finite rank representations*, Ph.D. dissertation, Harvard University, 1992; J. Funct. Anal. **129** (1995), 428–454.

[Z]   R. Zierau, *Unitarity of certain Dolbeault cohomology representations*, pp. 239–259 in *The Penrose Transform and Analytic Cohomology in Representation Theory* (M. Eastwood, J. Wolf, and R. Zierau, eds.), Contemporary Mathematics **154**, Amer. Math. Soc., Providence, Rhode Island, 1993.

Index

Perspectives in Mathematics